DIFFERENTIATION *IN VITRO*

THE FOURTH SYMPOSIUM OF
THE BRITISH SOCIETY FOR CELL BIOLOGY

DIFFERENTIATION
IN VITRO

EDITED BY

M. M. YEOMAN

Regius Professor of Botany, University of Edinburgh

D. E. S. TRUMAN

Senior Lecturer, Department of Genetics, University of Edinburgh

CAMBRIDGE UNIVERSITY PRESS

CAMBRIDGE

LONDON   NEW YORK   NEW ROCHELLE

MELBOURNE   SYDNEY

Published by the Press Syndicate of the University of Cambridge
The Pitt Building, Trumpington Street, Cambridge CB2 1RP
32 East 57th Street, New York, NY 10022, USA
296 Beaconsfield Parade, Middle Park, Melbourne 3206, Australia

First published 1982

Printed in Great Britain at the University Press, Cambridge

Library of Congress catalogue card number: 81-6094

British Library Cataloguing in Publication Data

Differentiation in vitro. – (Symposium of the British Society for Cell Biology; 4th)

1. Cell differentiation – Congresses
I. Yeoman, M. M. II. Truman, D. E. S.
III. Series
574.1'7   QH607

ISBN 0 521 23926 5

CONTENTS

Contents

PREFACE

Workers in the field of animal tissue culture probably have a limited acquaintance with the achievements of plant cell and tissue culture. They will know of the demonstrations of totipotency achieved by the culture of carrot phloem cells and may be vaguely aware that other species and cell types have been studied. The plant cell workers may be better acquainted with the longer history of cultured animal cells and cell lines. The meeting of the British Society for Cell Biology held in Edinburgh in September, 1980 devoted to studies of cell differentiation *in vitro* was held in the belief that the two groups of workers would benefit from mutual contact and the editors of the published papers hope that readers of this volume might derive a similar benefit.

Beliefs about the similarity of animal and plant cells fill the whole spectrum from those who would regard the two types as differing merely in the presence or absence of plastids and variation in the chemistry of the intercellular material, to those who would regard the differences as being so fundamental that nothing can be gained from acquaintance with the cells of the other kingdom.

The editors of this symposium volume would take a central position within these opinions. The properties of plant cells both *in vivo* and *in vitro* can be seen as a reflexion of the extent to which the living plant must adapt to its environment rather than seek an alternative environment. The open growth pattern of plants, with their persistent meristems ensures that some, at least, of the cells of a plant retain their totipotency and a capacity to respond by cell division and differentiation to new challenges from the environment. Another way of preserving the capacity to respond to the environment may be to maintain a limited activity in a wide range of metabolic pathways and to regulate overall metabolism by changes in the activity of certain key enzymes.

Such a possibility is illustrated by the work of Northcote. The question of whether the regulation of enzyme activity is achieved by control of gene transcription, translation or post-translational modification is not easily answered for any particular enzyme.

The classical view of cell differentiation held by those concerned with animal cells has been that specific cell types arise by selective transcription of limited portions of the genome to produce cell-specific messengers and proteins. This view is questioned by Clayton who has emphasised that, given sufficiently sensitive methods of assay, substances previously thought to be tissue specific may be found in a wide range of cell types.

On the other hand, substances previously regarded as widely distributed may be found by sufficiently sensitive analysis to be heterogeneous, with one type of molecule apparently specific to a particular tissue. Such seems to be the case with type II collagen which is characteristic of the extracellular matrix of cartilage. The importance of extracellular materials in the regulation of differentiation and morphogenesis is being increasingly recognised in animal systems, while in plants their importance has been long appreciated. The role of such extracellular matrix in cartilage induction was described by Lash.

The behaviour of cartilage cells during limb development is considered by Ede. The importance of the direction of cell movement during development emerges from this work. Cell movement by components of the immune system was also emphasised in the paper of Mitchison, who also stressed the biological function of the cell surface structures which originally revealed their existence to biologists by their part in histocompatibility. A growing knowledge of the interaction between the cell surface and extracellular material emerged in several other papers in the symposium, including the papers on myogenesis given by John and by Konigsberg. The elegance of this system of *in vitro* differentiation has permitted the investigation of various aspects of cellular behaviour during differentiation. The relationship between the cell cycle and the initiation of synthesis, or increase in synthetic rate, associated with differentiation remains controversial, but the application of new techniques or combinations of techniques may replace speculation with data. Konigsberg's combination of immunofluorescence and autoradiography has showed that myosin accumulation can be found in a significant proportion of myoblasts which are still synthesising DNA.

New techniques of identifying cells in a particular stage of the cell cycle were also used in the work described by Rifkind. Fractionation of friend cells according to size, and measurement of their DNA content has provided populations of synchronised cells and permitted a detailed investigation of the timing of induction and synthesis of haemoglobin in these cells.

It is clear from the proceedings of this Symposium that there is still considerable interest in the study of differentiation, and that cultured cells and tissues provide an important tool for further investigation of this process. It is, however, a complex field of research in which progress is slow and in which new approaches are needed.

The Society gratefully thanks the Principal of Edinburgh University, Dr J. H. Burnett, for his hospitality and scholarly introduction to the Symposium which was held in the University of Edinburgh from 24 to 25 September 1980. Thanks are also due to Dr C. E. Jeffree who performed his duties as local secretary with faultless efficiency and considerable enthusiasm. The editors would also like to thank the Cambridge University Press, especially Peter Silver, for their skilled assistance.

M. M. YEOMAN
D. E. S. TRUMAN

Induction of growth in the culture of pollen

N. SUNDERLAND

John Innes Institute, Colney Lane, Norwich NR4 7UH

INTRODUCTION

The greatest impact of plant tissue culture is in the field of morphogenesis. From cultures of many types of somatic cell and tissue in the Angiosperms, it is now possible to regenerate plants, and these play an increasing role in commercial practice and plant improvement. Rapid clonal propagation of elite and desirable genotypes, eradication of disease from economic plants and the development of new hybrid combinations are among the more important applications. The reproductive tissues are no exception and plants derived *in vitro* from pollen or embryo sacs provide a speedy route to pure lines and uniform F_1 hybrids.

There are two aspects to the expression of this latent morphogenic potency, namely induction and maintenance. In the case of somatic tissues, initiation of growth in a single cell or small group of cells may suffice to start a culture: the problem lies mainly with maintenance and the provision of an appropriate combination of nutrients and hormones to sustain growth. In the case of the reproductive tissues, however, the aim is to exploit the whole range of a plant's genetic resources and to give phenotypic expression to as many gene combinations as possible. Much more attention has to be given to the problem of induction. This chapter discusses induction in relation to the culture of pollen inside anthers and describes a method for obtaining large populations of induced pollen cells. Details are given of the mechanism of induction, the cellular changes that accompany induction, the cell lineages established and the patterns of division invoked. A final section deals with the culture conditions and describes a simple modification for greatly improved maintenance of the induced pollen populations.

POLLEN AS A TEST SYSTEM FOR INDUCTION

Produced inside the anthers in sacs created by an enveloping layer of polyploid, nutritive cells (tapetum), pollen provides an ideal tissue for study of induction. It develops in a highly characteristic manner in

AO

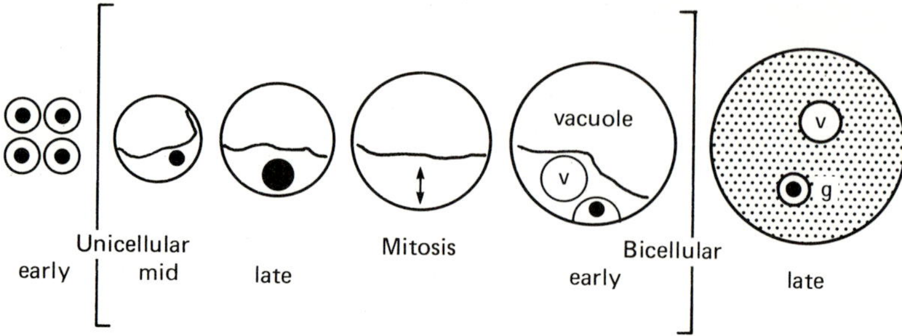

Fig. 1. Diagram indicating the period (within brackets) in pollen ontogeny that is critical for morphogenic expression *in vitro*. The term *microspore* (or spore) is used to denote the haploid cells formed by meiosis. After release from the pollen meiocytes, the spores enlarge and undergo an unequal mitotic division to give the generative (g) and vegetative (v) cells. These two cells comprise the male gametophyte or pollen grain.

the form of separate, relatively synchronous units. Each comprises a double-walled cell (the vegetative or tube cell) which at maturity has dense basophilic cytoplasm (Fig. 1). Free inside this cell is a smaller cell (the generative cell) which lacks a true wall but nevertheless has its own complement of light basophilic cytoplasm. In contrast, the respective nuclei show reciprocal staining affinities; the chromatin of the generative nucleus is highly condensed whereas that of the vegetative nucleus is diffuse. Any deviation from this unique cell-within-cell arrangement can easily be recognised in the light microscope by reference to suitably stained squash preparations of anthers. Likewise *in vitro*, changes that accompany induction and lead to division of the gametophytic cells can also be recognised and used to monitor effects of experimental treatments.

THE POLARISED MICROSPORE DIVISION

Induction is largely concerned with events that immediately precede and succeed the spore division which gives rise to the generative and vegetative cells. The division is highly polarised and unequal (Fig. 1). The small amount of cytoplasm apportioned to the generative cell contains few organelles and in some species may exclude plastids altogether. Little if any cytoplasmic synthesis takes place in the generative cell. Sauter (1969) attributes this to the presence of transcription-suppressing histones bound to the condensed chromatin. In contrast, the vegetative cell is the site of intense cytoplasmic synthesis. According to Mascarenhas (1975), rRNA genes and probably the tRNA genes are switched on shortly before the spore division and switched off shortly after it. These are genes specific to the gametophyte. The DNA-bound histones in the vegetative nucleus are considered to be in a form that permits transcription.

NATURALLY OCCURRING POPULATIONS OF MORPHOGENIC POLLEN

Variants found in some natural populations (Table 1) provide clues to the characterization and identification of the morphogenic pollens formed *in vitro*. Formal demonstration of the morphogenic potency of these variants was made on paeony anthers (Sunderland & Dunwell, 1974). The relationship has since been studied in barley (Dale, 1975) and a Burley cultivar of tobacco (Horner & Street, 1978). In the paeonies, there are marked differences in the numbers of variants produced between species and between cultivars of the same species. Some genotypes produce many, others few (Table 2). The variants often persist to anthesis, but in some genotypes, e.g. *P. emodi*, they are short-lived and mostly degenerate before anthesis. The phenomenon is a form of male sterility. In general, those genotypes that produce large numbers of variants *in vivo* are easy to culture and amongst the most productive.

The variants differ from the main population in respect of size, staining properties, cell or nuclear number and polarity. They are smaller and have only light basophilic cytoplasm. Growth is restricted by a delay in the onset of the microspore division. Some may be dividing or still in interphase as late as anthesis. Active cytoplasmic synthesis does not occur after the division. Instead, the gametophytic cells start to divide and may undergo at least one division before anthesis. Divisions of the vegetative cell occur in random planes and are equal, unequal or the divisions may be nuclear only (Fig. 2). The generative cell is often displaced from its usual position in relation to the spore axis and it divides either before or after detachment from the inner pollen wall. It may sometimes be larger than usual and occasionally the dividing wall is straight or sinuous instead of curved (Fig. 2). All the variants having different-sized cells are referred to as type A. In yet other variants (type B), polarity is so disturbed that a generative cell is not formed. The microspore division is equal in this case, incomplete

Table 1. *Species showing natural populations of pollen variants or populations induced* in vivo *by heat treatments*

	Species	Reference
Type A variants	*Amaryllis belladonna*	Upcott (1939)
	Eichhornia crassipes	Smith (1898)
	Hemerocallis fulva	Fullmer (1899)
	Lilium auratum	Chamberlain (1897)
	Scirpus palustris	Piech (1924)
	Secale cereale	Poddubnaja-Arnoldi (1936), Wenzel & Thomas (1974), Sunderland (1978)
Type B variants	*Anemone coronaria*	Sunderland (1977)
	Eremurus spp.	Charzynska, Maleszka & Hon (1977)
	Gasteria spp.	Geitler (1935)
	Hyacinthus orientalis	Němec (1898)
	Sambucus racemosa	Schüroff (1921)
	Sarcodes sanguinea	Oliver (1890)
	Scilla sibirica	Darlington & Mather (1949)
Types A and B	*Avena sativa*	Sunderland (1978)
	Cuscuta epithymum	Fedortschuk (1931)
	Hordeum vulgare	Sunderland (1974), Dale (1975)
	Lilium tigrinum	Chamberlain (1897)
	Nicotiana knightiana	Sunderland (1978)
	Nicotiana stocktonii	Sunderland (unpublished)
	Nicotiana tabacum	Horner & Street (1978)
	Paeonia hybrida	Sunderland (1974, 1977)
	Sorghum purpurea-sericeum	Darlington & Thomas (1941)
	Tradescantia sp.	Sax (1935)
	Tradescantia bracteata	La Cour (1949), Sunderland (1977)
Embryo-sac-like pollen grains	*Hyacinthus orientalis*	Němec (1989), De Mol (1921), Stow (1930), Naithani (1937)
Multicellular pollen grains	*Solanum tuberosum*	Ramanna (1974)

Table 2. *Percentage distribution of potentially morphogenic, mature and empty pollen grains in anthers of various species and hybrids of* Paeonia *growing in the gardens of the John Innes Institute; counts made about 24 h before anthesis*

	Morphogenic	Mature	Empty
Species			
P. delavayi	1	80	19
P. potanini trollioides	4	61	35
P. peregrina	9	43	48
P. triternata	9	87	4
P. emodi	14	4	82
P. officinalis	15	61	24
P. mlokosewitschii × *tenuifolia*	16	13	71
Cultivars			
P. emodi Lady Beatrice	2	92	6
P. lactiflora Elwes Hybrid	2	62	36
Charles Leveque	2	45	53
Victoria	8	61	31
The Bride	18	64	18
Dreadnought	23	69	8
Cleopatra	25	70	5
Victoire de la Marne	42	32	26
Darius	45	34	21

or nuclear only (Fig. 2). The two nuclei are similar in size and have diffuse chromatin. Additional divisions of the B variants occur in some species though not in paeony. The B variants constitute only a small fraction of most naturally occurring populations. There is no reason for attributing a greater inductive significance to them than to the more abundant A variants.

The variants are essentially cells in which gametophytic genes have not been activated. The lack of cytoplasmic synthesis after the microspore division suggests failure of the gametophytic transcription mechanism, possibly because the variants, as they develop, become so out of phase with the main population, and with the tapetum, that they miss the appropriate signals. Division of the gametophytic cells is thus an expression of sporophytic influences carried over from the diplophase through meiosis. Isolation of the variants from the main population could be due to retarded growth caused by positional effects within the

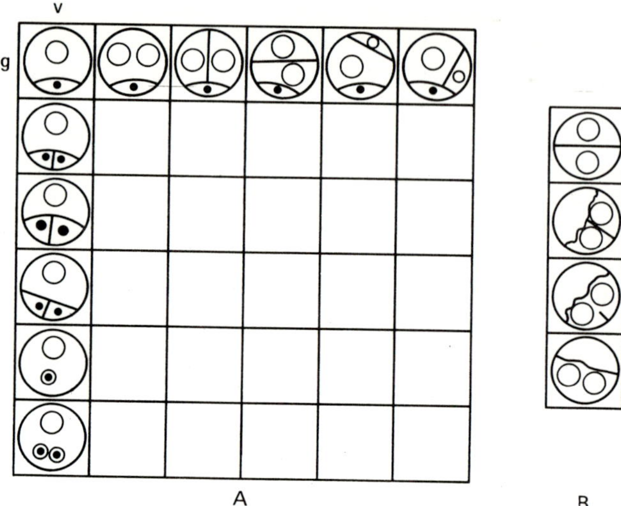

Fig. 2. Diagram of variant forms of spores and pollen grains found in paeonies and cereals (e.g. barley, rye and oats). The latin square represents the variants in which there is a recognisable generative cell (A). Variants without a generative cell (B) are shown separately.

anther or the variants may lag behind the rest from the start owing to asynchrony between meiocytes.

INDUCTION *IN VITRO*

The earliest studies on induction *in vitro* were carried out on anthers cultured direct from the plant and on species that do not exhibit variant (dimorphic) populations of pollen *in vivo*. These studies defined the best developmental stages of the anther for culture; they also revealed the similarity between the variants found in natural populations and the morphogenic pollens produced *in vitro*.

Developmental stage

Two stages are critical for induction. They are referred to, by cytological criteria, as the midunicellular (spore) stage and the early bicellular (gametophyte) stage (Fig. 1). Some species respond preferentially at the spore stage, others at the gametophyte stage. Henbane (*Hyoscyamus niger*) and Sabarlis barley (*Hordeum vulgare*) are used here as

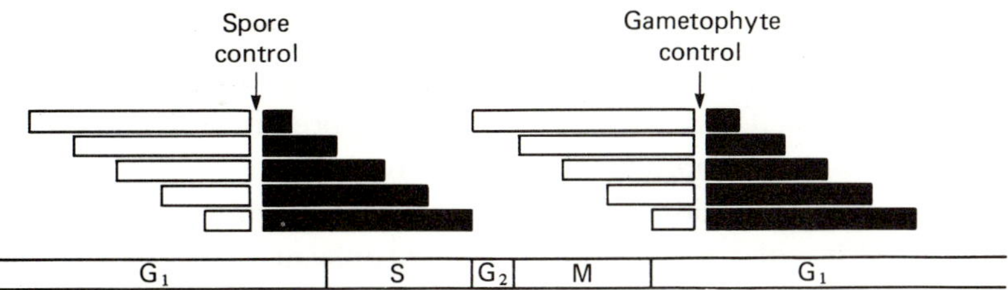

Fig. 3. Diagram illustrating the control point hypothesis. The inducible fraction (cells that have not passed the control point when the bud is removed from the plant) is represented by open rectangles for five anthers of a henbane (spore stage) or tobacco (gametophyte stage) bud. The non-inducible fraction is represented by solid rectangles. The phases of the cell cycle are purely arbitrary.

models for the spore stage, and the thornapple (*Datura innoxia*) and White Burley tobacco (*Nicotiana tabacum*), two of the most productive species in anther culture, as models for the gametophyte stage. Details given for Sabarlis barley actually refer to pretreated anthers (see later).

In terms of the cell cycle, the midunicellular stage coincides with the later part of the G_1 interphase of the spore cycle and the bicellular stage with the beginning of the next interphase after the spore division. It is envisaged that the pollen programme is diverted at specific control points (Fig. 3). As the respective cells pass the control their development may be checked, whereas those that have passed the control before culture continue the normal programme. A dimorphic population thus develops. In theory, the best time for culture would be before any cell has passed the control point, but in practice, culture of barley anthers in early G_1 of the microspore cycle results in rapid degeneration of the cells. Also, culture of tobacco anthers just before the onset of the microspore division, i.e. at the late unicellular stage (Fig. 1), is usually less productive. The anthers must be cultured as close to the control points as possible. Because neither the pollen within the anthers nor the anthers of the same bud are strictly in phase, the numbers of cells that can be checked (*inducible fraction*) must vary from one anther to another (Fig. 3).

Another feature of the culture of tobacco anthers at the late unicellular stage is that a dimorphic population again develops (Sunderland, 1979). Some cells are clearly more susceptible to check than others. The size of this *susceptible fraction* in relation to the size of the inducible fraction is indeterminate.

Cellular changes associated with induction

Spore stage. Induction at the spore stage concerns both the generative and vegetative cells. In Sabarlis barley, the spore division and the first division of the gametophytic cells occurs within 24–48 h (Sunderland, Roberts, Evans & Wildon, 1979). There is no obvious change in the stain intensity of the vegetative-cell cytoplasm before division; it increases as division continues. The diffuse nature of the chromatin is retained in the derived nuclei. In contrast, the stain intensity of the generative-cell cytoplasm, as shown by Raghavan (1976) in henbane, increases before division and the intensity increases as division continues. The condensed state of the chromatin is retained in the derived nuclei.

In barley, both the increase in cytoplasmic staining and the division of the generative cell begin before detachment of the cell from the inner pollen wall. The new cytoplasm is clearly of sporophytic origin. The control mechanism is thus associated with suppression of gametophytic transcription. If Sauter (1969) is correct about the transcription-suppressing activity of the DNA-bound histones, induction of the generative cell involves deactivation of the histones without any obvious relaxation of the chromatin structure. Alternatively, transcription of the sporophytic genes begins before the histones become active. The fact that the generative cell does not divide independently at the gametophyte stage suggests that once the histones are activated the fate of the cell may be irrevocably determined.

Gametophyte stage. Induction at the gametophyte stage is concerned mainly with the vegetative cell. At this stage, gametophytic RNAs and proteins are already accumulating in the cell. In tobacco, this gametophytic cytoplasm is degraded before the cell divides. In ultrastructural terms, the cell is almost entirely cleared out of ribosomes and other cytoplasmic organelles (Dunwell & Sunderland, 1974). The derived cells are subsequently repopulated with new cytoplasm (Dunwell & Sunderland, 1975). Removal of the gametophytic cytoplasm presents conditions which are more conducive to expression of the sporophytic genes.

The inductive changes in tobacco take at least 6 days at an incubation temperature of 25 °C (Sunderland & Wicks, 1971). In thornapple, on the other hand, the inductive period is much shorter and there is no obvious cytoplasmic degradation. This has been ascribed to the lower cytoplasmic content of the vegetative cell in thornapple after completion of the microspore division (Sunderland & Dunwell, 1974).

The evidence is consistent with the view that control resides in the switching on and off of gametophytic genes. In spore control the switch does not operate whereas in gametophyte control, it operates either prematurely (thornapple) or on time (tobacco), the latter species having the advantage of the cytoplasm clearing process.

There are two possibilities open to experimentation. Either the control points coincide with the actual times at which the switches operate *in vivo*, or the check occurs some time before the switches operate, the check to development in some way providing a signal for subsequent gene action. Vasil (1973) has suggested that the switching off of the rRNA genes (Mascarenhas, 1975) could be the time at which the fate of the vegetative cell is irrevocably determined.

INDUCTION BEFORE CULTURE

In the culture of anthers direct from the plant, induction frequencies are low. Moreover, only a few of the induced pollen grains survive to emerge from the anthers as macroscopic structures (Sunderland & Wicks, 1971). Assuming a 1% induction frequency and a recovery of 1%, the culture yield in tobacco amounts to a mean of four plants per anther. Because barley anthers are much smaller and contain fewer pollen grains, the same calculation gives a yield of one callus for every three anthers. Since relatively few calluses give plants in barley, the plant yield is considerably lower than 1 in 3. In practice, mean induction frequencies are usually somewhat larger than 1% in tobacco, but smaller in barley. Recovery values are also less than 1%.

The low frequencies are due to a high death-rate imposed by the culture conditions. After 3–4 days, the pollens that have not been diverted are mostly dead and have lost their contents, but many of the diverted pollens that have not started to divide or have divided only once or twice are also degenerating. To ensure induction and continued division of the whole susceptible fraction, this high death-rate must be circumvented. Preincubation of cultures at cold temperatures (Sunderland & Wildon, 1979) improves culture yields but the effect on induction is relatively small. A more effective way is to induce the pollen into division or into a state of incipient division before culture, that is, to induce a dimorphic population.

There has so far been little conscious attempt to impose physical constraints on whole plants in relation to the culture of anthers. Anthers cultured from nitrogen-starved plants of tobacco are known to give better culture yields than fed plants (Sunderland, 1978), but the effect on induction is not known. The formation of pollen variants in Sabarlis barley is more frequent in plants grown at 10 °C than at 15 °C (Sunderland, 1978). Much work has been done to define the best growth conditions of the plants before culture of the anthers, but again the data refer to culture yields rather than induction (but see Dunwell, 1976). Hormone treatments have been used but ineffectively (Nitzsche & Wenzel, 1977). Most progress has been made on the temperature-stress of excised floral parts, a procedure which derives from observations of Nitsch & Norreel (1973) and Nitsch (1974) on chilled buds of thornapple and tobacco.

The use of temperature-stress is best illustrated by reference once again to paeony anthers. If these are excised at the early bicellular stage, placed dry in a petri dish without medium and incubated, the pollen continues to develop and two populations diverge as *in vivo*. Survival of the anthers can be prolonged to 2–3 weeks at refrigerator temperatures and 4–5 weeks if the whole bud is used to prevent desiccation of the anthers. The rate of degeneration of the pollen is much less than it is in the presence of the culture medium. The induced pollens start to divide during the stress, the number doing so being dependent on the duration of stress (Sunderland, unpublished data).

Similar 'pretreatment' of tobacco buds also checks the pollen and this leads to the cytoplasmic degradation referred to earlier (Sunderland, 1979). In this instance the induced cells do not start to divide during the pretreatment. Since the inductive changes take 6 days in culture at 25 °C, it follows that much longer is needed at lower temperatures. The lowest temperature for tobacco appears to be about 7 °C; below this, degeneration is accelerated by chill damage (Sunderland & Roberts, 1979). Experience has shown that cold temperatures are not necessarily the best. Any temperature between 7 °C and 25 °C will suffice provided the time of stress is adjusted accordingly. This is illustrated by the data of Fig. 4. Timing is critical, more so at higher than at lower temperatures. A difference of a few hours at 20–25 °C can markedly alter the culture yield. The times used in the experiment depicted in Fig. 4 are known from experience to be near optimal for each temperature. It will be appreciated that a day either way in the time of the pretreatment, especially at the higher

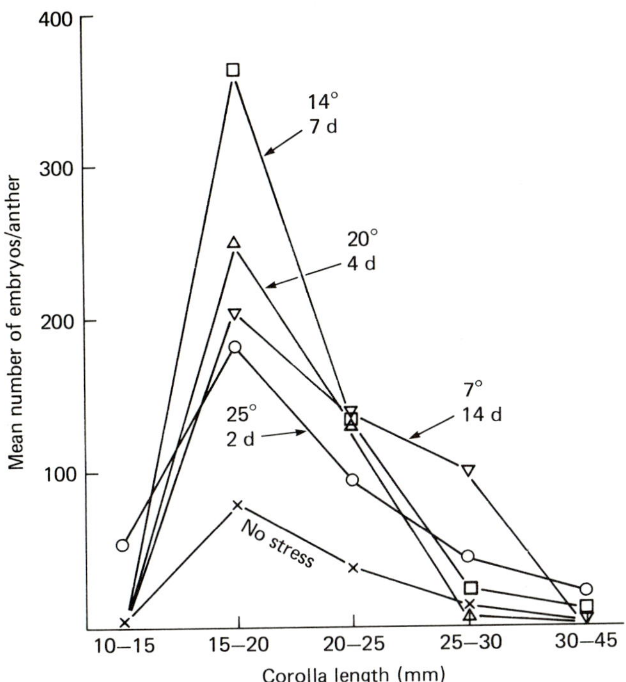

Fig. 4. Pollen embryo yields in float cultures of White Burley tobacco anthers stressed in excised buds at different temperatures before culture. The 15–20 mm category includes the mitotic to early bicellular stages. Embryos formed from anthers of older buds arise from lagging pollen grains. Twelve anthers in 5 ml liquid A medium (Sunderland & Roberts, 1977) in 50-mm petri dishes. Counts made after about 35 days at 25 °C (0–14 days in darkness).

temperatures would have given different results. The most productive treatment in Fig. 4 is 7 days at 14 °C. Pretreatment at 15 °C has also been found highly effective in henbane (Sunderland & Wildon, 1979).

To make the most of the temperature-stress therefore the strategy is to find for each species or cultivar the best combination of time and temperature. The data of Fig. 5 show the results of one such attempt with thornapple. A temperature of 7 °C is clearly better than either 4 °C, 14 °C or 20 °C. The results for 4 °C and 7 °C contrast sharply with those of other workers on this genus who insist that short treatments (2–3 days) at temperatures below 4 °C are essential (see e.g. Sangwan-Norreel, 1979; Gupta & Barbar, 1980).

In the case of Sabarlis barley, whole spikes removed from the tiller can be kept for as long as 28 days at 4 °C in sealed petri dishes before

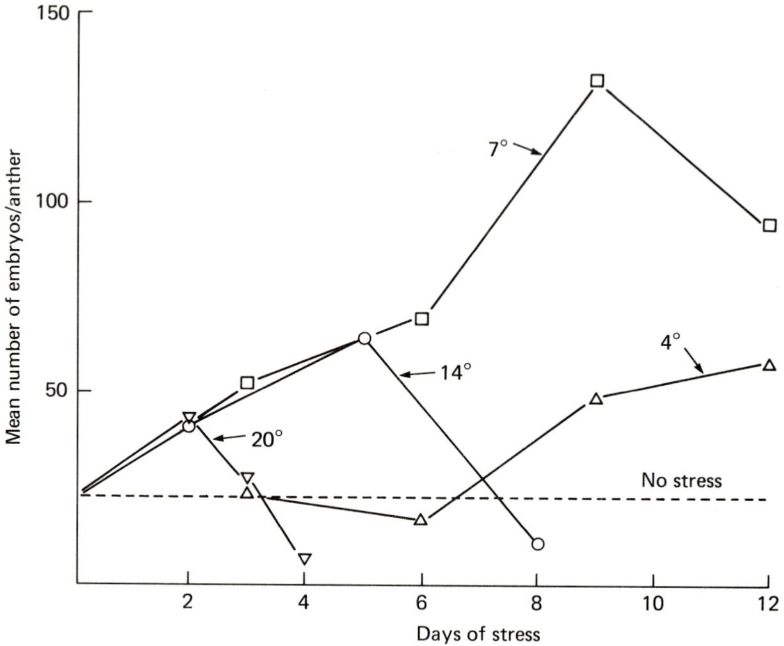

Fig. 5. Pollen embryo yields in float cultures of thornapple anthers stressed in excised buds (35–40 mm) at different temperatures before culture. Four anthers per dish as in the caption to Fig. 4.

serious deterioration sets in. During this period large populations of dividing pollen cells are induced provided the spikes are excised at the midunicellular stage. The number induced depends upon the size of the inducible fraction (Fig. 3) at the time of excision. The data of Table 3 indicate an induction frequency of 76% in a barley genotype that normally does not exhibit dimorphism *in vivo*. In this case, there is every indication that all the inducible cells have survived the treatment and been switched into morphogenesis.

There are two attitudes to the use of the pretreatment. The first, advocated by Nitsch & Norreel (1973) is to shock the microspores into an equal division (Fig. 2) by a short chill. Such shock treatments, which stem from the work of Sax, Stow and others (Table 1), have the disadvantage that much damage occurs during the process. The second, advocated here, is to use relatively long periods of stress. Excision of the buds from the plant followed by stress treatment initiates changes which lead to suppression of gametophytic transcrip-

Table 3. *Percentage of morphogenic, maturing and empty pollen grains in an anther of* Hordeum vulgare *var,* abyssinicum *after spike pre-treatment for 21 days at 4 °C followed by 24 h at 25 °C. Spike excised from the plant at the midunicellular (spore) stage*

	Morphogenic		Maturing	Empty	Unclassified
	76		3	19	2
g attached		g detached			
g + v	1	g + v	10		
g + v in mitosis	10	g + v in mitosis	8		
g + 2v	32	2g + v	2		
g + 3v	1	2g + v in mitosis	1		
2g + 2v	3	2g + 2v	4		
2g + 2v in mitosis	4				

g = generative cell; v = vegetative cell.

tion. The stress also affects polarity but does not necessarily lead to equal microspore divisions. Stress must also initiate changes in endogenous hormone levels and in the general metabolic activity of the cells. In henbane, the stress alters the optimal stage for culture of the anthers (Sunderland & Wildon, 1979) and this may be seen as a reflection of the changed metabolic status.

CELL LINEAGES AND PLOIDY LEVEL

The cell lineages established during the first few division cycles *in vitro* may be haploid, non-haploid or heteroploid (Table 4).

Spore stage

The evidence available suggests that in both henbane and barley, divisions of the generative cell are equal, and haploid, whereas those of the vegetative cell, at least in barley, are equal, unequal or the divisions may be nuclear only (See Fig. 2). In henbane, the vegetative cell is thought to function as a suspensor-like mother cell, undergoing only limited division, whereas the generative cell functions as an embryo-like mother cell. The pollen grains are thus segmented into haploid cells derived from the generative cell together with a small appendage

Table 4. *Cell lineages, ploidies and culture products in the four model species*

	Cell lineages	Ploidies	Culture products	Plants	References
Barley	A-G?, A-V, A-GV, B, C	n, non-n, mixed	callus	albino, variegated, green	Clapham (1977), Mix, Wilson & Foroughi-Wehr (1978), Sunderland *et al.* (1979), Sunderland & Evans (1980)
Henbane	A-G, A-V, A-GV, B	n, non-n	callus, bipolar embryos, multi-polar embryos, embryo clusters	green	Corduan (1975), Raghavan (1976, 1978), Sunderland *et al.* (1977)
Thornapple	A-V, B, C	n, $2n$, $3n$, $4n$, $6n$	bipolar embryos	green	Engvild *et al.* (1972), Sunderland (1974)
Tobacco	A-V, B rare	n, $2n$ occasionally	bipolar embryos	green	Sunderland (1970), Kasperbauer & Collins (1972), Engvild (1974)

A-G, generative-cell lineage only; A-V, vegetative-cell lineage only; A-GV, both lineages, partitioned units; B, spore lineage; C, lineages from fused generative and vegetative nuclei; n, haploid; non-n, non-haploid.

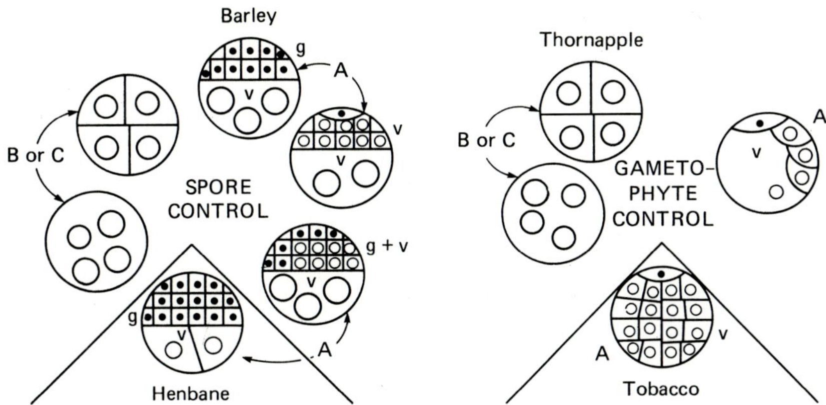

Fig. 6. Diagram representing the cell lineages established in the four model species. Generative-derived nuclei are indicated by solid circles, vegetative-derived nuclei, B and C nuclei, by open circles. Haploid and non-haploid nuclei are represented by different sized circles.

derived from the vegetative cell (Fig. 6). Other segmentation patterns occur in henbane (Raghavan, 1978; Sunderland, Wildon & Evans, 1977) but they have not been described in detail and are thus omitted from Fig. 6. In barley, by contrast, the vegetative cell is thought to function more like an endosperm mother cell (Sunderland *et al.*, 1979). There is an initial phase of free synchronous nuclear division during which nuclei fuse in varying numbers. When wall formation begins the mother cell is segmented into a series of non-haploid cells which are of the same or different ploidy. The generative cell forms a separate haploid embryo-like component (Fig. 6). Alternatively, the vegetative cell divides first by an unequal division (see Fig. 2). The larger derivative then divides by free nuclear division whereas the smaller derivative divides by haploid divisions, forming a small-celled embryo-like component. This latter may develop separate from, or in conjunction with, the generative-cell component. Sometimes the generative cell degenerates and only the vegetative-derived embryo component is represented (Fig. 6).

In barley, these chimerical units, which in some respects resemble embryo sacs, can only be observed in squash preparations between 48–72 h. After this the cell lineages cannot be distinguished one from the other.

Two other lineages are formed in barley in addition to those formed by independent division of the gametophytic cells (Sunderland &

Evans, 1980). They are referred to as the B and C lineages. The B lineage arises from the type B variants (Fig. 2), which when they occur in Sabarlis barley are wholly of the binucleate type. The two nuclei fuse during the first cycle and a diploid lineage is established. The C lineages arise by fusion of generative and vegetative nuclei in varying numbers before wall formation and regular division begin. The pollen grains are segmented into equal non-haploid cells in which all the nuclei are alike. Wall formation may also be delayed after the first fusion cycle. Further fusion occurs in succeeding cycles, generating still higher ploidies. After the first division, the B and C lineages cannot be distinguished cytologically one from the other.

Gametophyte stage

In the absence of an independent generative-cell lineage, partitioned units are not formed in tobacco or thornapple. The vegetative cell functions as a haploid mother cell dividing by equal or unequal divisions (Fig. 6). Both B and C lineages are formed in thornapple but not in White Burley tobacco. There is evidence for the formation of $2n$, $3n$, $4n$ and $6n$ C lineages in thornapple (Sunderland, Collins & Dunwell, 1974).

MORPHOGENIC COMPETENCE OF INDUCED POLLEN

Among the species discussed here, tobacco is the only one in which a single cell-lineage is represented in the induced pollen. The culture product consists entirely of bipolar embryos which give rise to green plants. The morphogenic competence of the vegetative-cell lineage is therefore assured; it has all the information, both nuclear and cyto-plasmic, for the development of a green plant.

In the other species the derivation of the end products of culture cannot be stated with such assurance. The difficulty is most acute in barley, in which the culture product is invariably callus. The plants regenerated from the calluses vary in ploidy and are albino, variegated or green (Clapham, 1977; Mix, Wilson & Foroughi-Wehr, 1978). Albinos predominate in Sabarlis barley. Devreux (1971) has suggested that albinism might be associated with the generative cell but the barley system is too complex to warrant such a conclusion. In henbane, not only are different lineages represented in the induced pollen, but the culture product is also diverse and the plants differ in ploidy (see Table 4). Raghavan (1976) gives prominence only to those units which show development of the generative cell and concludes that the green plants obtained are therefore derived from that cell. Because of the complexities at the start and end of culture, some measure of doubt exists. In thornapple, the culture product consists of bipolar embryos which give green plants. These vary in ploidy from *n* to 6*n*. Because C lineages are the only ones established initially with ploidies up to 6*n* (Sunderland *et al.*, 1974) there seems little reason for doubting the competence of the C lineages.

The B lineage is the one for which there is least hard evidence of competence. Yet this is taken as fact in much of the literature sometimes on the most slender of evidence. There are three main lines of argument:

(*a*) While the frequency of B pollens in a population may be low, there are sufficient in absolute terms to account for the small numbers of plants produced *in vitro* (Rashid & Street, 1974). This presupposes that the B lineage is not only competent but that it is more competent than the rest. An equal division of the spores for some reason endows them with special properties to withstand the rigours of the culture process. In a situation where there are many A forms and only an occasional B form, there doubtless does appear to be a high rate of degeneration of the A forms. Close scrutiny of the population, however, shows that the B pollens are just as subject to degeneration as the A forms.

(*b*) The few surviving multicellular units after several days of culture have cells that all look alike; therefore they must have arisen by an equal division of the spores (see e.g. Nitsch, 1974). Conclusions drawn from observations made only on the surviving units without reference to previous events are meaningless.

(*c*) Cold pretreatments of anthers for short periods shock the spores into equal divisions (Nitsch & Norreel, 1973). While this may possibly occur in some genotypes, an increase in the frequency of B pollens has not been observed in any of the experiments referred to in this paper in what are sometimes exceedingly long exposures to cold temperature (as in Table 3). This applies to paeony, tobacco and Sabarlis barley. Details for thornapple and henbane are unfortunately not yet available.

To date, only one attempt has been made to culture anthers of a plant that produces B pollens only. In this plant, a genotype of *Anemone coronaria*, the B pollens were induced to form multicellular pollen grains, but these did not survive the culture process (Sunderland, 1977).

The objection to the above arguments is the implication that an equal division of the spores is essential for morphogenic expression. As already indicated, the essential requirement is blockage of gametophytic transcription. There are perfectly valid reasons why an equal division *per se* may be desirable. For instance, in a system that lacks the restraints that are normally present in an embryo sac, an equal division in any lineage is more likely to lead to early establishment of polarity and organised growth.

THE ROLE OF THE CULTURE MEDIUM

The high induction frequencies that can now be obtained by stress pretreatments raise the question of whether culture has any role in the induction process. In tobacco, henbane and thornapple, the pretreatment induces the cells into a state of incipient division. Division begins within a few hours after culture of the anthers. The death-rate during the early stages is still the same as in the culture of anthers direct from the plant but because division is triggered earlier and in large numbers of cells the culture yield is enhanced (as in Fig. 4). This rapidity in the onset of division is most noticeable in tobacco, in which without the pretreatment the first divisions do not begin for at least 6 days. Hence a major role of the culture medium is in providing the trigger for division. For henbane, tobacco and thornapple, only a simple medium is required to generate high culture yields. It contains mineral salts and 2% sucrose (Sunderland & Roberts, 1977). Sucrose alone is sufficient to trigger division. It is generally supposed in the culture of somatic tissues that hormones provide the trigger for division. If hormones are involved in the anther system they must come from the anther tissues or the pollen itself. Addition of hormones to the medium may have slight enhancing effects on culture yields in thornapple (Sopory & Maheshwari, 1976) but these effects are more related to maintenance than to triggering of division. More often than not addition of hormones leads to callus, rather than embryo formation (see e.g. Corduan, 1975; Raghavan, 1978).

Conditioning factors are known to emanate from the anther tissues. Thus in tobacco, if optimally pretreated anthers are floated on liquid medium, they soon dehisce (Sunderland & Roberts, 1977). Pollen is shed into the medium and although it has probably only just started to divide, it will continue to grow if the anthers are removed from the culture vessel. The shed pollen will not develop, however, if it is

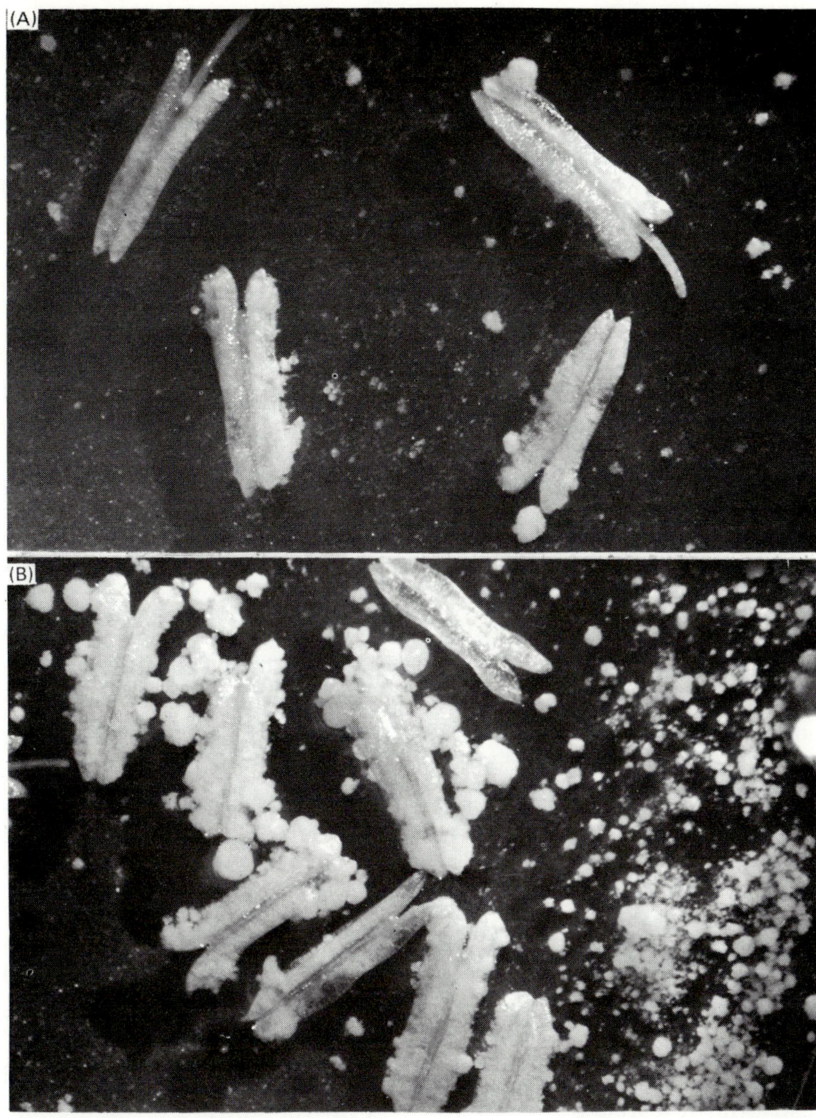

Fig. 7. Anthers of Sabarlis barley floating on (A) liquid potato medium (Sunderland *et al*. 1979) and (B) the same medium conditioned by Sabarlis anthers for 7 days. Ten anthers in 1 ml medium in 30-mm petri dishes for 28 days at 25 °C. Spikes pretreated at 7 °C for 14 days.

removed and put into fresh medium. The conditioning factor thus has an important role in the maintenance of growth.

Low culture yields are still a problem in barley despite the high induction frequencies. The best medium so far devised contains, in addition to the usual spectrum of minerals and vitamins, 9% sucrose (to maintain osmotic stability), hormones (to trigger division?) and an aqueous extract of boiled potato tuber (Chuang *et al.*, 1978). The active ingredient/s of the latter is/are not known. The observations on conditioned media prompted one of my colleagues, Xu Zhi-hong, to try conditioning with Sabarlis barley anthers. Fig. 7 indicates the spectacular results that can be obtained by a conditioning of the potato medium for only 7 days by Sabarlis anthers. Much of the pollen is shed into the conditioned medium within the first 24 h and it will continue to grow if the anthers are subsequently removed.

The conditioning effect is highly reproducible so long as Sabarlis anthers are used. A few other barley genotypes and cereals like maize, rye and oats have been tried as conditioning agents with varying degrees of success. Another colleague, Miss Huang Bin, finds that Sabarlis barley ovaries will also condition the medium whereas the rest of the spikelet is largely ineffective. The conditoning factor is thermolabile and thus probably unrelated to the potato factor. The role of the conditioning factor is now being evaluated with respect to both the triggering of division and the maintenance of growth.

The fact that the anther tissues contribute materials essential for growth of the morphogenic pollen grains and possibly also triggering agents may account for the inability of many species to respond to anther culture. Oats (*Avena sativa*) is a notable example of a species which produces pollen variants *in vivo* and in pretreated anthers. *Nicotiana stocktonii* is another. Anthers of neither species respond to the usual culture media. It may be that these species do not produce the essential wall factors.

It is a salutory thought that after more than 10 years of intensive and frustrating research by many workers on the culture of barley and other cereal anthers, the answer to both high induction frequencies and high culture yields lies with the anther itself and exploitation of the anther's own resources.

Acknowledgement

My thanks are due to Dr J. M. Dunwell for providing much of the literature survey given in Table 1.

REFERENCES

CHAMBERLAIN, C. J. (1897). Life history of *Lilium philadelphicum*. II. The pollen grain. *Botanical Gazette*, **23**, 423–30.

CHARZYNSKA, J., MALESZKA, J. & HON, B. (1977). Anomalous pollen grain development after nondifferentiating microspore division in *Eremurus*. *Acta Societatis Botanicorum Poloniae*, **46**, 459–69.

CHUANG, C-C, OUYANG, T-W., CHIA, H., CHOU, S-M. & CHING, C-K. (1978). A set of potato media for wheat anther culture. In *Proceedings of Symposium on Plant Tissue Culture*, pp. 51–6. Peking: The Science Press.

CLAPHAM, D. H. (1977). Haploid induction in cereals. In *Applied and Fundamental Aspects of Plant Cell, Tissue, and Organ Culture*, ed. J. Reinert & Y. P. S. Bajaj, pp. 279–98. Berlin: Springer-Verlag.

CORDUAN, G. (1975). Regeneration of anther-derived plants of *Hyoscyamus niger* L. *Planta*, **127**, 27–36.

DALE, P. J. (1975). Pollen dimorphism and anther culture in barley. *Planta*, **127**, 213–20.

DARLINGTON, C. D. & MATHER, K. (1949). In *The Elements of Genetics*, p. 196, London: Allen & Unwin.

DARLINGTON, C. D. & THOMAS, P. T. (1941). Morbid mitosis and the activity of inert chromosomes in *Sorghum*. *Proceedings of the Royal Society*, **B130**, 127–50.

DE MOL, W. E. (1921). On the influence of circumstances of culture on the habitus and partial sterility of the pollen grains of *Hyacinthus orientalis*. Proceedings, *Koninklijke Nederlanse Akademie van Wetenschappen*, **23**, 1–14.

DEVREUX, M. (1971). New possibilities for the *in vitro* cultivation of plant cells. *Eurospectra*, **9**, 105–10.

DUNWELL, J. M. (1976). A comparative study of environmental and developmental factors which influence embryo induction and growth in cultured anthers of *Nicotiana tabacum*. *Environmental and Experimental Botany*, **16**, 109–18.

DUNWELL, J. M. & SUNDERLAND, N. (1974). Pollen ultrastructure in anther cultures of *Nicotiana tabacum*. II. Changes associated with embryogenesis. *Journal of Experimental Botany*, **25**, 363–73.

DUNWELL, J. M. & SUNDERLAND, N. (1975). Pollen ultrastructure in anther cultures of *Nicotiana tabacum*. III. The first sporophytic division. *Journal of Experimental Botany*, **26**, 240–52.

ENGVILD, K. C. (1974). Plantlet ploidy and flower-bud size in tobacco anther culture. *Hereditas*, **76**, 320–2.

ENGVILD, K. C., LINDE-LAURSEN, I. & LUNDQVIST, A. (1972). Anther cultures of *Datura innoxia*: flower bud stage and embryoid level of ploidy. *Hereditas*, **72**, 331–2.

FEDORTSCHUK, W. (1931). Embryologische Untersuchung von *Cuscuta monogyna* Vahl und *Cuscuta epithymum* L. *Planta*, **14**, 94–111.

FULLMER, E. L. (1899). The development of the microsporangia and microspores of *Hemerocallis fulva*. *Botanical Gazette*, **28**, 81–8.

GEITLER, L. (1935). Beobachtungen über die erste Teilung im Pollenkorn der Angiospermen. *Planta*, **24**, 361–86.

GUPTA, S. C. & BARBAR, S. B. (1980). Enhancement of plantlet formation in anther culture of *Datura metel* L. by pre-chilling of buds. *Zeitschrift für*

Pflanzenphysiologie, **96**, 465–70.

HORNER, M. & STREET, H. E. (1978). Pollen dimorphism – origin and significance in pollen plant formation by anther culture. *Journal of Experimental Botany*, **42**, 763–71.

KASPERBAUER, M. J. & COLLINS, G. B. (1972). Reconstitution of diploids from leaf tissue of anther-derived haploids in tobacco. *Crop Science*, **12**, 98–101.

LA COUR, L. F. (1949). Nuclear differentiation in the pollen grain. *Heredity*, **3**, 319–37.

MASCARENHAS, J. P. (1975). The biochemistry of Angiosperm pollen development. *Botanical Reviews*, **41**, 259–314.

MIX, G., WILSON, H. M. & FOROUGHI-WEHR, B. (1978). The cytological status of plants of *Hordeum vulgare* L. regenerated from microspore callus. *Zeitschrift für Pflanzenzüchtung*, **80**, 89–99.

NAITHANI, S. P. (1937). Chromosome studies on *Hyacinthus orientalis* L. III. Reversal of sexual state in the anthers of *Hyacinthus orientalis* L., var. Yellow Hammer. *Annals of Botany*, **1**, 370–7.

NĚMEC, B. (1898). Über den Pollen der Petaloiden Antheren von *Hyacinthus orientalis* L. *Bulletin International de l'Acadèmie des Sciences Bohème*, **5**, 17–23.

NITSCH, C. (1974). Pollen culture – a new technique for mass production of haploid and homozygous plants. In *Haploids in Higher Plants: Advances and Potential*, ed. K. J. Kasha, pp. 123–35. Guelph: University of Guelph Press.

NITSCH, C. & NORREEL, B. (1973). Effet d'un choc thermique sur le pouvoir embryogène de pollen de *Datura innoxia* cultivé dans l'anthère ou isolé de l'anthère. *Comptes Rendu Hebdomadaire des Séances de l'Acadèmie des Sciences. Paris*, **276**, 303–6.

NITZSCHE, W. & WENZEL, G. (1977). Haploids in Plant Breeding. *Fortschritte der Pflanzenzüchtung*, Supplement No. 8, 1–80.

OLIVER, F. W. (1890). On *Sarcodes sanguinea*, Torr. *Annals of Botany*, **IV**, 303–26.

PIECH, K. (1924). Über die Teilung des primaren Pollenkorns und die Entstehung der Spermazellen bei *Scirpus paluster* L. *Bulletin International de l'Acadèmie Polonaise des Sciences et des Lettres, Serie B*, Botanique, 605–21.

PODDUBNAJA-ARNOLDI, V. (1936). Beobachtungen über die Keimung des Pollens einiger Pflanzen auf Künstlichem Nährboden. *Planta*, **25**, 502–29.

RAGHAVAN, V. (1976). Role of the generative cell in androgenesis in henbane. *Science*, **191**, 388–9.

RAGHAVAN, V. (1978). Origin and development of pollen embryoids and pollen calluses in cultured anther segments of *Hyoscyamus niger* (henbane). *American Journal of Botany*, **65**, 984–1002.

RAMANNA, M. S. (1974). The origin and *in vivo* development of embryoids in the anthers of *Solanum* hybrids. *Euphytica*, **23**, 623–32.

RASHID, A. & STREET, H. E. (1974). Segmentations of microspores of *Nicotiana sylvestris* and *N. tabacum* which lead to embryoid formation in anther culture. *Protoplasma*, **80**, 323–34.

SANGWAN-NORREEL, B. S. (1979). Données nouvelles sur l'androgenèse chez le *Datura* et le Tabac. *Bulletin Société Botanique de France. Lettres botaniques*, **126**, 413–43.

SAUTER, J. (1969). Cytochemische Untersuchung der Histone in Zellen mit unterschiedlicher RNA-und Protein-Synthese. *Zeitschrift für Pflanzenphysiologie*, **60**, 434–49.

SAX, K. (1935). The effect of temperature on nuclear differentiation in microspore development. *Journal of the Arnold Arboretum, Harvard University*, **16**, 301–10.

SCHÜROFF, P. N. (1921). Über die Teilung des generativen Kerns vor der Keimung des Pollenkorns. *Archiv für Zellforschung*, **15**, 145–59.

SMITH, W. R. (1898). A contribution to the life history of the Pontederiacae. *Botanical Gazette*, **25**, 324–37.

SOPORY, S. K. & MAHESHWARI, S. C. (1976). Development of pollen embryoids in anther cultures of *Datura innoxia*. II. Effects of growth hormones. *Journal of Experimental Botany*, **27**, 58–68.

STOW, I. (1930). Experimental studies on the formation of the embryo-sac-like giant pollen grains in the anther of *Hyacinthus orientalis*. *Cytologia*, **1**, 417–39.

SUNDERLAND, N. (1970). Pollen plants and their significance. *New Scientist*, **47**, 142–4.

SUNDERLAND, N. (1974). Anther culture as a means of haploid induction. In *Haploids in Higher Plants: Advances and Potential*, ed. K. J. Kasha, pp. 91–122. Guelph: University of Guelph Press.

SUNDERLAND, N. (1977). Observations on anther culture of ornamental plants. In *La Culture des Tissus et des Cellules des Végétaux: Travaux Dédiés à la Memoire de Georges Morel*, ed. R. J. Gautheret, pp. 34–46. Paris: Masson.

SUNDERLAND, N. (1978). Strategies in the improvement of yields in anther culture. In *Proceedings of Symposium on Plant Tissue Culture*, pp. 65–85. Peking: The Science Press.

SUNDERLAND, N. (1979). Comparative studies of anther and pollen culture. In *Plant Cell and Tissue Culture: Principles and Applications*, ed. W. R. Sharp, P. O. Larsen, E. F. Paddock & V. Raghavan, pp. 203–19. Columbus: Ohio State University Press.

SUNDERLAND, N. & DUNWELL, J. M. (1974). Pathways in pollen embryogenesis. In *Tissue Culture and Plant Science* 1974, ed. H. E. Street, pp. 141–67. London: Academic Press.

SUNDERLAND, N. & EVANS, L. J. (1980). Multicellular pollen formation in cultured barley anthers. II. The A, B, and C pathways. *Journal of Experimental Botany*, **31**, 501–14.

SUNDERLAND, N. & ROBERTS, M. (1977). New approach to pollen culture. *Nature*, **270**, 236–8.

SUNDERLAND, N. & ROBERTS, M. (1979). Cold-pretreatment of excised flower buds in float culture of tobacco anthers. *Annals of Botany*, **43**, 405–14.

SUNDERLAND, N. & WICKS, F. M. (1971). Embryoid formation in pollen grains of *Nicotiana tabacum*. *Journal of Experimental Botany*, **22**, 213–26.

SUNDERLAND, N. & WILDON, D. C. (1979). A note on the pretreatment of excised flower buds in the culture of *Hyoscyamus* anthers. *Plant Science Letters*, **15**, 169–75.

SUNDERLAND, N., COLLINS, G. B. & DUNWELL, J. M. (1974). The role of nuclear fusion in pollen embryogenesis of *Datura innoxia* Mill. *Planta*, **117**, 227–41.

SUNDERLAND, N., WILDON, D. C. & EVANS, L. J. (1977). The role of the generative cell in pollen embryogenesis. *John Innes Annual Report*, No. 68, 57–8.

SUNDERLAND, N., ROBERTS, M., EVANS, L. J. & WILDON, D. C. (1979). Multicellular pollen formation in cultured barley anthers. I. Independent division of the generative and vegetative cells. *Journal of Experimental Botany*, **30**, 1133–44.

UPCOTT, M. (1939). The external mechanics of the chromosomes. VII. Abnormal mitosis in the pollen grain. *Chromosoma*, **1**, 178–90.

VASIL, I. K. (1973). The new biology of pollen. *Naturwissenschaften*, **60**, 247–53.

WENZEL, G. & THOMAS, E. (1974). Observations on the growth in culture of anthers of *Secale cereale*. *Zeitschrift für Pflanzenzüchtung*, **72**, 89–94.

The long-term maintenance of differentiated tissues *in vitro*: paracetamol metabolism and toxicity in organ culture*

MICHAEL BALLS, MARIA ANA SANTOS† AND
RICHARD CLOTHIER

Department of Human Morphology, University of Nottingham Medical School,
Queen's Medical Centre, Nottingham NG7 2UH

The process of cellular differentiation involves not only the development of tissue-type specificity of structure and function in cell populations during embryogenesis and organogenesis, but also their maintenance in the mature organism. *In vitro* methods provide opportunities for studies on the control mechanisms involved in the control of cell differentiation and function in particular tissues in isolation from the systemic factors operating in the whole organism, and for investigations on the direct effects of hormones, drugs, toxins and other chemicals. However, although it is often said that tissue culture systems are simple in comparison with the *in vivo* situation, the setting-up of cultures, the design of meaningful experiments, and the critical evaluation and proper use of the results obtained in fact involve great conceptual and technical problems. In this article we shall discuss some of these problems in relation to the use of long-term organ culture in studies on drug metabolism and drug-induced tissue damage, with particular emphasis on interactions between paracetamol and adult amphibian liver cultures.

Organ culture involves 'the maintenance or growth (for more than 24 hours in a nutrient medium) of tissues, organ primordia, or the whole or parts of an organ *in vitro* in a way that may allow differentiation and preservation of architecture and/or function' (Federoff, 1967). The method has been especially useful in developmental biology (partly because embryonic tissues survive much better than adult tissues), for example, in studies on the development of cartilage and bone (Fitton-Jackson, 1976; Reynolds, 1976), the effects of vitamins (Fell, 1964), and on epithelio-mesenchymal interactions (Saxén & Karkinen-

* This work was supported by the Humane Research Trust and the Servico de Ciencia, Fundação Calouste Gulbenkian.
† Present address: Instituto de Ciências Biomédicas 'Abel Salazar', Praça da Escola Médica, Porto 4000, Portugal.

Jääskeläinen, 1975). Organ culture is especially useful in endocrinology (Fell, 1976), and is now an important complementary technique to cell culture in animal virology, since the replication of some viruses is strictly dependent on the specific features of certain host cells (Bucknall, 1980). Organ culture thus 'provides a convenient intermediate between the complexities of the whole organism and the simplicity of cell cultures' (Reed, 1976).

There have been few systematic attempts to use organ cultures in investigations on the pharmacokinetics, pharmacodynamics, or toxicity of particular drugs, largely because cultured fragments of such major adult mammalian organs as the liver, pancreas or kidney survive in culture for only very brief periods. The insuperable problem appears to be the provision of oxygen to the cells within cultured tissue fragments, where simple diffusion from the culture medium is inadequate (Hodges, 1976). By contrast, we have found that fragments of a range of organs from individuals of a number of amphibian species can survive and function in organ culture for several weeks (Balls, Brown & Fleming, 1976) – without the increased oxygen tension, high serum concentration, added insulin, gas–medium interface raft or platform culture, or complex circumfusion procedures which have been used in attempts to prolong the survival of mammalian cultures (Hodges, 1976). The longer survival of amphibian tissues probably results mainly from their lower metabolic rates and therefore lower oxygen requirement (Monnickendam & Balls, 1973). The potential value of long-term amphibian organ culture in biomedical research, including its use in studies in the general and specific effects of drugs, was recently reviewed by Balls & Rao (1980).

Since its first inclusion in the British Pharmacopoeia in 1963, the popularity of paracetamol (acetominophen; *N*-acetyl-*p*-aminophenol) as an analgesic has increased steadily, and by 1974 3000 million 500-mg tablets were used in the UK alone (Spooner & Harvey, 1976). Though the drug is safe and effective in normal therapeutic use, the number of hospital admissions resulting from overdosage increased from about 150 in 1968 to 7000 by 1973 (Volans, 1976). The most serious effect of overdosage is hepatic necrosis. At normal dose levels in man, paracetamol is overwhelmingly excreted in the urine either unchanged or as two metabolites of low toxicity, paracetamol sulphate (Fig. 1, reaction 1) and paracetamol glucuronide (reaction 2). As a result of the work of Brodie, Mitchell and their colleagues (Jollow *et al.*, 1973; Mitchell *et al.*, 1973a, b, 1974; Potter *et al.*, 1973), it is generally

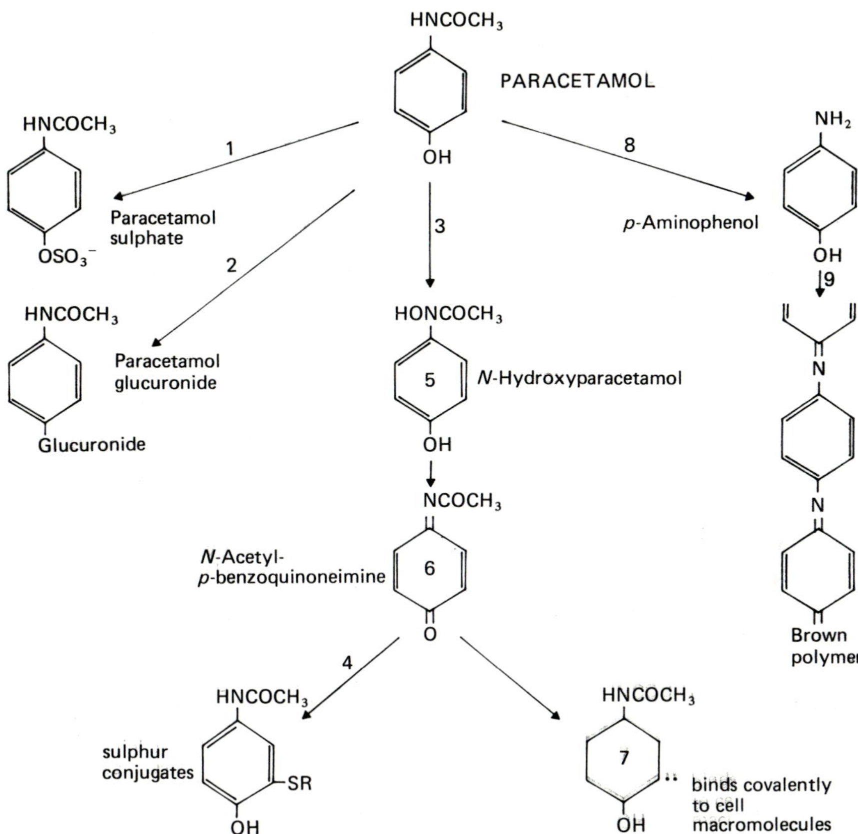

Fig. 1. Summary of paracetamol metabolism in mammals and urodele amphibians (Smith & Griffiths, 1976; Hinson, Nelson & Mitchell, 1977; Healey *et al.*, 1978; Clothier *et al.*, 1981).

considered that the oxidation of paracetamol via the monoxidase system (reaction 3) is more significant in the hepatotoxicity resulting from overdosage. Non-toxic glucuronides and sulphate conjugates can still be produced (e.g. reaction 4) from the intermediate metabolites produced in this way (5 and 6), but following overdosage glutathione, sulphate and glucuronide pools become exhausted so that a toxic metabolite (7) is formed, which binds covalently to various macromolecules in the hepatocyte, resulting in damage.

We have recently found that various urodele amphibian tissues, including liver, pancreas and kidney fragments *in vitro*, deacetylate paracetamol to *p*-aminophenol (reaction 8), which then spontaneously polymerises to form a brown substance (reaction 9) (Clothier *et al.*,

1981). Since significant amounts of *p*-aminophenol are produced from paracetamol in rats (Smith & Griffiths, 1976) and since *p*-aminophenol causes tissue damage (Calder *et al.*, 1971, 1975; Crowe *et al.*, 1977), this metabolite may also be involved in the toxic effects resulting from paracetamol overdosage. The aim of the experiments described in this article was to establish whether long-term amphibian organ cultures could be used to further understanding of the basis of the toxicity of high doses of paracetamol, and, in particular, whether *p*-aminophenol induced changes in cultured tissues, and whether tissues exposed to paracetamol could subsequently respond to other drugs or hormones.

ORGAN CULTURE MEDIA AND METHODS

Removal of the organ to be cultured and its subdivision into fragments of a size suitable for culture must be carried out as quickly and carefully as possible, in order to reduce to a minimum the damage which inevitably results. A single *Amphiuma means* (Congo-eel) liver can provide about 500 replicate cultures, each containing 10 fragments. In our experience this setting-up takes four workers about three hours, but the last cultures to be completed are no worse than the first. Organ culture on this scale could not be carried out with a mammalian liver, since the preparation time would be too great a part of the total survival time, and the viability of the last cultures to be established would be considerably reduced.

Importance of size of cultured fragments

Fragment size is important; when cultured fragments are much too large, central necrosis occurs, but more subtle changes resulting in confusing metabolic variations within the fragment can also occur. For example, the newt liver fragment shown in Fig. 2 consists of an outer region of cells containing glycogen, and a central region of healthy cells which have broken down the glycogen, presumably to provide glucose for energy production by glycolysis.

The penetration of drugs or other agents must also be taken into account, particularly if results are calculated on the assumption that all cells in the fragment were equally affected. Roller tube cultures were set up containing approximately the same amount of *A. means* liver, but cut into 4 large, 8 normal-size, or 16 small fragments; 250 µg/ml paracetamol was added and media were sampled at intervals up to 96 h and assayed for paracetamol concentration. After 24 h (Table 1), less

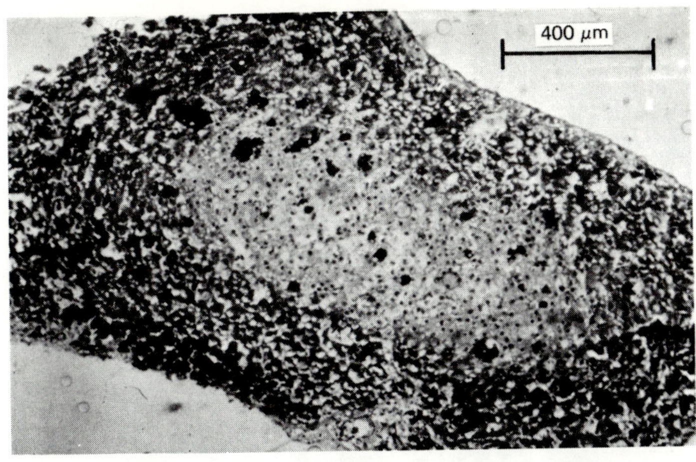

Fig. 2. Liver fragment from *Triturus cristatus carnifex* after 72 h in culture, stained by the Periodic Acid/Schiff (PAS) method to show an outer zone of cells containing glycogen and a central, glycogen-free area.

Table 1. *Paracetamol metabolism by* A. means *liver cultures of similar total weight but different fragment size*

Size	Time (h)	Paracetamol metabolised Total (μg)	μg/mg	p-Aminophenol polymer produced $(\Delta OD_{610} \times 10^3/mg)$
Large	24	298 ± 12	4.3 ± 0.2	8.4 ± 0.5
Normal		371 ± 20	5.8 ± 0.1	13.4 ± 0.4
Small		434 ± 17	7.5 ± 0.2	16.4 ± 0.8
Large	96	862 ± 7	13.3 ± 0.7	25.4 ± 0.8
Normal		857 ± 15	12.7 ± 0.3	27.0 ± 1.1
Small		853 ± 4	14.8 ± 0.1	31.5 ± 2.0

As elsewhere in this chapter, unless otherwise stated, tissues were cultured according to the methods of Balls *et al.* (1976), at 25 °C in roller tubes containing 4 ml medium (50% MEM, 40% water, 10% fetal calf serum, with 15 mM HEPES, and antibiotics). Results are given as the mean ± standard error of the mean ($n = 4$). Medium paracetamol concentration was determined by the ABTS method and p-aminophenol polymer production by a spectrophotometric method of Clothier *et al.* (1981). Calculations were based on wet weight of tissue.

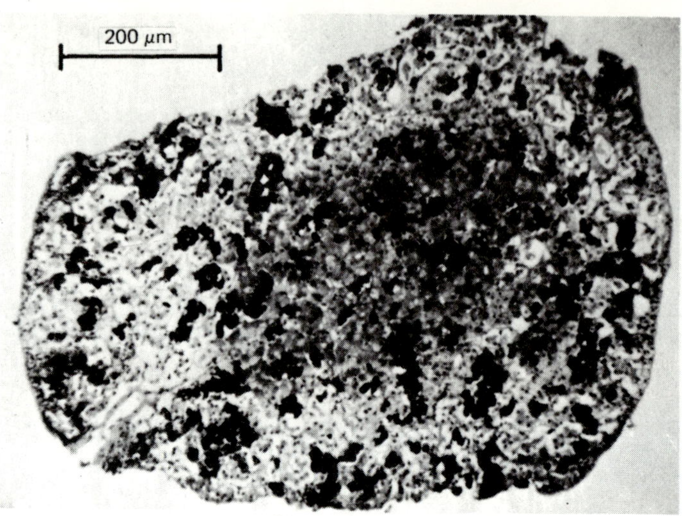

Fig. 3. Large *A. means* liver fragment after 96 h in medium initially containing 250 µg/ml paracetamol, stained by the PAS method, and showing glycogen loss from the outer region and a central area rich in glycogen. Glycogen loss occurred throughout normal-sized and small fragments.

paracetamol had been metabolised in tubes containing large fragments, but by 96 h almost all the available drug had been metabolised in all the culture tubes. When histological sections of the cultured liver were stained by the PAS method, it was found that the small fragments contained no glycogen and almost all the glycogen had been lost from the normal-sized fragments, whereas large amounts of glycogen were still present at the centres of the larger pieces (Fig. 3). Since paracetamol rapidly induces glycogen breakdown *in vitro* (Gater, Brown & Balls, 1976), these results suggest that the proportion of cells in the cultures active in paracetamol metabolism varied according to fragment size. Further indications of the importance of drug penetration and the effects of fragment size were obtained from an experiment with kidney cultures. The brown polymer of *p*-aminophenol accumulates in the cytoplasm of the tubules of *A. means* kidney cultures exposed to paracetamol, but tubules toward the centres of larger fragments remained free of such accumulations.

Effects of culture agitation

Mammalian organ culture methods usually involve the maintenance of tissues at the air–medium interface on solid substrates or on rafts or

Table 2. *Metabolism of paracetamol by agitated and stationary rat liver preparations*

Group	Mean weight of tissue (µg)	Paracetamol metabolised after 6 h	
		Total (µg)	µg/mg
Agitated	86 ± 6	154 ± 14	1.8 ± 0.2
Stationary	94 ± 8	87 ± 8	0.9 ± 0.1

Tissues were maintained in 4 ml medium (90% MEM, 10% fetal calf serum, 15mM HEPES) containing 250 µg/ml paracetamol, in 25 ml conical flasks ($n = 6$), in a water bath at 37 °C, with a 5% CO_2/95% O_2 gas phase. Agitated cultures were shaken linearly at 2 cycles/s.

platforms above a fluid medium (Hodges, 1976). We recently designed a simple apparatus for the short-term maintenance of rat liver fragments in submerged culture in Erlenmeyer flasks in a shaking water bath. We were surprised to find that agitation of the flasks had such a dramatic effect on the amount of paracetamol metabolised during a 6-h experimental period (Table 2), presumably due to improved medium circulation and gaseous exchange (Hodges, 1976).

Choice of basic culture medium

Chemically defined basic culture media consist of a buffered salt solution, amino acids, vitamins, and an energy source, usually glucose. Such media were originally designed to promote survival and growth in cell monolayer cultures, rather than for the maintenance of specific functions in organ cultures. For example, L15 medium (Leibovitz, 1963) appears to be ideal for amphibian cell culture (Balls & Ruben, 1965); it contains large quantities of free-base amino acids which provide an effective zwitterionic buffering system, allowing cultures to be maintained in free gaseous exchange with the atmosphere. The need for the 5% CO_2/95% air gas phase required for the more conventional CO_2-bicarbonate buffering system can also be avoided by using minimum essential medium (MEM) buffered with HEPES (Fleming, Brown & Balls, 1975). Since hepatocytes monitor the circulating blood and regulate the levels of many factors, including amino acids, it is not surprising that total nitrogenous excretion is much higher in L15 than in MEM (Fig. 4). *A. means* liver fragments show adaptability in long-term culture, since when transferred from L15 to MEM after 4 or 8 days

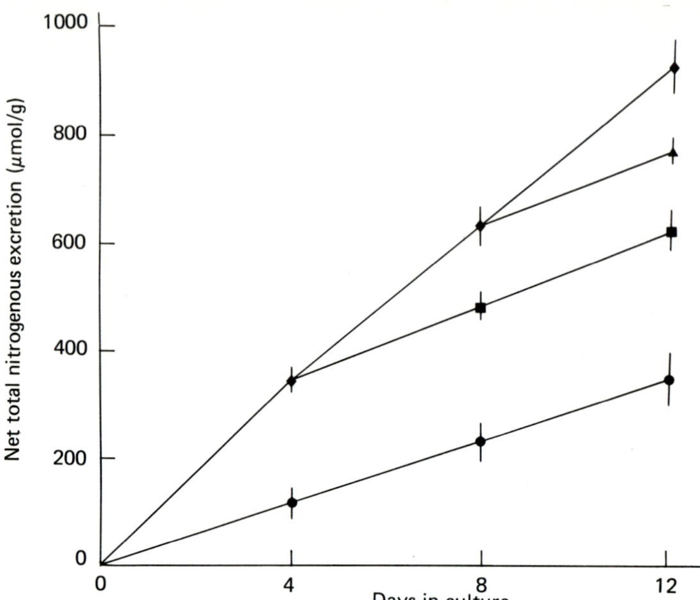

Fig. 4. Net total nitrogenous excretion by *A. means* liver cultures maintained for 12 days in MEM (●) or L15 (◆), or cultured for 4 (■) or 8 (▲) days in L15 then in MEM (*n* = 4). Media were changed after 4 and 8 days. Similar curves were obtained for net total urea and free ammonia production.

they reverted to the rate of nitrogenous excretion shown by other cultures in MEM.

To confirm that this increase in nitrogenous excretion was due to the high amino acid concentration in L15, *A. means* liver fragments were maintained in glucose-free MEM containing additional non-essential amino acids at final concentrations equivalent to those in L15. Phenformin (an oral hypoglycaemic agent, which reduces urea and total nitrogenous excretion while increasing free ammonia release – Table 3) was added to some groups, to illustrate the complexity which can result from combined medium and drug effects. Nitrogenous excretion and urea production were significantly higher in media containing extra amino acids, but reduced in media containing phenformin (Table 4). Intermediate values were obtained for cultures containing both extra amino acids and phenformin. Both factors increased lactate output, and these effects appear to have been additive.

The role of serum

In attempts to improve the viability and quality of cell and organ cultures, many defined and undefined additives have been included in

Table 3. *Effects of 10^{-6}M phenformin on nitrogenous excretion by* A. means *liver cultures*

	Total excretion during a 16 day period (µmol/g)		
Medium	Total nitrogen	Free ammonia	Urea
L15	957 ± 17	187 ± 25	385 ± 18
L15 + phenformin	756 ± 28	363 ± 9	189 ± 10

Culture media (50% L15, 40% water, 10% fetal calf serum, plus antibiotics) were changed on days 4, 8 and 12 ($n = 4$). Medium urea and ammonia concentrations were determined with a Boehringer test combination.

Table 4. *Effects of extra amino acids on nitrogenous excretion and lactate production by* A. means *liver cultures with and without* 10^{-6}M *phenformin*

Group	Total nitrogen production (µmol/g)	Free ammonia release (µmol/g)	Urea production (µmol/g)	Lactate production (µg/mg)
GFMEM	148 ± 13	60 ± 4	44 ± 6	32 ± 3
+AA	351 ± 37	69 ± 4	141 ± 18	55 ± 10
+PHE	115 ± 8	77 ± 9	19 ± 1	61 ± 7
+AA+PHE	257 ± 26	75 ± 8	91 ± 10	90 ± 9

Groups of 4 cultures were maintained for 8 days in glucose-free MEM (plus serum, HEPES and antibiotics) with or without added non-essential amino acids (AA-Flow Laboratories Ltd, Irvine, Scotland) and phenformin (PHE). Media were changed on day 4. Medium lactate concentration was estimated with a Boehringer test combination.

culture media, including insulin and serum (Hodges, 1976). The main problems related to the use of sera result from their variability and, more recently in the case of fetal calf serum, availability. *A. means* liver cultures can survive for weeks in serum-free media, though glycogen loss is more rapid than in media containing serum (Balls *et al.* 1976), and the absence of serum (like the presence of phenformin) results in a reduction in urea and total nitrogenous excretion, but an increase in free ammonia release (Table 5).

Culture media and paracetamol metabolism

Our preliminary results indicate that paracetamol metabolism also appears to be significantly affected by the choice of basic culture

Table 5. *Nitrogenous excretion by* A. means *liver cultured in L15 or MEM with or without 10% fetal calf serum*

	Total nitrogenous excretion during a 24-day period (μmol/g)		
	Total nitrogen	Free ammonia	Urea
L15 + serum	1470 ± 89	295 ± 11	589 ± 48
L15 − serum	1203 ± 68	379 ± 22	381 ± 9
MEM + serum	653 ± 10	184 ± 15	235 ± 11
MEM − serum	438 ± 16	242 ± 7	98 ± 7

Media were changed on days 4, 8, 12, 16 and 20 ($n = 4$).

Table 6. *Effects of basic medium and serum on paracetamol metabolism by* T. c. carnifex *liver cultures*

	Paracetamol metabolised		p-Aminophenol polymer produced ($\Delta OD_{610} \times 10^3/mg$)
Medium	Total (μg)	μg/mg	
L15 + S	223 ± 9	6.7 ± 0.2	20.7 ± 0.5
MEM + S	379 ± 7	13.0 ± 1.0	30.9 ± 1.8
MEM − S	370 ± 10	12.6 ± 0.6	19.0 ± 1.0

Cultures were maintained for 48 h in media containing 250 μg/ml paracetamol ($n = 4$; + S = with 10% fetal calf serum).

medium and the presence or absence of serum (Table 6). *T. c. carnifex* (Italian great crested newt) liver cultures metabolised less paracetamol in L15 (with 10% fetal calf serum) than in MEM (with or without serum). However, though the amounts of paracetamol metabolised by cultures in MEM with or without serum were similar, less *p*-aminophenol polymer was produced in serum-free medium. Since non-polymerised *p*-aminophenol, but not the polymer, appears to cause tissue damage (p. 38) slower polymerisation would be expected to result in increased tissue damage.

VIABILITY, NORMALITY AND TOXICITY

The removal of organs from the body, their dissection into fragments, and their culture in artificial media under imperfect conditions inevitably leads to changes from the *in vivo* state and to slow (*A. means* liver) or rapid (rat liver) degeneration. Since the cells within fragments cannot be seen, as can cells in monolayer culture, it is very difficult to assess their *viability*. Sample cultures can be subjected to structural, ultrastructural, biochemical or physiological analysis, but this takes time and uses tissue. Similar problems relate to the *normality* of particular cultures and their *comparability* with other cultures of the same type and with the equivalent tissues *in vivo*.

Microanalysis of culture media provides for the relatively rapid assessment of tissue condition – for example, urea, ammonia and lactate production, glucose uptake or release, and the release of lactate dehydrogenase (LDH) or transaminases can provide a base-line of general information about the state of liver cultures, against which the results of specific experiments can be judged. Indeed, some of the analytical methods concerned were specifically designed for assessing liver function and damage *in vivo*. For example, damage to only 1% of the hepatocytes will markedly increase the plasma level of glutamate pyruvate transaminase (APT) (Iber, 1974). The kidneys are not merely organs of excretion but are also major homeostatic regulators, endocrine organs and sites of major metabolic activity. Some of these functions (e.g. renin production, gluconeogenesis) continue for long periods in amphibian organ cultures. Like the liver, the kidneys can also be damaged by drugs, including phenacetin and paracetamol overdosage (Crowe *et al.*, 1977; Healey *et al.*, 1978). A number of routine tests give indications of the integrity of renal function and the state of renal tissues (Foulkes & Hammond, 1975). For example, increases in urine alkaline phosphatase, acid phosphatase and LDH indicate renal damage.

Any evidence of synthesis *in vitro* (e.g. insulin and amylase synthesis in pancreas cultures, renin synthesis in kidney cultures, glycogenesis in liver cultures) strengthens the culturist's faith in the general validity of the system being used. However, the experimental toxicologist has the problem of distinguishing between the effects of general degeneration and the specific toxic effects of the chemical being studied. For example, glycogen loss is a good indicator of liver damage, which is induced in *A. means* liver cultures by ethanol, allyl alcohol and carbon tetrachloride, as well as by paracetamol (Gater, 1976); but, glycogenolysis can also result from donor- and culture-related factors, such as health of the animal used, high initial glycogen content, overlarge fragment size, and quality of serum (Balls & Rao, 1980).

Table 7. *Effects on* A. means *liver cultures of single and repeated exposure to media containing paracetamol*

Group	Paracetamol metabolised (μg/mg)	p-Aminophenol polymer produced ($\Delta OD_{610} \times 10^3$/mg)	Tissue glycogen content (%)	Tissue LDH content (mU/mg)
Control	0	0	6.1 $\pm$ 0.1	14.2 $\pm$ 0.2
*S-1 day	7.0 $\pm$ 0.4	12.0 $\pm$ 1.1	3.0 $\pm$ 0.4	8.0 $\pm$ 1.0
S-2 days	10.0 $\pm$ 1.0	17.0 $\pm$ 1.3	4.0 $\pm$ 1.3	8.5 $\pm$ 1.0
S-4 days	15.1 $\pm$ 2.2	25.3 $\pm$ 3.8	2.1 $\pm$ 0.1	6.4 $\pm$ 2.0
Control	0	0	5.6 $\pm$ 0.4	14.0 $\pm$ 0.3
†1-48 h	10.4 $\pm$ 1.4	14.7 $\pm$ 2.2	3.7 $\pm$ 0.6	7.9 $\pm$ 1.3
2-48 h	16.2 $\pm$ 1.4	23.8 $\pm$ 1.9	2.2 $\pm$ 0.1	7.8 $\pm$ 0.7
3-48 h	25.2 $\pm$ 2.2	40.6 $\pm$ 3.8	0.8 $\pm$ 0.2	4.0 $\pm$ 1.4
4-48 h	31.7 $\pm$ 1.7	62.8 $\pm$ 3.4	0.2 $\pm$ 0.1	3.4 $\pm$ 0.5

* Exposure to a single initial dose of 250 μg/ml paracetamol for 1, 2, or 4 days.

† Exposure to media initially containing 250 μg/ml paracetamol for 1, 2, 3 or 4 48-h periods.

Tissue glycogen was estimated by the method of Monnickendam, Brown & Balls (1974), lactate dehydrogenase (LDH) activity with a Boehringer test combination.

Further questions to be considered are related to calculation of physiological drug concentration, binding of drugs to medium constituents or culture vessels, effective half-life of the drug *in vitro*, numbers of treatments, intervals between repeated treatments, intervals between treatment and assessments of effects, and intervals between culture medium changes and the effects of such changes.

Toxicological studies would, in general, be expected to require longer intervals between exposure and the assessment of effects than in the case of, for example, the effect of drugs on blood pressure or rate of heart beat. This is because chemicals may have to be metabolised to an active metabolite, the toxic effects of which are dependent on concentration, time, and levels and activities of enzymes involved in detoxification. Paracetamol and *A. means* cultures are well suited to this pilot study, since the concentration and metabolism of the former can be readily and rapidly monitored (Clothier *et al.*, 1981) and the latter survive and function *in vitro* for relatively long periods.

Table 8. *Effects of a series of paracetamol concentrations on* A. means *kidney cultures after 48 h*

Paracetamol concentration (μg/ml)	Paracetamol metabolised (μg/mg)	p-Aminophenol polymer produced ($\Delta OD_{610} \times 10^3$/mg)	Lactate output (μg/mg)	Glucose uptake (μg/mg)	Tissue LDH content (mU/mg)
0	0	0	28.3 $\pm$ 3.1	27.4 $\pm$ 1.9	17.9 $\pm$ 1.4
50	7.0 $\pm$ 0.8	7.2 $\pm$ 0.6	25.2 $\pm$ 1.8	25.3 $\pm$ 1.2	19.4 $\pm$ 1.1
150	14.6 $\pm$ 1.4	21.9 $\pm$ 1.3	15.6 $\pm$ 0.4	13.2 $\pm$ 1.4	5.0 $\pm$ 0.7
250	24.1 $\pm$ 1.6	31.6 $\pm$ 2.8	15.8 $\pm$ 1.2	9.9 $\pm$ 0.8	2.2 $\pm$ 0.1

$n = 4$; medium glucose concentration was determined by the Sigma GOD/POD method.

Table 9. *Effects of paracetamol-derived p-aminophenol polymer on A.* means *liver and kidney cultures*

Parts of media combined*					
Fresh medium	Control culture medium	Medium from paracetamol cultures	Liver glycogen content (%)	Liver LDH content (mU/mg)	Kidney LDH content (mU/mg)
4	0	0	5.0 ± 0.4	12.7 ± 1.1	19.9 ± 0.6
2	0	2	5.7 ± 0.4	11.4 ± 0.8	19.8 ± 1.3
3	0	1	6.3 ± 0.5	11.5 ± 0.5	20.5 ± 1.0
2	2	0	4.5 ± 0.2	11.2 ± 0.8	19.1 ± 0.7
3	1	0	4.2 ± 0.4	11.7 ± 0.8	21.7 ± 0.8

$n = 4$; experimental time = 4 days;
* Liver cultures were exposed to 250 µg/ml paracetamol until all the paracetamol had been metabolished and the maximum OD_{610} reached; this medium was then combined with fresh medium, as was medium from control liver cultures.

Paracetamol-related damage to liver and kidney cultures

Gater *et al.* (1976) found that paracetamol treatment resulted in a dose-dependent increase in APT, glutamate oxaloacetate transaminase (AST) and lactate release from *A. means* liver cultures. From the results given in Table 7, we can conclude that reductions in liver glycogen and LDH were related to time following a single dose and to number of exposures to media containing paracetamol. Reductions in lactate output, glucose uptake, and tissue LDH content were related to paracetamol dose in kidney cultures (Table 8).

p-Aminophenol and tissue damage

Since paracetamol is overwhelmingly metabolised by urodele amphibians to *p*-aminophenol which spontaneously polymerises to form a brown-coloured substance, tissue damage could be caused by unmetabolised paracetamol, unpolymerised *p*-aminophenol, or polymerised *p*-aminophenol. Liver and kidney cultures were maintained for four days in media containing different proportions of the polymerised *p*-aminophenol product of paracetamol metabolism. The results (Table 9) provided no evidence that the polymer damaged the cultured tissues, since liver glycogen and liver and kidney LDH levels were not significantly different from those of the control cultures.

Table 10. *Effects on glycogen content of* A. means *and* X. laevis *liver cultures of exposure to media initially containing paracetamol (250 µg/ml) or freshly-dissolved p-aminophenol (180 µg/ml)*

| | A. *means* cultures (48 h) | | X. *laevis* cultures (24 h) | |
| | Polymer | Tissue | Polymer | Tissue |
Group	produced*	glycogen (%)	produced*	glycogen (%)
Control	0	3.9 ± 0.2	0	2.4 ± 0.2
Paracetamol	23.5 ± 1.8	2.8 ± 0.4	0	2.2 ± 0.1
p-Aminophenol	23.2 ± 0.9	1.2 ± 0.1	23.1 ± 0.4	1.4 ± 0.1

$n = 4$; $^*\Delta OD_{610} \times 10^3/mg$.

Liver cultures from *A. means* and from *Xenopus laevis* (the South African clawed toad – an anuran amphibian which does not metabolise paracetamol) were therefore cultured for 48 h and 24 h, respectively, in media containing paracetamol or an equivalent amount of freshly-dissolved *p*-aminophenol. The final levels of glycogen were significantly lowered in the *A. means* liver fragments, but were further significantly reduced in the cultures directly exposed to *p*-aminophenol (Table 10). The glycogen levels of *X. laevis* cultures in medium containing paracetamol were not different from the control culture levels, but the glycogen content of liver fragments directly exposed to *p*-aminophenol was significantly lowered. These results suggest that unpolymerised *p*-aminophenol induces damage in amphibian organ cultures.

SEQUENTIAL TREATMENTS AND RESPONSES

Among the reasons for studying the toxicity of chemicals are the search for a better understanding of the cellular and biochemical bases of their toxic effects, and attempts to find other agents which will prevent tissue damage and/or promote recovery from damage. Our current work on the inhibition of paracetamol metabolism *in vitro* will be discussed elsewhere, but we have found, for example, that the addition of *N*-acetylcysteine (currently the preferred treatment for paracetamol overdosage in man, Prescott *et al.*, 1979) reduces paracetamol metabolism and *p*-aminophenol production by *T. c. carnifex* and rat liver fragments *in vitro*.

The long steady-state period available with many amphibian organ

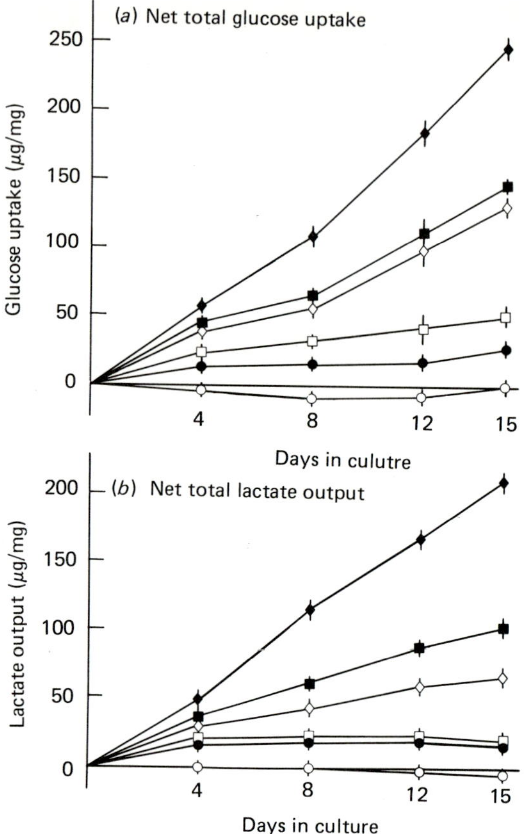

Fig. 5. (a) Net total glucose uptake and (b) net total lactate output by A. *means* kidney cultures maintained for 15 days in media with or without phenformin after previously being cultured for 0, 1 or 4 days in the presence of paracetamol. Groups of 4 cultures without paracetamol (◇◆) or for 24 h (□■) or 96 h (○●) in media initially containing 250 µg/ml paracetamol. These culture media were then replaced with media with (closed symbols) or without (open symbols) 10^{-5}M phenformin. These media were changed on days 4, 8 and 12.

cultures permits experiments involving repeated and intermittent treatment with the same agent, sequential exposure to different agents, and the study of recovery after exposure and damage. In the two experiments which will be discussed, paracetamol was used to change the original state of cultured A. *means* liver and kidney fragments, then other agents were added, to see whether the altered cultures could still respond to them and progress to new states comparable with those attained by control cultures not exposed to paracetamol.

Responses of kidney cultures to phenformin

Paracetamol reduces glucose uptake and lactate release by kidney fragments *in vitro*, whereas phenformin has the opposite effect. Groups of four kidney cultures were exposed to media initially containing 250 µg/ml paracetamol for 24 h or 96 h, then transferred to medium containing 10^{-5}M phenformin. Glucose uptake and lactate output were then monitored during a further 15 days (with medium renewal on days 4, 8 and 12). Kidney cultures treated with phenformin took in more glucose and produced more lactate than cultures in medium without phenformin (Fig. 5); however, cultures previously exposed to paracetamol took in less glucose and produced less lactate than the control cultures, and cultures exposed to paracetamol (and therefore *p*-aminophenol) for 96 h took in and produced less than those exposed to paracetamol for 24 h, and showed a reduced response to the presence of phenformin.

Thus, exposure to paracetamol did not remove the ability of cultured tissues to respond to phenformin, but did affect the basal levels from which responses were made and the magnitude of these responses.

Responses of liver cultures to adrenaline and insulin

Although there are many ways in which glycogenolysis can be induced, glycogen synthesis requires much more specific conditions and events. To investigate the effects of previous exposure to paracetamol on glycogenolysis in response to adrenaline and glycogenesis in response to high glucose/insulin (HGI) medium, 14 groups of 4 cultures were treated in various combinations (Table 11). The results were as follows.

1. Adrenaline induced glycogen breakdown and glucose release into the culture medium (compare groups D and A), but subsequent exposure to HGI medium resulted in increased glucose uptake and an increase in glycogen content (F and E).

2. Exposure to paracetamol for 24 h began to induce a fall in glycogen level, which continued after the cultures were placed in fresh medium (G/H and A/B). Transfer to HGI medium resulted in a significantly higher glucose uptake and glycogen content than would have occurred (I and H).

3. When cultures were exposed to adrenaline after 24 h in paracetamol-containing medium, glycogen level was further reduced and glucose output increased (J and G/H), but subsequent culture for

Table 11. *Effects of adrenaline and insulin on liver cultures previously exposed to paracetamol*

Group	Treatment	Tissue glycogen (%)	Glucose uptake $(+)$ or release $(-)$ (μg/mg)
A	Initial control (ended with D, G, L)	4.5 ± 0.1	-2.3 ± 0.5
B	Final control (ended 72 h later with other groups)	5.2 ± 0.2	$+16.0 \pm 0.6$
C	In HGI medium (final 48 h)	5.8 ± 0.3	$+30.7 \pm 2.3$
D	Adrenaline (24 h)	3.6 ± 0.2	-13.7 ± 1.3
E	Adrenaline then ordinary medium (72 h)	3.4 ± 0.3	$+7.1 \pm 1.4$
F	Adrenaline then HGI medium (48 h)	4.4 ± 0.4	$+30.2 \pm 4.9$
G	Paracetamol (24 h)	4.1 ± 0.2	-4.6 ± 0.7
H	Paracetamol then ordinary medium (72 h)	2.6 ± 0.1	$+1.6 \pm 0.9$
I	Paracetamol then HGI medium (48 h)	4.4 ± 9.2	$+27.7 \pm 0.3$
J	Paracetamol (24 h), adrenaline (24 h), ordinary medium (48 h)	2.3 ± 0.2	-3.1 ± 1.7
K	Paracetamol, adrenaline (24 h), HGI medium (48 h)	3.3 ± 0.2	$+25.3 \pm 2.4$
L	Paracetamol (96 h)	2.1 ± 0.2	-8.4 ± 0.8
M	Paracetamol (96 h) then ordinary medium (72 h)	1.0 ± 0.1	$+4.3 \pm 0.3$
N	Paracetamol (96 h) then HGI medium (48 h)	1.3 ± 0.1	$+2.8 \pm 3.5$

Initial paracetamol concentration, 250 μg/ml; adrenaline concentration, 5×10^{-6}M; HGI medium, 2 mg/ml glucose and 10 mU/ml insulin; cultures in groups F, I and N were maintained in ordinary medium for 24 h before the final period in HGI medium.

48 h in HGI medium resulted in significant increases in tissue glycogen and in glucose uptake (K and J).

4. Exposure to paracetamol for 96 h resulted in a greater fall in glycogen level (L and A/G), which continued after transfer to ordinary medium (M). When such cultures were transferred to HGI medium there were no significant increases in glucose uptake or in glycogen content (N and M).

These results clearly show that cultures could respond to adrenaline and HGI medium after a 24 h exposure to paracetamol. However, cultures exposed to paracetamol and its metabolites for 96 h did not respond to HGI medium.

DISCUSSION

The aim of organ culture is 'to provide an environment that will permit differentiated tissues to exercise their normal functions under the closely controlled conditions obtainable in an *in vitro* system', and 'once this aim has been achieved, many kinds of experiments will become possible which could not be done *in vivo*' (Dame Honor Fell, 1976, in the first BSCB symposium volume). This aim has not been achieved for most adult mammalian tissues, because it has not yet been possible to develop adequate techniques for providing their basic physiological requirements (Franks, 1976). Hodges (1976) stressed the need for the development of better media for organ culture, and the experiments described here illustrate what great effects the basic culture medium and serum can have on the performance of cultured tissues. Media containing high concentrations of amino acids to provide a zwitterionic buffering system, such as L15, though well-suited for cell cultures, clearly should not be used for *in vitro* experiments on glucose metabolism with tissues capable of deamination and gluconeogenesis. Amino acid breakdown on the scale we have found would be expected to have marked effects on major metabolic pathways.

Organ cultures have three phases (Biggers, 1965) – the shock phase (resulting from the trauma of removal from the body and setting-up *in vitro*), the stable phase (when meaningful experiments can be carried out), and the hydration phase (when wet weight increases in relation to dry weight as the tissue declines). Successful long-term organ culture requires minimisation of the effects of the first phase, prolongation of the second, and delay of the onset of the third. Once such a method has been developed for a particular tissue, its vailidity as a model system with relevance to what happens *in vivo* must be considered. For a model to be valid, there must be a considerable parallel between what can reasonably be expected of it and what occurs *in vivo*; responses to metabolites, hormones and drugs can provide useful indications of such parallels. For example, *A. means* liver cultures take in glucose and synthesise glycogen in high glucose medium containing insulin, and the glycogenolytic response to β adrenoreceptor agonists is blocked by β adrenoreceptor antagonists (Brown, Pryor & Balls, 1976).

The length of the stable period is particularly important when repeated treatments with the same agent, sequential treatments with different agents, or recovery from such treatment are being studied. In addition, the effects of some agents, including toxic chemicals, may not become apparent for some time after the dose. This may be because the chemical needs to be metabolised to the active compound, because, as in the case of paracetamol, the enzymes and substrates of protective detoxication pathways take time to become exhausted, or because the development of signs of damage itself takes time. For example, lethal doses of most hepatotoxins (carbon tetrachloride, pyrrolizidine alkaloids, dimethylnitrosamine) cause no obvious cell injury in the rat before about 12 h after the dose, and often do not kill before 48 h (McLean & Nuttall, 1978), which limits the usefulness of *in vitro* preparations which survive for only 2–6 h. One approach to this problem is to set up liver preparations at intervals after *in vivo* dosage, another is to use active metabolites rather than the parent compound. However, many drugs and some toxic chemicals exert their effects more rapidly. Oral dosage with paracetamol leads to signs of liver injury in rats within 8 h, or less in animals pretreated with phenobarbital, so McLean & Nuttall (1978) were able to develop a valid *in vitro* model for paracetamol-induced liver injury. As *in vivo*, injury was dependent on paracetamol dose and on phenobarbital, was preceded by a fall in glutathione levels, and took several hours to develop.

One advantage of *in vitro* systems is that, merely by changing the culture medium, exposure of a tissue to a drug can be brought to an end much more quickly than is possible *in vivo*. This is likely to be very useful as studies on recovery from toxic effects become more possible. However, it must also be added that, *in vivo*, drugs tend to be removed from the tissues via the circulation and excreted from the body, whereas in closed *in vitro* systems they could remain in contact with cultured tissues until the next change of medium. Perfusion and circumfusion methods would alleviate this problem (and some of the other problems which currently limit the survival of adult mammalian organ cultures), but they require such complex apparatus (Murrell, 1976) that it is unlikely that a large number of replicate cultures could be simultaneously maintained and studied.

The long stable period available with adult amphibian organ cultures and the simple techniques involved make them ideal for studies on the direct effects of drugs, toxins and other chemicals on isolated tissues. We have established that the toxic effects resulting from the exposure of liver and kidney cultures to paracetamol probably result from the unpolymerised form of the *p*-aminophenol produced from paracetamol by deacetylation. We also found that the ability of the cultured tissues to respond to insulin and phenformin was affected by the time of maintenance in culture media containing paracetamol.

During the course of this work, and while carrying out experiments on the basis of the hyperglycaemic effect of diazoxide, an antihypertensive drug, and on the basis of the lactate acidosis associated with the use of phenformin (Balls & Rao, 1980), we have repeatedly been made aware that, once techniques for maintaining differentiated tissues *in vitro* have been developed, their proper use in meaningful experiments presents a wide range of new problems. We have tried to show in this chapter that, in the absence of methods for the long-term maintenance of the equivalent adult mammalian tissues, amphibian organ culture is well-suited to a general consideration of these problems, as well as for original work on the effects of particular agents *in vitro*.

REFERENCES

BALLS, M., BROWN, D. & FLEMING, N. (1976). Long-term amphibian organ culture. *Methods in Cell Biology*, **13**, 213–38.

BALLS, M. & RAO, R. R. (1980). Organ culture in pharmacology. In *The Use of Alternatives in Drug Research*, ed. A. N. Rowan & C. J. Stratmann, pp. 49–69. London: The Macmillan Press.

BALLS, M. & RUBEN, L. N. (1965). Cultivation *in vitro* of normal and neoplastic cells of *Xenopus laevis*. *Experimental Cell Research*, **43**, 694–5.

BIGGERS, J. D. (1965). Cartilage and bone. In *Cells and Tissues in Culture*, vol. 2, ed. E. N. Willmer, pp. 197–260. New York & London: Academic Press.

BROWN, D., PRYOR, J. S. & BALLS, M. (1976). The effects of drugs on cultured amphibian tissues: activation and blockade of adrenoceptors in *Amphiuma means* liver. In *Organ Culture in Biomedical Research*, ed. M. Balls & M. A. Monnickendam, pp. 481–501. Cambridge University Press.

BUCKNALL, R. A. (1980). The use of cultured cells and tissues in the development of antiviral drugs. In *The Use of Alternatives in Drug Research*, ed. A. N. Rowan & C. J. Stratmann, pp. 15–27. London: The Macmillan Press.

CALDER, I. C., FUNDER, C. C., GREEN, C. R., HAM, K. N. & TANGE, J. D. (1971). Comparative nephrotoxicity of aspirin and phenacetin derivatives. *British Medical Journal*, **4**, 518–21.

CALDER, I. C., WILLIAMS, P. J., WOODS, R. A., FUNDER, C. C., GREEN, C. R., HAM, K. N. & TANGE, J. D. (1975). Nephrotoxicity and molecular structure. *Xenobiotica*, **5**, 303–7.

CLOTHIER, R. H., DEWAR, J. R., SANTOS, M. A., NORTH, A. D., FOSTER, S. & BALLS, M. (1981). A comparative study of the deacetylation of paracetamol by urodele and anuran amphibian organ cultures. *Xenobiotica*, **11**, 149–57.

CROWE, C. A., CALDER, I. C., MADSEN, N. P., FUNDER, C. C., GREEN, C. R., HAM, R. N.

& TANGE, J. D. (1977). An experimental model of analgesic-induced renal damage – some effects of *p*-aminophenol on rat kidney (mitochondria). *Xenobiotica*, **7**, 345–56.

FEDEROFF, S. (1967). Proposed usage of animal tissue culture terms. *Experimental Cell Research*, **46**, 642–8.

FELL, H. B. (1964). The role of organ cultures in the study of vitamins and hormones. *Vitamins and Hormones*, **22**, 81–127.

FELL, H. B. (1976). The development of organ culture. In *Organ Culture in Biomedical Research*, ed. M. Balls & M. A. Monnickendam, pp. 1–13. Cambridge University Press.

FITTON JACKSON, S. (1976). The maintenance of differentiation in skeletal tissues. In *Organ Culture in Biomedical Research*, ed. M. Balls & M. A. Monnickendam, pp. 165–77. Cambridge University Press.

FLEMING, N., BROWN, D. & BALLS, M. (1975). Hepatocyte function in long-term organ culture of *Amphiuma means* liver. *Journal of Cell Science*, **18**, 533–44.

FOULKES, E. C. & HAMMOND, P. B. (1975). Toxicology of the kidney. In *Toxicology: the Basic Science of Poisons*, ed. L. J. Casaret & J. Dorell, pp. 190–200. New York: Macmillan.

FRANKS, L. M. (1976). Summary and future developments. In *Organ Culture in Biomedical Research*, ed. M. Balls & M. A. Monnickendam, pp. 549–56. Cambridge University Press.

GATER, S. (1976). Some aspects of amphibian pancreas and liver function in organ culture. Ph. D. thesis, University of East Anglia, Norwich, UK.

GATER, S., BROWN, D. & BALLS, M. (1976). The use of amphibian organ cultures in toxicological investigations. In *Organ Culture in Biomedical Research*, ed. M. Balls & M. A. Monnickendam, pp. 502–14. Cambridge University Press.

HEALEY, K., CALDER, I. C., YONG, A. C., CROWE, C. A., FUNDER, C. C., HAM, K. N. & TANGE, J. D. (1978). Liver and kidney damage induced by N-hydroxy-paracetamol. *Xenobiotica*, **7**, 403–11.

HINSON, J. A., NELSON, S. D. & MITCHELL, J. R. (1977). Studies on the microsomal formation of arylating metabolites of acetaminophen and phenacetin. *Molecular Pharmacology*, **13**, 625–33.

HODGES, G. M. (1976). A review of methodology in organ culture. In *Organ Culture in Biomedical Research*, ed. M. Balls & M. A. Monnickendam, pp. 15–59. Cambridge University Press.

IBER, F. L. (1974). Normal and pathologic physiology of the liver. In *Pathologic Physiology: Mechanisms of Disease*, 5th edn, ed. W. A. Sodeman, Jr & W. A. Sodeman, pp. 790–817. Philadelphia: W. B. Saunders.

JOLLOW, D. J., MITCHELL, J. R., POTTER, W. Z., DAVIS, D. C., GILLETTE, J. N. & BRODIE, B. B. (1973). Acetaminophen-induced hepatic necrosis. II. Role of covalent binding *in vitro*. *Journal of Pharmacology and Experimental Therapeutics*, **187**, 195–202.

LEIBOVITZ, A. (1963). The growth and maintenance of tissue cultures in free gaseous exchange with the atmosphere. *American Journal of Hygiene*, **78**, 173–80.

McLEAN, A. E. M. & NUTTALL, L. (1978). An *in vitro* model of liver injury using paracetamol treatment of liver slices and prevention of injury by some antioxidants. *Biochemical Pharmacology*, **27**, 425–30.

MITCHELL, J. R., JOLLOW, D. J., POTTER, W. Z., DAVIS, D. C., GILLETTE, J. R. & BRODIE, B. B. (1973a). Acetaminophen-induced hepatic necrosis. I. Role of drug metabolism. *Journal of Pharmacology and Experimental Therapeutics*, **187**, 185–94.

MITCHELL, J. R., JOLLOW, D. J., POTTER, W. Z., GILLETTE, J. R. & BRODIE, B. B. (1973b). Acetaminophen-induced hepatic necrosis. IV. Protective role of glutathione. *Journal of Pharmacology and Experimental Therapeutics*, **187**, 211–17.

MITCHELL, J. R., THORGEIRSON, S. S., POTTER, W. Z., JOLLOW, D. J. & KEISER, H. (1974). Acetaminophen-induced hepatic injury: protective role of glutathione in man as rationale for therapy. *Clinical Pharmacology and Therapeutics*, **16**, 676–84.

MONNICKENDAM, M. A. & BALLS, M. (1973). The relationship between cell sizes, respiration rates and survival of amphibian tissues in long-term organ cultures. *Comparative Biochemistry and Physiology*, **44A**, 871–80.

MONNICKENDAM, M. A., BROWN, D. & BALLS, M. (1974). Organ culture of *Amphiuma means* liver: control of glycogen content. *Comparative Biochemistry and Physiology*, **47A**, 567–72.

MURRELL, L. R. (1976). Circumfusion organ culture of rat acinar pancreas with chemically defined medium. In *Organ Culture in Biomedical Research*, ed. M. Balls & M. A. Monnickendam, pp. 273–92. Cambridge University Press.

POTTER, W. Z., DAVIS, D. C., MITCHELL, J. R., JOLLOW, D. J., GILLETTE, J. R. & BRODIE, B. B. (1973). Acetaminophen-induced hepatic necrosis. III. Cytochrome P_{450}-mediated covalent binding *in vitro*. *Journal of Pharmacology and Experimental Therapeutics*, **187**, 203–9.

PRESCOTT, L. F., ILLINGWORTH, R. N., CRITCHLEY, J. A. J. H., STEWART, M. J., ADAM, R. D. & PROUDFOOT, A. T. (1979). Intravenous N-acetylcysteine: the treatment of choice for paracetamol poisoning. *British Medical Journal*, **2**, 1097–100.

REED, S. E. (1976). Organ cultures for studies of viruses and mycoplasmas. In *Organ Culture in Biomedical Research*, ed. M. Balls & M. A. Monnickendam, pp. 513–32. Cambridge University Press.

REYNOLDS, J. J. (1976). Organ cultures of bone: studies on the physiology and pathology of resorption. In *Organ Culture in Biomedical Research*, ed. M. Balls & M. A. Monnickendam, pp. 355–66. Cambridge University Press.

SAXÉN, L. & KARKINEN-JÄÄSKELÄINEN, M. (1975). Inductive interactions in morphogenesis. In *The Early Development of Mammals*, ed. M. Balls & A. E. Wild, pp. 319–34. Cambridge University Press.

SMITH, G. E. & GRIFFITHS, L. A. (1976). Comparative metabolic studies of phenacetin and structurally-related compounds in the rat. *Xenobiotica*, **6**, 217–36.

SPOONER, J. B. & HARVEY, J. G. (1976). The history and usage of paracetamol. *Journal of Internal Medicine, Research Supplement*, **4**, 1–6.

VOLANS, G. N. (1976). Self-poisoning and suicide due to paracetamol. *Journal of Internal Medicine, Research Supplement*, **4**, 7–13.

Control of enzyme activity during plant cell development

D. H. NORTHCOTE

Department of Biochemistry, University of Cambridge, Cambridge, CB2 1QW

INTRODUCTION

In an intact stem, the vascular cambium gives rise to the xylem, and the major chemical changes that occur in the walls of the cells during this development are known (Thornber & Northcote, 1961). The sequence of events takes place from the formation of the cell plate at telophase to the final lignification of the secondary thickened wall. At the initial stages pectic substances, hemicellulose and cellulose are laid down while during secondary thickening the deposition of pectin ceases, the amount of cellulose and hemicellulose (mainly 4-OMe glucuronoxylan in dicotyledons) formed increases and the process of lignification begins. These changes in the synthetic capabilities of the cell are accomplished by the activation and inactivation of certain key enzymes, the induction and repression of other enzymes and the control of transport and organisation processes.

It is possible to limit the consideration of the enzymic sites that are involved in the control of differentiation by knowing the chemical changes that take place and the mechanism of transport of the materials involved. Some of these enzymic steps are the effectors whereby the control is monitored and they will respond to signals such as hormonal supply or changes in environment of the cell or tissue.

Polysaccharides and lignin

The donors for all the polysaccharides are nucleoside diphosphate sugars and the main flow of carbohydrate is from UDP glucose to UDP glucuronic acid and UDP xylose by means of a dehydrogenase and a decarboxylase, to form the donors of hemicellulose synthesis, or via epimerases from each of these three uridine diphosphate sugars to form UDP galactose, UDP galacturonic acid and UDP arabinose (Fig. 1) (Hassid, Neufeld & Feingold, 1959; Northcote, 1962). These last three UDP sugars form the donors for the pectic substances. There are, therefore, two series of uridine diphosphate sugars, one based on glucose which forms the hemicelluloses and which can be derived from

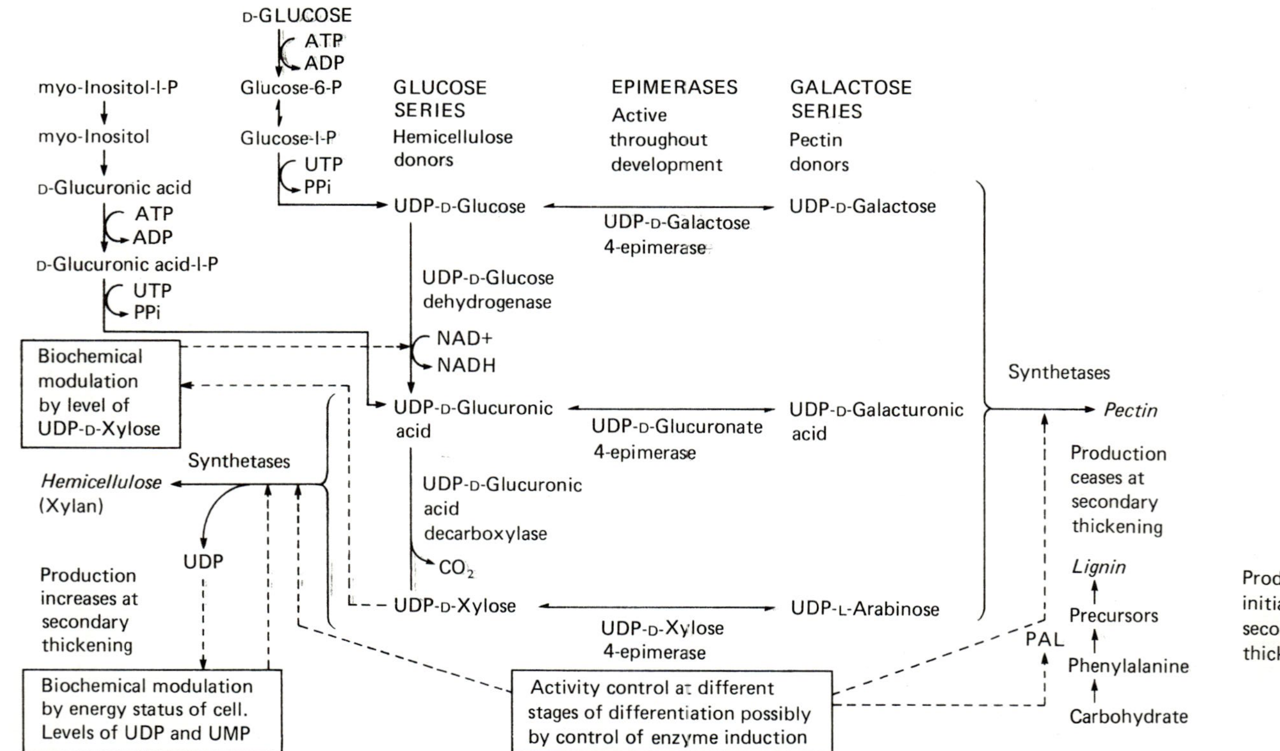

Fig. 1. Control of synthesis of wall components during cell development. The solid lines represent metabolic routes; the controls are indicated by broken lines. The main wall components are in italics.

UDP glucose, and the other series which is based on galactose forms the pectic substances. The galactose series arises from the corresponding uridine diphosphate sugars of the glucose series by the epimerases.

The onset of lignification occurs at the very early stage of secondary thickening (Wooding & Northcote, 1964). It starts at the middle lamella (outer region of the wall), progresses inwards towards the developing secondary thickenings and between them along the middle lamella (Kerr & Bailey, 1934; Bailey, 1936). The precursors of lignin are phenolic alcohols produced in the cytoplasm of differentiating cells (Freudenberg, 1959; Rubery & Northcote, 1968). These precursors, such as coniferyl alcohol, arise from cinnamic acid which is formed from phenylalanine by phenylalanine ammonia lyase and the aromatic ring can be further hydroxylated and the phenolic groups methylated by hydroxylases and O-methylases (Neish, 1961; Koukol & Conn, 1961; Freudenberg, 1965; Towers, 1974). In the wall, the final stage of polymerisation takes place by an oxidation of the phenolic alcohols which produces free radicals. This oxidation is probably brought about by a peroxidase (Harkin & Obst, 1973).

The enzyme activities concerned with the production of polysaccharides and lignin are shown in Fig. 1 and it is the changes in activities of these that we have studied during cell differentiation. We have used two experimental systems for the investigation, either the stems of growing trees or the induced differentiation of plant tissue cultures.

EXPERIMENTAL SYSTEMS

With tree stems, in spring, the tissues from a developing row of cells from cambium to xylem may be separated by a gentle and simple scraping technique into three fractions (Thornber & Northcote, 1962; Dalessandro & Northcote, 1977a, b). Each of these is considerably enriched in cambial cells or differentiating xylem cells or differentiated xylem cells. The variation in enzymic activity associated with the changes in pectin and hemicellulose synthesis during differentiation of xylem can be determined by measurement of the activities in the three fractions and thus the possible control sites may be identified.

Plant tissue cultures either as a solid callus or in suspension culture can be induced to differentiate by subculturing the cells into media with varying ratios of auxin to kinetin and increased levels of sucrose (Wetmore & Rier, 1963; Jeffs & Northcote, 1966, 1967; Haddon & Northcote, 1976c). The extent of differentiation can be measured by counting the appearance of xylem and phloem tissue in nodules arising within the culture and also by measuring the activities of appropriate enzymes (Haddon & Northcote, 1975).

ENZYMES FOR POLYSACCHARIDE FORMATION

The epimerases

It would be expected that the epimerase activities would be shut down during secondary thickening so that no pectic precursors would be formed. However, when these enzymes are extracted from cells undergoing either primary or secondary growth of their walls, although the activities are found to vary throughout growth, there is little difference between them and they are always maintained at levels similar to those in the very young tissues when relatively large amounts of pectin are being formed, so that unless there is a biochemical modulation of the epimerases there is no control at these sites (Dalessandro & Northcote, 1977a, b). The cells at all stages of growth are capable of establishing a pool of uridine diphosphate sugars even if some of these are not used to any extent for wall formation at secondary thickening. Thus the main control for the shut-down of pectin synthesis must take place at another enzymic site and it is probably exerted at some aspect of the synthetase system.

During active primary growth the pectin composition alters and it is possible that these relatively minor changes in composition are partly controlled by the changes in flux of carbohydrate into the donors of pectin synthesis (UDP galactose, UDP galacturonic acid and UDP arabinose) brought about by the small variations in epimerase activities that occur at this time (Dalessandro & Northcote, 1977c). The pectin is composed of chains of polygalacturonic acid to which are attached arabinogalactans (Barrett & Northcote, 1965). This substitution alters the physical and chemical properties of the polysaccharide which can vary with the degree of substitution. Thus the composition of the pectin, and especially its interaction with water, determines the physical nature of the wall, and it undergoes these changes during growth as the area of the wall increases (Northcote, 1972). That the composition of the pectin varies during the initial stages of growth can be seen from the results of at least three experiments. (1) Protoplasts prepared from a suspension culture will regenerate a cell wall and during this process large amounts of pectin, which are formed very early, are secreted into the medium in which the protoplasts are incubated. The pectin can therefore be easily monitored and analysed during the early stages of wall reformation. It is found that its composition changes in that there is a variation in the amount of arabinogalactan relative to the poly-galacturonic acid (Hanke & Northcote, 1974). (2) The plasmolysis of tobacco leaf discs results in a change in the patterns of formation of wall material. As soon as plasmolysis occurs a change in the synthesis of pectin takes place. This is shown by incubation of the discs in radioac-tive glucose when a variation in the incorporation of the glucose into the arabinose of the pectin polymers occurs (Boffey & Northcote, 1975; Northcote, 1977). (3) The composition of the pectin in a sycamore culture cell can be directly influenced by the application of the growth substance, 2,4-dichlorophenoxyacetic acid. This is indicated by the difference in incorporation of radioactive glucose or arabinose into the pectin of the wall of cells grown in the absence or presence of the growth factor. The growth factor increases the incorporation into the arabinose of the pectin polymers (Rubery & Northcote, 1971).

Pectin composition varies therefore according to growth conditions and by the application of exogenous growth hormones. This change can be, in part, correlated with the changes in the epimerase activities of the young cells during initial growth while the cells are dividing and the walls are increasing in surface area.

UDP glucose dehydrogenase and UDP glucuronic acid decarboxylase

In angiosperms the main hemicellulose is xylan (Northcote, 1969). During differentiation of a xylem cell of a dicotyledon such as sy-camore or poplar, the activities of UDP glucuronic acid decarboxylase and UDP glucose dehydrogenase that produce UDP xylose from UDP glucose, increase significantly (Table 1) (Dalessandro & Northcote, 1977a). Thus the flow of glucose towards the nucleoside diphosphate donor for xylan synthesis increases. The UDP dehydrogenase has always a lower activity than the decarboxylase and is probably a rate limiting step for the flow to UDP xylose. This enzyme is inhibited by an increased concentration of UDP xylose so that the dehydrogenase can monitor the flow to the xylan precursor by a biochemical feed-back mechanism (Neufeld & Hall, 1965; Dalessandro & Northcote, 1977a). UDP glucuronic acid can also arise from glucose via myo-inositol (Loewus, Chen & Loewus, 1973) and this route will bypass the modu-lation of the dehydrogenase so that an effective control of UDP xylose production by the decarboxylase can occur.

Table 1. *Specific activities and units of enzyme activities per cell of the enzymes of UDP-sugar interconversions of the glucose series during differentiation in sycamore stems*

Enzyme	Cambial cells		Differentiating xylem cells		Differentiated xylem cells	
	specific activity (nmol/min/ mg protein)	units of enzyme activity per cell (nmol/min/cell)	specific activity (nmol/min/ mg protein)	units of enzyme activity per cell (nmol/min/cell)	specific activity (nmol/min/ mg protein)	units of enzyme activity per cell (nmol/min/cell)
UDP-D-glucose dehydrogenase	6.4	7.9×10^{-7}	10.0	15.8×10^{-7}	18.0	38.8×10^{-7}
UDP-D-glucuronic acid decarboxylase	132.8	16.5×10^{-6}	313.5	49.7×10^{-6}	303.7	60.4×10^{-6}

Table 2. *Xylan synthetase activity in cambial cells, differentiating and differentiated xylem cells in sycamore and poplar trees*

Source of enzyme fraction	Sycamore xylan synthetase activity (nmol/min/mg of protein)	Poplar xylan synthetase activity (nmol/min/mg of protein)
Cambial cells	0.7	2.3
Differentiating xylem cells	1.6	5.1
Differentiated xylem cells	4.2	6.3

The synthetase systems

Most of the polysaccharides of the hemicelluloses and pectin are built up of polymers containing more than one type of monosaccharide, e.g. the xylans contain 4-OMe-glucuronic acid side branches, the mannans are galactoglucomannans, the pectin consists of chains of polygalacturonic acid substituted with rhamnose units and carries side chains of arabinogalactans (Northcote, 1963, 1969). The transfer of the branches or the various minor sugars to the main backbone of the polymer depends on the formation of the main chain as an acceptor. It is possible therefore that control of the polymerisation of these substituent sugars is only expressed in the presence of the main chain as an acceptor for the transglycosylation reactions. The control of the polymerisation of these minor sugars could therefore be secondary and depend on the primary control of the synthesis of the backbone chain.

The control during differentiation of the synthesis of the main chain of the xylan in the hemicellulose of sycamore and poplar trees can be clearly seen when the activity of the synthetase is measured in cells which are differentiating into xylem, compared with that from the cambium initials. The activity of the xylan synthetase increases 3–6 fold during the period of the maturation of the cells (Table 2) (Dalessandro & Northcote, 1981).

Xylan synthetase, like all polysaccharide synthetases, is membrane bound. It occurs within some parts of the endomembrane system (endoplasmic reticulum, Golgi apparatus and plasma membrane) of the cells (Northcote, 1974). The details of the intermediates in xylan synthetases are not known but by comparison with some other synthetases of polysaccharides and glycoproteins at least two transglycosylases may possibly be involved. One from UDP-D-xylose to an acceptor such as a lipid (e.g. polyprenyl phosphate) or a protein and a second transfer from the acceptor to give the growing chain of the polysaccharide (Northcote, 1979). The increase in activity of the system that occurs during secondary thickening may therefore be due to the increased activity of one or all of the several possible transglycosylases involved. These increases in activity are probably due to an increase in the amount of enzyme and therefore the control is exerted at the stage of transcription and translation of the mRNAs for the enzymes and the insertion of the proteins into the membrane system.

A biochemical feed-back control is also possible at the xylan synthetase since UDP and UMP inhibit it (Dalessandro & Northcote, 1981). This type of control would relate the synthetase activity to the available energy of the cell in the form of ATP. UDP is produced from UDP xylose after the incorporation of the xylose into the polymer by transglycosylase. The UDP must be converted to UTP and then react with sugar-1-phosphate by a pyrophosphorylase reaction before more UDP xylose as substrate for the synthetase is formed. At the level of the xylan synthetase therefore there is both a long-term control operated by the level of the synthetase system which changes during differentiation probably by control of protein synthesis either at transcription or translation and a constant immediate biochemical modulation which relates the synthesis of the polysaccharide to the energy supply and demands of the cell.

The xylan formed in the secondary wall of sycamore xylem is not a glucoxylan and thus no stimulation by the presence of UDP glucose was expected. However, even with the cambial tissue the xylan synthetase activity was not stimulated by the addition of UDP glucose as found with the primary wall system of pea shoots (Ray, 1980). On the contrary UDP glucose was found to inhibit the enzyme at concentrations of 1 mM. It is not known whether this inhibition has any biological significance.

Increases in synthetase activities also occur in callus tissue of bean, which is induced to differentiate in a medium containing 1-naphthylacetic acid and kinetin in a ratio of 5 : 1 w/w. Both xylem and phloem are formed and in addition to an increase in xylan synthetase activity, a specific marker for the phloem differentiation also increases. The phloem formation is characterised by the synthesis of callose, a β 1,3 glucan, which is formed at the pores of the sieve tubes (Aspinall & Kessler, 1957; Kessler, 1958; Esau, Cheadle & Risley, 1962; Wooding &

Northcote, 1965). Callose synthetase activity increases during the induction of the differentiation and thus the control of the synthesis of this particular polysaccharide is associated with the induction of the synthetase activity (Haddon & Northcote, 1975).

PHENYLALANINE AMMONIA LYASE AND THE ACTION OF PLANT GROWTH HORMONES

The plant growth hormones, indole acetic acid (or artificial compounds such as 2,4-dichlorophenoxyacetic acid), cytokinins, gibberellic acid and abscisic acid together with sucrose, when applied to plant tissue cultures influence the differentiation of the callus cells to xylem and phloem (Skoog & Miller, 1957; Wetmore & Rier, 1963; Jeffs & Northcote, 1966, 1967; Wright & Northcote, 1972; Haddon & Northcote, 1976a). These hormones and possibly sucrose therefore act in part to influence the induction of particular enzymes for polysaccharide synthesis such as xylan synthetase and callose synthetase, already discussed, and in addition they must induce the activities of enzymes for lignin synthesis which are necessary for the formation of xylem.

The induction of differentiation of a bean callus transferred to an induction medium is measured by the increase in activity of phenylalanine ammonia lyase (PAL) over a period of about 12 days (Haddon & Northcote, 1975). Gibberellic acid in the induction medium (one containing the required ratio of auxin to kinetin) will influence the type of differentiation and may slightly delay its onset and the induction of the enzyme activities, while the presence of abscisic acid will inhibit both the enzyme induction and the differentiation (Haddon & Northcote, 1976a). In a series of experiments the induction of PAL activity can be directly related to the induction of the differentiation of xylem (Haddon & Northcote, 1975, 1976a, b). It is difficult to prove that the induced activity is due to a *de novo* increase or onset of synthesis of enzyme rather than an activation of enzyme previously synthesised but inactive. However, inhibitors of transcription (actinomycin D) and translation (D-2-(4-methyl-2, 6-dinitroanilino)-*N*-methylpropionamide, MDMP) applied to bean suspension culture cells grown in induction media inhibit the rising phase of PAL activity (Jones & Northcote, 1981). This indicates that both transcription and translation are required for the increase in PAL activity.

Superinduction by actinomycin D during the falling phase of PAL activity suggests that availability of PAL mRNA for translation may be controlled to produce the PAL activity response to plant hormones (Jones & Northcote, 1981). If this is correct then the induction medium has brought about an increase, or an initiation in some cases, of the transcription and translation of the mRNA from the plant cell genome and the control exerted by the application of the exogenous plant growth factors is operated at the stage of transcription, RNA processing, translation or post translational modification of the enzyme.

With bean tissue, the increase of phenylalanine ammonia lyase activity that is associated with cell differentiation depends upon the presence of 1-naphthylacetic acid (NAA) and kinetin. The NAA needs to be present at least 48 h prior to the increase in PAL activity while the kinetin can be applied immediately prior to the expected rise in PAL activity (Bevan & Northcote, 1979a). Therefore it is probable that the hormones are acting at two different sites to control the production of the enzyme.

The bean tissue loses its potential for differentiation. When it is transferred four times to maintenance medium and then once to induction medium, differentiation and induction of enzyme activity occurs. However, as the number of transfers on maintenance medium is increased above four, there is a gradual reduction in the amount of differentiation that occurs upon transference to induction medium (Haddon & Northcote, 1975). As the number of transfers to maintenance medium increases so the potential for differentiation is gradually lost as measured by the histological changes and induction of enzyme activities.

There are various possibilities for this failure to differentiate, which occurs both on solid and in liquid media (Bevan & Northcote, 1979b). One is that there may be two kinds of cells in the tissue, one capable of differentiation, the other not. If this were so then perhaps selection of non-differentiating cells might occur with continued propagation. This possibility can be investigated by cloning the cells. When this was done there was no evidence to show that selective growth of non-inducible clones occurred on the maintenance medium which caused loss in inducibility, and thus such selective growth could not be responsible for the loss of morphogenetic potential. Cell selection as a factor therefore does not seem to be very probable, and in addition after about thirty generations in maintenance medium the ability to differentiate was restored quite suddenly over one transfer (average

3.7 generations per transfer). The change occurred so rapidly that cell selection could not be responsible (Bevan & Northcote, 1979b).

Another possibility was that genetic material was being lost during culture and its loss would prevent induction of necessary enzymes and thus prevent the induction of differentiation. We have used the phenomenon which Hahlbrock and his colleagues have shown with cultures of parsley cells (Hahlbrock & Grisebach, 1979), that PAL activity is induced on dilution of the cultures. When the bean cell cultures were diluted and transferred to induction medium there was a peak of PAL activity after 10 h, due to the dilution, followed by a second peak of PAL activity after 4–5 days which was due to the hormones, NAA and kinetin, in the medium. When cultures were diluted onto a medium without the growth hormones there was only the one dilution peak (Bevan & Northcote, 1979b). The failure of bean tissues to differentiate after continued transfer to maintenance medium was coupled with the inability of these tissues to show the hormone-induced production of PAL on transfer to an induction medium. On the other hand there was still an induction of PAL by these cultures by the dilution technique. So all the mechanism for the induction of PAL activity was present in the cells; they just did not respond to the growth substances (Bevan & Northcote, 1979b).

This seems to rule out the loss of genetic information as a reason for the loss of potential to differentiate and the corresponding inability of the cells to synthesise the various enzymes in response to the application of growth factors. The cultured tissue no longer responded to the ratio of exogenous hormones because of a change in its metabolism (Bevan & Northcote, 1979b), such as the rate of formation and level of endogenous hormones; brought about by a change in their synthesis or rate of degradation monitored by a biochemical feed-back control by the pool-size of the growth factors within the cell. This pool-size is in part dependent on the external supply.

The work with the tissue cultures indicated that the ratio of the various growth factors at the cell together with nutrients such as sucrose determined the path of differentiation. The plant cell *in situ* receives these growth factors by two routes, one by synthesis within the cell and secondly by transport from other cells. In the intact plant differentiation and the establishment of tissues and organs depends upon the synthesis, degradation and transport of the various growth factors such as auxin, cytokinin, gibberellic acid and nutrients such as sucrose.

POSSIBLE CONTROL SIGNALS FROM THE WALL BACK TO THE CYTOPLASM

The work on the plasmolysis of tobacco leaf discs, previously mentioned, has shown that plasmolysis alters the type of pectin synthesised. The pectins and hemicellulose polysaccharides are synthesised within the cytoplasm at the membrane system and especially at the Golgi apparatus. The polysaccharides are packed within vesicles and these vesicles are incorporated into the plasmamembrane so that the polysaccharides are transferred into the wall. There is a membrane fusion mechanism between the vesicles and the plasmamembrane so that material which is inside the vesicle becomes outside the plasmamembrane and incorporated into the cell wall (Northcote, 1974, 1979).

The effects of plasmolysis which alter the pattern of pectin synthesis were partially reversed when the plasmolysed tissue was placed in a solution which brought about deplasmolysis. During plasmolysis of the cells the change in pectin synthesis was brought about immediately the cells were placed in solutions just above the isotonic point, that is, as the plasmalemma begins to be withdrawn from the wall and the plasmadesmata were not yet broken (Northcote, 1977). This immediate effect was much greater than the subsequent response to decreasing the plasmalemma surface area as the osmotic pressure of the plasmolysing solution was progressively increased above the isotonic point. Thus the loss of contact between much of the cell wall and the plasmalemma occurs simultaneously with a change in the synthesis occurring within the cytoplasm. It is possible that this change is stimulated because the ionic atmosphere at the outer surface of the membrane has changed. The removal, because of the plasmolysis, of the acidic pectin molecules of the cell wall, from the membrane surface, could alter the charge distribution across the membrane and the distribution and orientation of the membrane constituents. In this way fusion of the vesicles from the Golgi bodies might be altered and control of membrane flow from the endomembrane system could be influenced.

A possible control point in the secretion of polysaccharides by the endomembrane system is therefore the rate of fusion of the vesicles with the plasmamembrane. There is evidence to suggest that one of the factors which initiate membrane fusion is the level of Ca^{2+} at the membrane surface.

Membrane fractions enriched in Golgi apparatus or plasmamembrane (Baydoun & Northcote, 1980a) prepared from maize root tips are made radioactive if the roots are preincubated with radioactive glucose or choline. When radioactive and non-radioactive membrane preparations are mixed, the amount of membrane fusion is related to the transfer of radioactivity between them (Baydoun & Northcote, 1980b); a quantitative assay of membrane fusion can therefore be made. Ca^{2+} ions are necessary for the fusion and a specific, integral protein of the membranes is necessary for the Ca^{2+}-dependent fusion. This protein is removed from the membrane by treatment with trypsin and it is soluble in deoxycholate; it is a Ca^{2+}-activated ATPase (Baydoun & Northcote, 1980b, 1981).

Suspension-cultured sycamore cells secrete soluble polysaccharides and the steady-state rate of secretion is substantially and very rapidly increased by addition of various electrolytes to the medium (Morris & Northcote, 1977). The increased secretion is induced, from within the cell, by the presence of cations, primarily at the outer surface of the plasmamembrane. The increase in rate of secretion is brought about by a stimulation of the normal mechanism of polysaccharide secretion, that is, it is the fusion of membrane-bounded vesicles from the Golgi apparatus to the plasmamembrane that is activated, so that this step is a rate-limiting process in polysaccharide secretion and a potential control point. Control of this event can occur at the outer surface of the plasmamembrane and the immediate effect is on the last stage of the mechanism of polysaccharide secretion. A further control of the biochemical synthetic steps within the membrane system at the level of the endoplasmic reticulum and Golgi apparatus must be co-ordinated with this control point (Morris & Northcote, 1977).

CONCLUSIONS

It is possible therefore that the operation of a control at the outer surface of a cell might modulate the mechanism of metabolic activity within the cytoplasm. The cell can thus respond to its environment and in an intact tissue the environment will be made up of other cells and their metabolic products. A co-ordinated development that involves cell–cell interaction for the differentiation of the tissue during its growth becomes possible. The results are consistent with the idea that there are control points for wall metabolism and deposition at the membrane surface; these could possibly be influenced by the ionic environment of the membrane which is in part controlled by the type of polysaccharide which is present in the wall. The ionic atmosphere of the membrane could also be affected by the growth factors, especially if these altered the permeability of the membrane, but in addition, although an initial response might be at the cell surface, subsequent effectors of the response, which may be the growth factors, operate at some internal control point, for example the synthetic sites especially the polysaccharide synthetases of the membrane system, within the cytoplasm of the cell.

REFERENCES

ASPINALL, G. O. & KESSLER, G. (1957). The structure of callose from the grape vine. *Chemistry and Industry*, 1296.

BAILEY, A. J. (1936). Lignin in Douglas fir. Composition of the middle lamella. *Industrial and Engineering Chemistry (Analytical edition)*, **8**, 52–5.

BARRETT, A. J. & NORTHCOTE, D. H. (1965). Apple fruit pectic substances. *Biochemical Journal*, **94**, 617–27.

BAYDOUN, E. A-H. & NORTHCOTE, D. H. (1980a). Isolation and characterisation of membranes from the cells of maize root tips. *Journal of Cell Science*, **45**, 147–67.

BAYDOUN, E. A-H. & NORTHCOTE, D. H. (1980b). Measurement and characteristics of fusion of isolated membrane fractions from maize root tips. *Journal of Cell Science*, **45**, 169–86.

BAYDOUN, E. A-H. & NORTHCOTE, D. H. (1981). The extraction from maize root cells of membrane-bound protein with Ca^{2+}-ATPase activity and its possible role in membrane fusion *in vitro*. *Biochemical Journal*, **193**, 781–92.

BEVAN, M. & NORTHCOTE, D. H. (1979a). The interaction of auxin and cytokinin in the induction of phenylalanine ammonia lyase in suspension cultures of *Phaseolus vulgaris*. *Planta*, **147**, 77–81.

BEVAN, M. & NORTHCOTE, D. H. (1979b). The loss of morphogenetic potential and induction of phenylalanine ammonia lyase in suspension cultures of *Phaseolus vulgaris*. *Journal of Cell Science*, **39**, 339–53.

BOFFEY, S. A. & NORTHCOTE, D. H. (1975). Pectin synthesis during the wall regeneration of plasmolysed tobacco leaf cells. *Biochemical Journal*, **150**, 433–40.

DALESSANDRO, G. & NORTHCOTE, D. H. (1977a). Changes in enzymic activities of nucleoside diphosphate sugar interconversions during differentiation of cambium to xylem in sycamore and poplar. *Biochemical Journal*, **162**, 267–79.

DALESSANDRO, G. & NORTHCOTE, D. H. (1977b). Changes in enzymic activities of nucleoside diphosphate sugar interconversions during differentiation of cambium to xylem in pine and fir. *Biochemical Journal*, **162**, 281–8.

DALESSANDRO, G. & NORTHCOTE, D. H. (1977c). Possible control sites of polysaccharide synthesis during cell growth and wall expansion of pea seedlings (*Pisum sativum* L.). *Planta*, **134**, 39–44.

DALESSANDRO, G. & NORTHCOTE, D. H. (1981). Increase of xylan synthetase activity during xylem differentiation of the vascular cambium of sycamore and poplar trees. *Planta*, **151**, 61–7.

ESAU, K., CHEADLE, V. I. & RISLEY, E. B. (1962). Development of sieve-plate pores. *Botanical Gazette*, **123**, 233–43.

FREUDENBERG, K. (1959). Biosynthesis and constitution of lignin. *Nature*, **183**, 1152–5.

FREUDENBERG, K. (1965). Lignin: its constitution and formation from p-hydroxy-cinnamyl alcohols. *Science*, **148**, 595–600.

HADDON, L. E. & NORTHCOTE, D. H. (1975). Quantitative measurement of the course of bean callus differentiation. *Journal of Cell Science*, **17**, 11–26.

HADDON, L. & NORTHCOTE, D. H. (1976a). The influence of gibberellic acid and abscisic acid on cell and tissue differentiation of bean callus. *Journal of Cell Science*, **20**, 47–55.

HADDON, L. & NORTHCOTE, D. H. (1976b). Correlation of the induction of various enzymes concerned with phenylpropanoid and lignin synthesis during differentiation of bean callus (*Phaseolus vulgaris* L). *Planta*, **128**, 255–62.

HADDON, L. & NORTHCOTE, D. H. (1976c). The effect of growth conditions and origin of tissue on the ploidy and morphogenetic potential of tissue cultures of bean (*Phaseolus vulgaris* L). *Journal of Experimental Botany*, **27**, 1031–51.

HAHLBROCK, K. & GRISEBACH, H. (1979). Enzymic controls in the biosynthesis of lignin and flavonoids. *Annual Review of Plant Physiology*, **30**, 105–30.

HANKE, D. E. & NORTHCOTE, D. H. (1974). Cell wall formation by soybean callus protoplasts. *Journal of Cell Science*, **14**, 29–50.

HARKIN, J. M. & OBST, J. R. (1973). Lignification in trees; indication of exclusive peroxidase participation. *Science*, **180**, 296–7.

HASSID, W. Z., NEUFELD, E. F. & FEINGOLD, D. S. (1959). Sugar nucleotides in the interconversion of carbohydrates in higher plants. *Proceedings of the National Academy of Sciences, USA*, **45**, 905–15.

JEFFS, R. A. & NORTHCOTE, D. H. (1966). Experimental induction of vascular tissue in an undifferentiated plant callus. *Biochemical Journal*, **101**, 146–52.

JEFFS, R. A. & NORTHCOTE, D. H. (1967). The influence of indole-3yl acetic acid and sugar on the pattern of induced differentiation in plant tissue culture. *Journal of Cell Science*, **2**, 77–88.

JONES, D. H. & NORTHCOTE, D. H. (1981). Induction by hormones of phenylalanine ammonia-lyase in bean cell suspension cultures: inhibition and superinduction by actinomycin D. *European Journal of Biochemistry*, **116**, 117–25.

KERR, T. & BAILEY, I. W. (1934). Structure, optical properties and chemical composition of the so-called middle lamella. *Journal of the Arnold Arboretum*, **15**, 327–49.

KESSLER, G. (1958). Zur Charakterisierung der Silbrohrenkallose. *Berichte der Schweizerische Botanische Gesellschaft*, **68**, 5–43.

KOUKOL, J. & CONN, E. E. (1961). The metabolism of aromatic compounds in higher plants. IV. Purification and properties of the phenylalanine deaminase of *Hordeum vulgare*. *Journal of Biological Chemistry*, **236**, 2692–8.

LOEWUS, F., CHEN, M-S. & LOEWUS, M. F. (1973). In *Biogenesis of Plant Cell Wall Polysaccharides*, ed. F. Loewus, pp. 1–27. New York: Academic Press.

MORRIS, M. R. & NORTHCOTE, D. H. (1977). Influence of cations at the plasmamembrane in controlling polysaccharide secretion from sycamore suspension cells. *Biochemical Journal*, **166**, 603–18.

NEISH, A. C. (1961). Formation m- and p-coumaric acids by enzymatic deamination of the corresponding isomers of tyrosine. *Phytochemistry*, **1**, 1–24.

NEUFELD, E. F. & HALL, C. W. (1965). Inhibition of UDP-D-glucose dehydrogenase by UDP-D-xylose: a possible regulatory mechanism. *Biochemical and Biophysical Research Communications*, **19**, 456–61.

NORTHCOTE, D. H. (1962). The nature of plant cell surfaces. *Biochemical Society Symposia*, **22**, 105–25.

NORTHCOTE, D. H. (1963). Polysaccharides. *Annual Review of Biochemistry*, **33**, 51–74.

NORTHCOTE, D. H. (1969). The synthesis and metabolic control of polysaccharides and lignin during the differentiation of plant cells. *Essays in Biochemistry*, **5**, 90–137.

NORTHCOTE, D. H. (1972). Chemistry of the plant cell wall. *Annual Review of Plant Physiology*, **23**, 113–32.

NORTHCOTE, D. H. (1974). Sites of synthesis of the polysaccharides of the cell wall. In *Plant Carbohydrate Biochemistry*, ed. J. P. Pridham, pp. 165–81. London, New York, San Francisco: Academic Press.

NORTHCOTE, D. H. (1977). The synthesis and assembly of plant cell walls; possible control mechanisms. In *The Synthesis, Assembly and Turnover of Cell Surface Components*, vol. 4, ed. G. Poste & G. L. Nicholson, pp. 717–39. Elsevier/North-Holland Biomedical Press.

NORTHCOTE, D. H. (1979). The involvement of the Golgi apparatus in the biosynthesis and secretion of glycoproteins and polysaccharides. *Biomembranes*, vol. 10, ed. L. A. Manson, pp. 51–76. New York: Plenum Publishing Company.

RAY, P. M. (1980). Co-operative action of β-glucan synthetase and UDP-xylose xylosyl tranferase of Golgi membranes in the synthesis of xyloglucan-like polysaccharide. *Biochimica et Biophysica Acta*, **629**, 431–44.

RUBERY, P. H. & NORTHCOTE, D. H. (1968). Site of phenylalanine ammonia lyase activity and synthesis of lignin during xylem differentiation. *Nature*, **219**, 1230–4.

RUBERY, P. H. & NORTHCOTE, D. H. (1971). The effect of auxin (2, 4-dichlorophenoxy-acetic acid) on the synthesis of cell wall polysaccharides in cultured sycamore cells. *Biochimica et Biophysica Acta*, **222**, 95–108.

SKOOG, F. & MILLER, C. O. (1957). Chemical regulation of growth and organ formation in plant tissue cultured *in vitro*. *Symposia of the Society of Experimental Biology*, **11**, 118–31.

THORNBER, J. P. & NORTHCOTE, D. H. (1961). Changes in the chemical composition of a cambial cell during its differentiation into xylem and phloem tissue in trees. 1. Main components. *Biochemical Journal*, **81**, 449–55.

THORNBER, J. P. & NORTHCOTE, D. H. (1962). Changes in the chemical composition of a cambial cell during its differentiation into xylem and phloem tissue in trees. 3. Xylan, glucomannan and α-cellulose fractions. *Biochemical Journal*, **82**, 340–6.

TOWERS, G. H. N. (1974). Enzymological aspects of flavonoid and lignin biosynthesis. In *Plant Biochemistry: MTP International Review of Science*, vol 11, ed. D. H. Northcote, pp. 247–76. London: Butterworths & Co. Ltd.

WETMORE, R. H. & RIER, J. P. (1963). Experimental induction of vascular tissues in callus of angiosperms. *American Journal of Botany*, **50**, 418–30.

WOODING, F. B. P. & NORTHCOTE, D. H. (1964). The development of the secondary wall of the xylem in *Acer pseudoplatanus*. *Journal of Cell Biology*, **23**, 327–37.

WOODING, R. H. & NORTHCOTE, D. H. (1965). The fine structure and development of the companion cell of the phloem of *Acer pseudoplatanus*. *Journal of Cell Biology*, **23**, 327–37.

WRIGHT, K. & NORTHCOTE, D. H. (1972). Induced root differentiation in sycamore callus. *Journal of Cell Science*, **11**, 319–37.

Accumulation of secondary products as a facet of differentiation in plant cell and tissue cultures

M. M. YEOMAN, K. LINDSEY, M. B. MIEDZYBRODZKA AND W. R. McLAUCHLAN

Department of Botany, University of Edinburgh EH9 3JH

INTRODUCTION

The complex metabolic changes which accompany the differentiation of cells in plants bring into operation, or initiate, biosynthetic pathways which result in the accumulation of new compounds. For example, the synthesis and accumulation of lignin by differentiating tracheidal and vessel elements of the xylem (Northcote, 1974) marks the termination of a differentiation process which began much earlier in the expanding cells of the procambium and which prepares the differentiated xylem for water and salt transport. Many of the substances accumulated, such as lignin and chlorophyll, are common, with few exceptions, to all higher plants. In contrast, other compounds appear to be characteristic of a species, or a genus, for example, ricin in *Ricinus communis*, the castor bean and diosgenin in certain species of the genus *Dioscorea*, the yams. Both groups of compounds, the general and the more or less specific, are the products of secondary metabolism and are described as secondary to distinguish them from primary compounds which are the products of primary metabolism. Although it is difficult to distinguish with certainty between primary and secondary metabolism it would appear that it is the products of secondary metabolism which are largely associated with the process of differentiation (Luckner, 1972; Böhm, 1980). Indeed, secondary products are normally found in highly differentiated parts of plants and constitute an important facet of differentiation. Secondary products may be formed as a consequence of differentiation or vice versa, so that a study of secondary metabolism should be expected to yield essential facts about the process of differentiation.

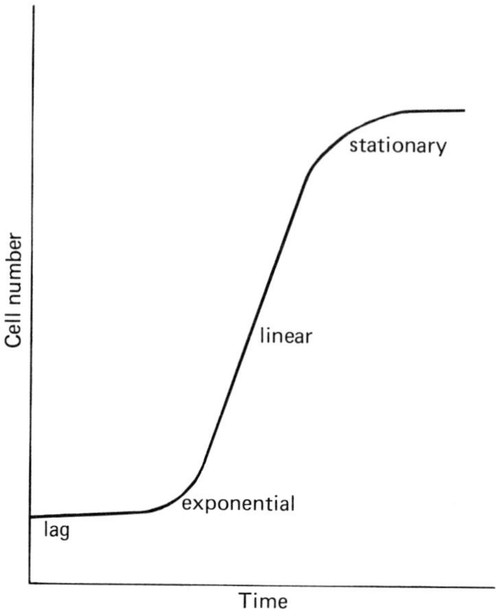

Fig. 1. The generalised growth of a callus or cell culture (shown here in terms of an increase in cell number). Differentiation and the accumulation of secondary metabolites tend to occur maximally when the growth rate decreases, at or approaching the stationary phase.

USE OF CELL CULTURES FOR THE STUDY OF SECONDARY METABOLISM

Cell cultures from higher plants would appear to be good systems for the study of secondary metabolism because of the ease with which the physical and chemical environment of the cultured cells can be closely controlled and monitored (Barz, Reinhard & Zenk, 1977). Generally, however, it has been found difficult to encourage cultures of plant cells to mimic the synthetic capabilities of the plants from which the culture originated, although there are exceptions (Butcher, 1977; Aitchison & Yeoman, 1977; Staba, 1980; Yeoman, Miedzybrodzka, Lindsey & McLauchlan, 1980). The basis of all studies on the accumulation of secondary products by tissue cultures is that the callus induced from a tissue fragment is sub-cultured to provide either more callus or a cell suspension culture. Although such cultures may appear to be reasonably homogeneous, they are not, and represent mixtures of cells of different shapes, sizes and synthetic capabilities (Yeoman & Forche, 1980). During each period of sub-culture the heterogeneous cell population proliferates and the products differentiate freely, producing a variety of cells with different morphologies. If simple growth kinetics are applied to any tissue in culture it can be clearly seen that differentiation follows division in a regular sequence (see Fig. 1) so that the greatest number of differentiated cells is present towards the end of the culture period when cell division has ceased. The transfer of a fragment of this callus to a fresh medium induces a new round of division and differentiation and this constitutes a growth cycle. The growth rate, as measured by the increase in cell number, is high at the beginning of sub-culture and slows down during the growth cycle reaching a low value when differentiation is occurring most extensively. It is during this phase of decreasing growth rate and increasing differentiation that accumulation of secondary products normally takes place. This point can be illustrated from work in this laboratory in which cells of *Dioscorea composita* were grown in suspension culture. It can be seen clearly from Fig. 2 that growth of the cells precedes the accumulation of diosgenin. This agrees well with the situation in many intact plants where accumulation occurs in differentiated structures in which growth is slow or has been terminated. However, here it is relevant to point out an important difference between the culture and the intact plant, which concerns the cytogenetical stability of the constituent cells. The instability of plant cells in culture is well known (Bayliss, 1980) and contrasts with the general stability of the constituent cells of the intact plant. During repeated sub-culture various changes occur in cells which modify the ability of these cells to differentiate. Often sub-culture over an extended period leads to a loss in the ability of a particular cell culture to differentiate in a particular way and may diminish the ability of the cells to synthesise and accumulate a particular secondary product.

A detailed consideration of the literature (see Butcher, 1977; King & Street, 1977; Staba, 1980) shows that rapidly growing cultures tend to accumulate cells and not secondary products, and conversely, mature, highly differentiated cells tend to accumulate secondary products. This suggests quite strongly that an inverse relationship exists between the accumulation of a secondary product (differentiation) and growth (accumulation of cells and primary cellular materials). If an intimate relationship between growth and the accumulation of a secondary product, which the theory predicts, actually exists, then it should be possible to terminate cell division and cell growth prematurely and initiate the accumulation of a particular secondary compound earlier

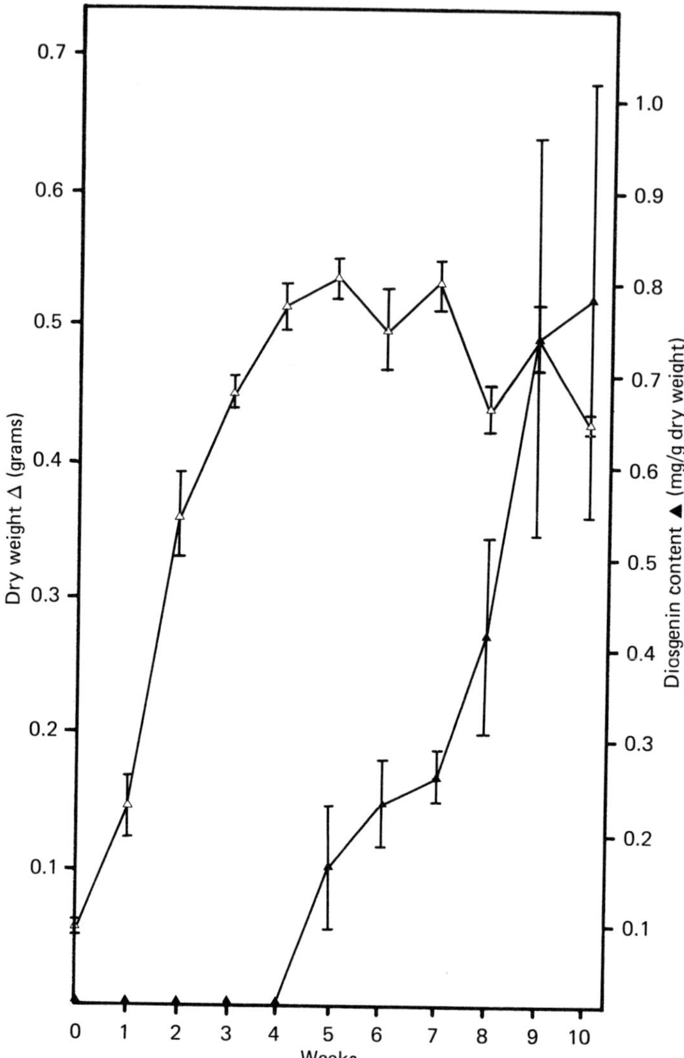

Fig. 2. The pattern of growth (measured as dry weight △) and accumulation of diosgenin (▲) during a growth cycle of *Dioscorea composita* cell suspension cultures. Cells were grown in 60 ml MS medium supplemented with 1.4×10^{-5} M 2,4-D, in 250 ml Erlenmeyer flasks, which were rotated at 24 ± 1 °C in continuous white light (approximately 200 lux). The vertical bars represent standard errors.

Table 1. *Relationship between alkaloid content, appearance
and growth rate of the callus*

Species	Appearance	Growth rate	Alkaloid content mg/g dry weight
Datura innoxia	pale cream, friable	fast	0.10
Datura innoxia	green, compact	slow	0.40
Datura stramonium	pale green	moderate	0.50
Datura stramonium	green/brown	slow	1.00
Datura clorantha	dark brown, friable	moderate	0.10
Datura clorantha	green compact	slow	1.20
Solanum dulcamara	brown	very slow	0.25
Solanum dulcamara	green compact	slow	1.00

Callus sampled at the end of a 28 day culture period.

than it would have occurred in a normal growth situation. It should also be possible to follow the switch from primary to secondary metabolism by the use of a radioactive precursor for the product. Phillips & Henshaw (1977) have successfully switched from protein synthesis (PRIMARY METABOLISM) to the synthesis of polyphenols (SECONDARY METABOLISM) by manipulating the phosphate levels in suspension cultures of sycamore (*Acer pseudoplatanus* L.) and have followed this switch using labelled phenylalanine. They have also shown that this switch can be influenced by other nutrients such as nitrogen, or growth substances.

RELATIONSHIP BETWEEN ALKALOID CONTENT, APPEARANCE AND GROWTH OF CALLUS

Lindsey (1979, unpublished data) in an extended investigation has studied the relationships which exist between alkaloid content, appearance and growth characteristics of callus cultures of a range of Solanaceous species (see Table 1). It is clear from the data presented that green calluses display a higher alkaloid content than cultures devoid of chlorophyll. This is also true for a range of leguminous species including *Trigonella balansae, T. corniculata, Lupinus angustifolius, L. luteus* and *L. mutabilis*. An additional point which provides further support for the conclusion that secondary product accumulation is inversely proportional to growth can also be seen from Table 1. Pale friable calluses which are actively growing and poorly differentiated show low levels of alkaloids even in the post-proliferation stage; in contrast slow-growing green cultures accumulate the highest levels of alkaloids.

EFFECTS OF INHIBITORS ON SECONDARY PRODUCT ACCUMULATION, PROTEIN SYNTHESIS AND GROWTH OF CALLUS

Further evidence in support of the general conclusion that secondary product accumulation is inversely proportional to growth is presented in Table 2. Mizukami, Konoshima & Tabata (1977) have demonstrated that cell cultures of *Lithospermum* can be stimulated to accumulate shikonin derivatives by the addition of streptomycin sulphate, an inhibitor of RNA and protein synthesis. Similar effects have been reported by Neumann & Müller (1971) who have shown that cycloheximide inhibits protein synthesis and therefore growth, but does not affect alkaloid accumulation by callus and suspension cultures of *Nicotiana tabacum*. D-threo-chloramphenicol inhibited growth of the culture, and stimulated free amino acid content as well as alkaloid production. In a more recent paper Neumann & Müller (1974) have shown that actinomycin D promotes alkaloid accumulation in callus cultures of *Macleaya cordata* while cycloheximide inhibits growth without stimulating alkaloid accumulation. Both of these antibiotics inhibit protein synthesis in callus cells of *Macleaya*. Aitchison (1977, unpublished data) working with callus cultures of *Capsicum frutescens* has shown that cycloheximide, at concentrations which inhibit protein synthesis, prevented the incorporation of radioactive phenylalanine and valine into callus cells, but stimulated the incorporation of these amino acids into capsaicin, a secondary product characteristic of the fruit of the plant (Chilli pepper) (see Table 3). From these somewhat limited data (Table 2) it may be tentatively concluded that inhibitors of RNA synthesis tend to stimulate secondary product accumulation while inhibitors of protein synthesis may stimulate secondary product formation or have no effect, certainly at the range of concentrations employed. In all cases the inhibitors of RNA and protein synthesis prevent growth although the long-term effects of substances such as actinomycin D, cycloheximide and D-threo-chloramphenicol are difficult to interpret and these results must be approached with some caution. Despite these reservations it seems reasonable to conclude that

Table 2. *Effects of inhibitors on secondary product accumulation, protein synthesis and growth of callus*

Species	Inhibitor	Compound	Effect			Reference
			Compound	Protein synthesis	Growth	
Lithospermum erythrorhizon	Streptomycin sulphate	shikonin derivatives	↑	↓	↓	Mizukami *et al.* (1977)
Macleaya cordata	Actinomycin D	alkaloid	↑	↓	↓	Neumann & Müller (1974)
Macleaya cordata	Cycloheximide	alkaloid	→	↓	↓	Neumann & Müller (1974)
Nicotiana tabacum	D-threo-chloramphenicol	alkaloid	↑	↓	↓	Neumann & Müller (1971)
Nicotiana tabacum	L-threo-chloramphenicol	alkaloid	→	→	→	Neumann & Müller (1971)
Nicotiana tabacum	Cycloheximide	alkaloid	→	↓	↓	Neumann & Müller (1971)
Capsicum frutescens	Cycloheximide	capsaicin	↑	↓	↓	Aitchison 1977 (unpublished data)

↑ indicates an increase in the level or a stimulation.
↓ indicates a decrease in the level or a depression.
→indicates no effect.

Table 3. *A summary of salient points relating to the incorporation of* ^{14}C *valine and* ^{14}C *phenylalanine into the capsaicin isolated from callus of* Capsicum frutescens *(see Yeoman et al., 1980)*

1   ^{14}C valine and ^{14}C phenylalanine when applied together or separately to actively growing cultures are incorporated into capsaicin at *very* low levels. Most of the recovered radioactivity is in protein.

2   ^{14}C valine and ^{14}C phenylalanine when applied together or separately to callus cultures in which the rate of growth has been *slowed down* by reducing the level of nitrogen and/or phosphorus in the medium are incorporated into capsaicin at *much higher* levels.

3   Application of cycloheximide to actively growing cultures blocks protein synthesis and increases the level of incorporation of ^{14}C valine and ^{14}C phenylalanine into capsaicin.

in several instances the accumulation of a particular compound may be stimulated by conditions in which protein synthesis is prevented, which suggests that the metabolic machinery essential for the synthesis of that compound is present in the cells at the time of application of the inhibitor and is sufficient to support synthesis of the designated secondary compound.

DO CELLS CONTAIN THE ENZYMIC MACHINERY TO SYNTHESISE SECONDARY PRODUCTS PRIOR TO ACCUMULATION?

The accumulation of a particular compound occurs when synthesis exceeds degradation. An increased rate of synthesis accompanied by a smaller increase or a cessation in degradation will lead to an accumulation; similarly in a condition where the rate of synthesis remains unchanged, but the rate of degradation is reduced, accumulation will also be observed. In cells the product may be transported away from the site of synthesis to a vacuole or exported to other cells or outside the organism. Here removal may be considered as synonymous with degradation. Clearly, the particular compound will not be produced in the absence of the biosynthetic pathway. From this it would appear that cultured cells which do not accumulate a particular compound during the growth phase will do so only when growth has slowed down or stopped completely; presumably because the growing cells do not possess the appropriate biosynthetic pathway. Further, it would

Table 4. *A summary of the evidence from the work of Banthorpe on monoterpene synthesis (1979, 1980, unpublished data) supporting the general supposition that biosynthetic pathways for designated secondary metabolites are present in the cells of callus cultures which are not accumulating these compounds (see text for full explanation)*

1. Callus cultures of *Ocimum basilicum*, *Tanacetum vulgare*, *Rosmarinus officinalis* and *Lavandula spica* do not accumulate monoterpenes or terpene glycosides.

2. Plants of the above-named species do accumulate monoterpenes and terpene glycosides.

3. Cell-free extracts from both the plant and the callus cultures have the ability to sustain the synthesis of monoterpenes (geraniol, nerol, and α pinene, β pinene, 1,8-cineole and carvone) from isopentenyl pyrophosphate (IPP) which supports the concept that the complete enzyme machinery is present in the callus for monoterpene synthesis.

follow that as growth slows down and the cells differentiate there is a controlled sequential read-out of the information in the genome which leads to the synthesis of enzymes which are organised into a biosynthetic pathway and the compound is synthesised and accumulated. It is therefore important before proceeding further to determine from the available evidence whether the complete biosynthetic pathways are present or absent from cells not accumulating a particular compound.

The simplest and perhaps the best approach is to discover whether cell-free preparations from cultured cells, which do not accumulate a particular compound, can incorporate a radioactive precursor, far removed from the product, into the designated compound. Here, the work of Banthorpe at the Department of Chemistry at University College, London University, is conclusive. Banthorpe (1979, 1980, unpublished data) (see Table 4) has for some time been examining the ability of established callus cultures from *Ocimum basilicum* (basil), *Tanacetum vulgare* (tansy), *Rosmarinus officinalis* (rosemary), and *Lavandula spica* (lavender) to synthesise and accumulate monoterpenes. None of these cultures accumulated or secreted detectable amounts of monoterpenes, or as far is known, terpene glycosides. However, cell-free extracts of each of these cultures, prepared by three different techniques (see Banthorpe, Bucknall, Doonan, Doonan & Rowan, 1976), could sustain the biosynthesis of monoterpenes (geraniol, nerol, α pinene, β pinene, 1,8-cineole and carvone) from ^{14}C isopentenyl pyrophosphate (IPP). Also the observed enzyme levels of the steps of the biosynthetic pathway were *20–4000* fold greater than the enzyme activities that could be extracted, using the same techniques, from the intact plant which displayed the same pattern of products. Additionally in experiments with *Rosa* and Jasmine species Banthorpe prepared active cell-free extracts from calluses which sustained similar reactions to the intact plant but the ratio of enzyme activity, callus/plant was only 0.5–1.5. In all of these studies with cell-free extracts it is important to realise that the lower activities observed with material from the intact plant may reflect a destruction of activity or deactivation of enzymes by complexing of proteins to phenols and quinones. Here it is relevant to point out that the callus extracts contained a lower concentration of phenolic compounds than the intact plant. Recent experiments in this laboratory on alkaloid accumulation in cell suspension cultures of *Solanum nigrum* (Perez-Frances 1980, unpublished data) have demonstrated that cell-free extracts of a culture in the exponential phase of growth not accumulating hyoscyamine, will sustain the conversion of ^{14}C ornithine to hyoscyamine. Perez-Frances also showed that exponentially growing cell cultures of *S. nigrum* would incorporate ^{14}C ornithine into hyoscyamine demonstrating the operation of the biosynthetic pathway in these cells.

The results obtained in investigations with callus and cell cultures of *Capsicum frutescens* in this laboratory (Yeoman *et al.*, 1980) also support the general idea that the enzymic machinery for the synthesis of a particular secondary compound, in this case capsaicin, is present within cells not accumulating capsaicin. Callus tissue, either freshly isolated or which has been in culture for four years, will not accumulate capsaicin during the course of a growth cycle on an agar growth medium. Although capsaicin cannot be detected in the culture by sensitive analytical procedures (e.g. GLC) the compound can be labelled by feeding the callus with ^{3}H or ^{14}C phenylalanine or valine (see Table 3). Extracts of Chilli pepper callus from experiments with these labelled amino acids contain radioactively marked capsaicin and dihydrocapsaicin. These compounds can be detected as radioactive areas on TLC plates and will co-chromatograph with pure samples of these compounds in a number of solvent systems. It is estimated, but this must be taken as very approximate, that the level of capsaicin in the cells is of the order of 10^{-10}M. However, the level of label in capsaicin can be raised substantially by manipulating the conditions in which the cultures are maintained (Yeoman *et al.*, 1980) or by the use of inhibitors (Table 3). In conditions of 'nutrient stress' in which most of the phosphate has been utilised and precursors are present (the cells are growing very slowly) the yield of capsaicin can be increased spectacularly to milligram levels which is comparable with the concentration present in the mature fruit of the Chilli pepper (2–3 mg).

The results of an experiment in which phenylalanine, valine, vanillylamine and iso-capric acid were fed in various combinations to callus cultures of *Capsicum frutescens* are shown in Table 5. The experiments were conducted using a 'flat-bed' culture apparatus. The principal feature of this culture system is the movement of a liquid nutrient medium across stationary cells. The medium drips from a reservoir into a culture vessel containing cells seated on a synthetic fabric substratum, moves by capillary action across the fabric, and is thence pumped from the vessel back into the reservoir. Such a system allows sequential chemical treatments and additions of relatively large quantities of precursors at low concentration. This culture situation is one in which cell differentiation takes place at rates comparable to that of callus on an agar medium. It can be seen from the data summarised in Table 5 that capsaicin cannot be detected in either callus or medium unless precursors are supplied to the cells. The highest yields were obtained in the presence of iso-capric acid, a precursor close to the end of the biosynthetic pathway (see Yeoman *et al.*, 1980). The presence of vanillylamine which might be expected to further increase the yield decreased the overall amount of capsaicin. This may be due to its toxicity or some antagonistic effect. The best yield was comparable to that found in the mature fruit, 2.3–2.5 mg of capsaicin/g dry weight. The general conclusion from this work with the Chilli pepper is that the slowing down of growth (phosphate is depleted within a few days) and the provision of precursors, results in the accumulation of capsaicin in the medium. In this case the induction of the biosynthetic pathway by the addition of precursors is possible but unlikely. Two pieces of evidence support this interpretation. The first is that in earlier experiments with *Capsicum* callus (Yeoman *et al.*, 1980) very small amounts of ^{3}H or ^{14}C valine or phenylalanine are rapidly incorporated into capsaicin; although enzyme induction is possible, it seems unlikely at the concentrations used. Secondly, in experiments with cultures from *Solanum nigrum* (Prez-Frances 1980, unpublished data), ^{14}C ornithine fed to intact cells, or a cell-free extract, was rapidly incorporated into

Table 5. *Levels of capsaicin accumulated in callus cultures of* Capsicum frutescens

	Source	Vanillylamine	Iso-capric acid	Phenylalanine	Valine	Incubation time (weeks)	µg per dish	% maximum theoretical yield
1	Callus	—	—	—	—	2	ND	—
	Medium							
2	Callus	—	5 mM	5 mM	—	2	ND	—
	Medium						75	0.14
3	Callus	—	5 mM	—	—	3	ND	—
	Medium						2371	4.20
4	Callus	10 mM	10 mM	—	—	3	ND	—
	Medium						597	0.47
5	Callus	5 mM	—	—	—	3	ND	—
	Medium						171	0.26
6	Callus	5 mM	—	—	5 mM	2	ND	—
	Medium						11	0.02

ND, Not Detectable.

hyoscyamine; this alkaloid could not be detected in these cultures without ornithine using conventional analytical procedures. Here the biosynthetic pathway from ornithine to hyoscyamine is present in cultures not accumulating the alkaloid. It is also interesting, and perhaps important, to note that if the detached fruits of *Capsicum annuum* var. grossum, the sweet pepper, which does *not* naturally accumulate capsaicin or dihydrocapsaicin, are supplied with iso-capric acid and vanillylamine, dihydrocapsaicin and related capsaicinoids can be detected at a level of 7.2 µg/fruit after six days. It is also highly significant that cell-free extracts of these sweet pepper fruits will synthesise dihydrocapsaicin from a reaction mixture containing vanillylamine and iso-capric acid (Iwai, Suzuki, Lee, Kobashi & Oka, 1977). Clearly here is another example where at least part of the biosynthetic pathway for capsaicin is present in a structure which does not naturally accumulate the compound.

A WORKING HYPOTHESIS

The evidence presented, although fragmentary, suggests that the inability of at least some cultures of plant cells to accumulate a designated compound or compounds characteristic of the species may *not* be due to the absence of the complete biosynthetic pathway(s) for that compound(s) from the cells. Although there may be exceptions the general rule appears to be that if the chemical milieu encourages growth then growth will occur and when growth ceases or is stopped, because of a shortage or withdrawal of an essential nutrient or nutrients, a metabolic shunt will occur in which biosynthetic pathways are modified, brought into play and new compounds are produced. This accumulation occurs, as in the intact plant at the expense of growth and proliferation, and the addition of precursors for the particular metabolite may encourage synthesis and accumulation, particularly at the 'down-turn' of growth.

At first, we (see Yeoman *et al.*, 1980) were in favour of a 'plumbing analogy' in which the primary and secondary pathways were always present and switching off or reducing the flow of metabolites through the primary pathways leading to protein synthesis and growth would divert certain essential key metabolites into the secondary pathways. However, the argument can be raised that if the flow of metabolites through the primary pathways is proceeding at a *maximum* rate then the provision of a block (slowing down or cessation of growth), would not increase the rate of flow through the 'narrower pipes' of the secondary pathways. Clearly the hypothesis lacks switches or perhaps more appropriate to the plumbing analogy 'valves'.

From the evidence available it would appear that at least in some culture systems the complete biosynthetic pathway for the particular metabolite in question is present in the cells at times when the culture is *not* accumulating the secondary product and that the addition of a precursor to a cell-free extract will result in the sustained production of the product. This, together with the suggestion that growth is inversely proportional to secondary product accumulation, would suggest that the pathways are present but inoperative and that a switch brings the existing pathway(s) into operation. If we can indulge ourselves with a further analogy it is rather like the switching of points on a railway system which diverts the train from the main to the branch line! Both sets of tracks are present and it is the setting of the points which determines the destination of the train.

Clearly the nature and extent of the switching is crucial. The simplest concept would be to invoke a small number of key enzymes which control the flux of a few important metabolites through various secondary pathways. One candidate immediately springs to mind, phenylalanine as the important metabolite and phenylalanine ammonia lyase (PAL) as the key enzyme. Phenylalanine is a molecule which stands at an important cross-roads in metabolism; it is used extensively in protein synthesis and for a wide range of polyphenolic compounds. Labelled phenylalanine fed to actively growing cultures of *Capsicum frutescens* is rapidly incorporated into protein; however the same compound fed to cultures which have ceased to grow but begun to differentiate will be incorporated into a range of polyphenolic compounds and capsaicin. It has been observed by Davies (1971, 1972) that PAL activity increases very considerably in cell cultures of a *Rosa* sp. immediately before the accumulation of polyphenolics begins. Similar observations have been made by Hahlbrock, Kuhlen & Lindl (1971) who have shown a large increase in PAL activity in dark grown batch cell cultures of *Glycine max* at the end of the growth period. These have been interpreted as indicative of the opening up of aspects of secondary metabolism. PAL is an enzyme which can be induced by light and an increase in the activity of PAL is usually followed by an increase in the formation of a particular range of polyphenolic compounds. Isoenzymes of PAL have not yet been found in cell cultures (Zaprometov, 1978). The synthesis and accumulation of many flavonoid compounds are light-dependent, particularly in cultured cells. The light-dependent regulation of biosynthesis of cinnamic acids and some classes of flavonoids at the enzymic level has been studied extensively by Hahlbrock (1977). He has shown that in cell cultures of parsley (*Petroselinum hortense*) and soybean (*Glycine max*), there are two integrated enzyme groups which respond to illumination in different ways. One of the groups is represented by the enzymes of the phenylpropanoid pathway, the other of the flavonoid pathway. Together the two groups of enzymes catalyse the conversion of phenylalanine and several other substrates derived from intermediary metabolism to a variety of flavone and flavonol glycosides (Hahlbrock, 1977; Ebel & Hahlbrock, 1977; Hahlbrock, Betz, Gardiner, Kreuzaler, Matern, Ragg, Schäfer & Schröder, 1978). The first group comprises the three enzymes of general phenylpropanoid metabolism beginning with PAL, whereas the enzymes of the flavonoid glycoside pathway are members of the second group. Clearly, PAL fills the role of a key enzyme which can be switched quite easily and divert an essential metabolite into secondary metabolism.

The essential features of the ideas developed in this article are summarised in Fig. 3. Schemes 1 and 2 represent two possible explanations. The pathways concerned with primary metabolism are lightly stippled, those with secondary metabolism heavily shaded. In both schemes A represents no synthesis of a designated secondary compound and B synthesis of a designated secondary compound. In scheme 1 the option is presented where the biosynthetic pathways are present but inoperative until switched by a key enzyme, on the left A is present but closed and inoperative, on the right B is present, switched on and operative. In scheme 2 the option is presented in which the biosynthetic pathway for a designated compound is absent, and therefore inoperative (A) and subsequently the complete pathway is produced *de novo* and when present is operative (B). The evidence presented in this article although incomplete is consistent with scheme 1; however considerable caution must be exercised in applying this model to all situations in which secondary products are accumulated, especially where the inverse relationship between the accumulation of the designated secondary product and growth is unclear. Like most models it is simplistic and is put forward only as a working hypothesis to encourage further experimentation in a difficult and complex area of research. The central idea of the control of some aspects of differentiation in plants by key switch enzymes which can bring biosynthetic

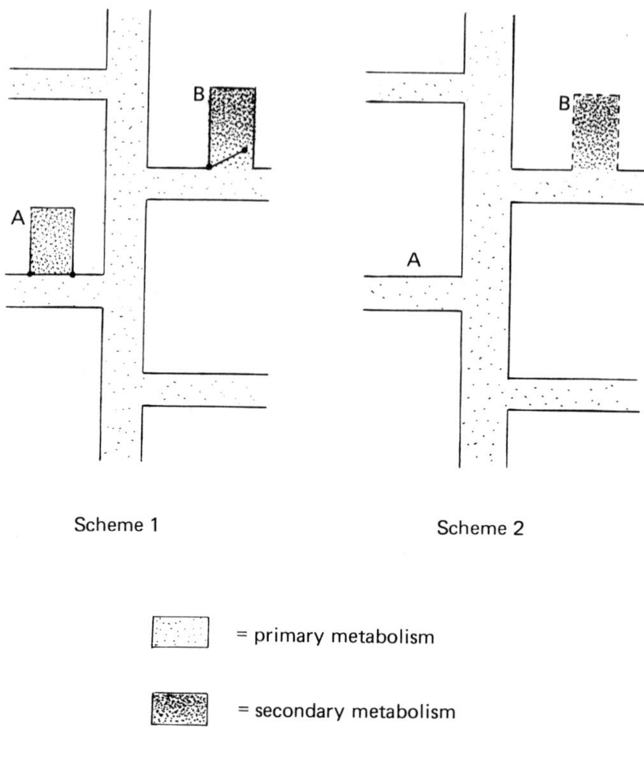

Fig. 3. A diagrammatic representation of two schemes providing an explanation of how a switch may occur between primary and secondary metabolism in cultured plant cells.

In both schemes A represents a state in which there is no synthesis of a designated secondary compound and B a state in which synthesis occurs. The lightly stippled areas are pathways of primary metabolism and the shaded areas pathways of secondary metabolism. (See text for full explanation of model.)

pathways already present rapidly into operation has some merit. Plants are organisms which are unable to move about, they have evolved against a changing environment and the successful individuals are those which respond most effectively to the prevailing physical conditions. For example, it is the nature of the metabolic machinery within the green plant which determines the speed of response and therefore a control system in which a limited number of key regulatory enzymes are involved would be of selective advantage. In contrast, the *de novo* production of a complete biosynthetic pathway would delay the response and jeopardise survival.

Acknowledgments

We are extremely grateful to Dr D. V. Banthorpe and Dr J. Perez-Frances for permission to use some of their unpublished data. We also wish to thank Mrs Marion Thomas for typing this manuscript and to Albright and Wilson Ltd for providing financial support.

REFERENCES

AITCHISON, P. A. & YEOMAN, M. M. (1977). Growth patterns in tissue (callus) cultures. In *Plant Tissue and Cell Culture*, ed. H. E. Street, pp. 267–306, *Botanical Monographs 11*. Oxford: Blackwell.

BANTHORPE, D. V., BUCKNALL, G. A., DOONAN, H. J., DOONAN, S. & ROWAN, M. G. (1976). Biosynthesis of geraniol and nerol in cell-free extracts of *Tanacetum vulgare*. *Phytochemistry*, **15**, 91–100.

BARZ, W., REINHARD, E. & ZENK, M. H. (1977). *Plant Tissue Culture and its Biotechnological Application*. Berlin: Springer-Verlag.

BAYLISS, M. W. (1980). Chromosomal variation in plant tissues in culture. In *Perspectives in Plant Cell and Tissue Culture*, ed. I. K. Vasil, pp. 113–44, *International Review of Cytology, Supplement IIA*. New York: Academic Press.

BÖHM, H. (1980). The formation of secondary metabolites in plant tissue and cell cultures. In *Perspectives in Plant Cell and Tissue Culture*, ed. I. K. Vasil, pp. 183–208, *International Review of Cytology, Supplement IIB*. New York: Academic Press.

BUTCHER, D. N. (1977). Secondary products in tissue cultures. In *Applied and Fundamental Aspects of Plant Cell, Tissue and Organ Culture*, ed. J. Reinert & Y. P. S. Bajaj, pp. 668–93. Berlin: Springer-Verlag.

DAVIES, M. (1971). Multisample enzyme extraction from cultured plant cell suspensions. *Plant Physiology, (Lancaster)*, **47**, 38–42.

DAVIES, M. (1972). Polyphenol synthesis in cell suspension cultures of Paul's Scarlet Rose. *Planta*, **104**, 50–65.

EBEL, J. & HAHLBROCK, K. (1977). Enzymes of flavone and flavonol-glycoside biosynthesis. *European Journal of Biochemistry*, **75**, 201–9.

HAHLBROCK, K. (1977). Regulatory aspects of phenylpropanoid biosynthesis in cell cultures. In *Plant Tissue Culture and its Biotechnological Application*, ed. W. Barz, E. Reinhard & M. H. Zenk, pp. 95–111. Berlin: Springer-Verlag.

HAHLBROCK, K., BETZ, B., GARDINER, S. E., KREUZALER, F., MATERN, U., RAGG, H., SCHÄFER, E. & SCHRÖDER, J. (1978). Enzyme induction in cultured cells. In *Frontiers of Plant Tissue Culture*, ed. T. A. Thorpe, pp. 317–24. Calgary: IAPTC.

HAHLBROCK, K., KUHLEN, E. & LINDL, T. (1971). Änderungen von Enzymaktivitaten während des Wachstums von Zellsuspensionskulturen von *Glycine max*: phenylalanin ammonium und p-cumarat: CoA Ligase. *Planta*, **99**, 311–18.

IWAI, K., SUZUKI, T., LEE, K., KOBASHI, M. & OKA, S. (1977). *In vivo* and *in vitro* formation of dihydrocapsaicin in sweet pepper fruits, *Capsicum annuum* L. var. grossum. *Agricultural and Biological Chemistry*, **41**, 1877–82.

KING, P. J. & STREET, H. E. (1977). Growth patterns in cell cultures. In *Plant Tissue and Cell Culture*, ed. H. E. Street, pp. 307–87, *Botanical Monographs 11*. Oxford: Blackwell.

LUCKNER, M. (1972). *Secondary Metabolism in Plants and Animals*. London: Chapman & Hall.

MIZUKAMI, H., KONOSHIMA, M. & TABATA, M. (1977). Effect of nutritional factors on shikonin derivative formation in *Lithospermum* callus cultures. *Phytochemistry*, **16**, 1183–6.

NEUMANN, VON D. & MÜLLER, E. (1971). Beiträge zur Physiologie der Alkaloide. *V* Alkaloidbildung in Kallus und Suspensionskulturen von *Nicotiana tabacum* L. *Biochemie und Physiologie der Pflanzen*, **162**, 503–13.

NEUMANN, VON D. & MÜLLER, E. (1974). Beiträge zur Physiologie der Alkaloide. *IV* Alkaloidbildung in Kalluskulturen von *Macleaya*. *Biochemie und Physiologie der Pflanzen*, **165**, 271–82.

NORTHCOTE, D. H. (1974). *Differentiation in Higher Plants. Oxford Biology readers, 44*, ed. J. J. Head. Oxford University Press.

PHILLIPS, R. & HENSHAW, G. G. (1977). The regulation of synthesis of phenolics in stationary phase cell cultures of *Acer pseudoplatanus* L. *Journal of Experimental Botany*, **28**, 785–94.

STABA, E. J. (1980). *Plant Tissue Culture as a Source of Biochemicals*. Florida: CRC Press.

YEOMAN, M. M. & FORCHE, E. (1980). Cell proliferation and growth in callus cultures. In *Perspectives in Plant Cell and Tissue Culture*, ed. I. K. Vasil, pp. 1–24, *International Review of Cytology, Supplement IIA*. New York: Academic Press.

YEOMAN, M. M., MIEDZYBRODZKA, M. B., LINDSEY, K. & McLAUCHLAN, W. R. (1980). The synthetic potential of cultured plant cells. In *Plant Cell Cultures: Results and Perspectives*, ed. F. Sala, B. Parisi, R. Cella & O. Ciferri, pp. 327–43. Amsterdam: Elsevier/North-Holland.

ZAPROMETOV, M. N. (1978). Enzymology and regulation of the synthesis of polyphenols in cultured cells. In *Frontiers of Plant Tissue Culture*, ed. T. A. Thorpe, pp. 335–43. Calgary: IAPTC.

Cellular and molecular aspects of differentiation and transdifferentiation of ocular tissues *in vitro*

R. M. CLAYTON

Department of Genetics, University of Edinburgh, West Mains Road, Edinburgh EH9 3JN

INTRODUCTION

Tissue type interconversion in the developing ocular system

The process of cell differentiation is effected by differential gene expression, but the status of the cell at any moment defines both the external signals to which it may respond and the patterns of this response. The concept of competence is long established in developmental biology. Manipulative procedures may disclose that an area or type of embryonic tissue has the potential to differentiate into several cell types, only one of which may be exhibited in normal development. Normally, as differentiation proceeds, competence to form other cell types is progressively restricted and finally lost.

The vertebrate embryonic eye offers exceptional advantages for investigations of the molecular basis of competence, since it exhibits a wide repertoire of possible tissue type interconversions *in vivo* and *in vitro*. Cells of neural retina or retinal pigmented epithelium each have the potential to re-differentiate into lens cells, a tissue of quite different developmental origin. (Other pathways are shown in Fig. 2.) This phenomenon is termed transdifferentiation: (reviewed Okada, 1976, 1980; Eguchi, 1976, 1979; Clayton, 1978).

The lens fibre cell is a terminally differentiated cell with a characteristic ultrastructure and a high content of several specific proteins, collectively called crystallins. These comprise several members of each of three antigenically unrelated gene families. α and β crystallins are universal; γ crystallin in fish, amphibians and mammals is replaced by δ crystallin in birds and reptiles. The lens continues to form fibre cells from its epithelium throughout life, and there is a regular ontogenic change in the representation of crystallin polypeptides in successively formed cells (reviewed Clayton, 1974). A terminally differentiated cell is therefore not defined by a unique representation of the possible crystallin gene products.

Table 1. *Superabundant mRNAs or proteins found at low levels in other tissues*

Protein	Major tissue source	Other tissues with low levels	Comments	Source
Crystallins	Lens	Crystallins reported in: iris, retina, embryo, adenohypophysis, cornea; lower levels in brain, liver, kidney, skin. Crystallin mRNA reported in: embryo eye cup, neural retina and pigmented retina layers	Iris, neural retina, pigment retina, embryo adenohypophysis, cornea can all transdifferentiate into lens cells with high levels of lens proteins, *in vitro*	Proteins: Clayton *et al.*, 1968; Clayton, 1970; Bours *et al.*, 1972; other references to transdifferentiation and to crystallin mRNA, in text
Chick ovalbumin	Oviduct	Ovalbumin mRNA sequences found in nuclear RNA of liver, spleen, brain, heart, and polysomes of liver, brain	Responsive to oestrogen stimulation in all tissues	Tsai *et al.*, 1979
Insulin	Pancreas	Liver, brain, cultured lymphocyte, cultured fibroblasts. Produced in some non-endocrine neoplasms	Levels unaffected by insulin depletion	Rosenzweig *et al.*, 1980
Actin, myosin, collagen	Muscle, connective tissue	Large numbers of cell types	Actins represent a gene family and each member may be restricted to certain cell types	In the lens system: see Ramaekers *et al.*, 1980 for recent review
ACTH	Pituitary	Carcinomata, mainly of lung (oat cell carcinoma), thymus, pancreas, phaochromocytoma, some ovarian tumours	Levels of ACTH and related peptides can be sufficient in these carcinomata to produce clinical symptoms	Reviewed: Bertagna *et al.*, 1979; Jeffcoate & Rees, 1978

Chorionic gonadotrophin	Placenta	Pituitary, liver, colon, ovary, testis, cultured fibroblasts. Also carcinomata of spleen, liver, pancreas, breast, kidney, melanoma, myeloma, lung, bladder, some urogenital cancers, teratomas	Not glycosylated in normal non-placental tissues. Appreciable levels synthesised by some carcinomas	Kawamura *et al.*, 1978; Braunstein *et al.*, 1979; Rosen, Kaminska, Calver & Aaronson, 1979; Yoshimoto, Wolfsen & Odell, 1977; Yoshimoto, Wolfsen, Hirose & Odell, 1979
Thy 1 (cell surface antigen)	Thymocyte, brain subpopulation of peripheral lymphocytes	Transient appearance in embryonic skeletal muscle, skin fibroblasts, on mammary tumour cell line	Thy 1 glycoproteins on thymocyte and brain may have different carbohydrate. Muscle cells have been obtained from thymus cells in culture	Lesley & Lennon, 1977; Lennon, Unger & Dulbecca, 1978; Barclay, Letarte-Muirhead, Williams & Faulkes, 1976; Wekerle, Paterson, Ketelson & Feldman, 1975
Ia (cell surface antigen)	lymphocytes B. cells	Transient on erythroid cells and granulocytes, Langerhans cells of skin, and on malignant melanoma, epidermoid carcinoma		Winchester *et al.*, 1977, 1978
Globin	Red cell	RNA sequences found in RSV transformed chick fibroblasts liver, brain, uninduced Friend leukaemia cells		Groudine & Weintraub, 1975; Gilmour *et al.*, 1974; Humphries *et al.*, 1976; Ono & Cutler, 1978

The problems offered for investigation by the vertebrate eye include, therefore, not only the conditions of the tissues which permit some of the tissue interconversions, but also, in the case of transdifferentiation into the lens pathway, the conditions regulating the crystallin representation in the transdifferentiated fibre cells (Clayton, 1978).

THE NATURE OF TISSUE SPECIFICITY

Tissue specificity in differentiated tissues

The concept that many differentiated tissues could be distinguished by their synthesis of characteristic and highly tissue-specific proteins has long been generally accepted as a basis for directing ontogenic and molecular studies of differentiation. Myosin and actin, for example, were long regarded as restricted to muscle, insulin to the pancreas, haemoglobin as exclusive to the red cell and crystallins as specific to the lens.

The application of immunological and other protein recognition techniques of increasing sophistication, and of molecular hybridisation techniques for the detection of specific species of mRNA have made it possible to detect such proteins or their mRNAs at very low levels in certain other tissues of different functions and histiotypic appearance. The locus of such low-level distributions varies according to the protein in question, but the concept that there are exclusively tissue-specific gene products is being undermined, and it may now be more appropriate to describe such proteins as super-abundant in a particular tissue, but expressed also in a certain range of other tissues (Table 1).

Both high and low levels of expression must be regulable, but attention has focussed largely on the regulatory factors involved in the synthesis of a superabundant product in the tissue it characterises.

Possible mechanisms of quantitative regulation must include transcription, processing, mRNA stability and availability for translation, translational controls and post-translational changes. Any or several of these may be affected by external signals.

In the case of crystallin synthesis in the lens cell, the initial signal for increase in crystallin mRNA is received at induction by the eye cup, subsequent signals evidently emanate from the retina and the precise balance of crystallin products expressed is affected by numerous other factors (reviewed Coulombre, 1965; Piatigorsky *et al.*, 1976; Clayton, 1978, 1979a).

Little is known so far of the regulation of the presence and the levels of superabundant ('tissue specific') gene products in the tissues in which they are present at low levels. These low levels may however also be responsive to various regulatory factors. For example, the levels of chick ovalbumin mRNA in the oviduct are regulated by oestrogen (reviewed Chan, Means & O'Malley, 1978) but levels of ovalbumin mRNA in those tissues where it is present in trace amounts also respond to oestrogen although they still remain low (Tsai, Bai, Lin & O'Malley, 1979).

In some cases, low levels of a heterologous product may decline as the tissue containing it pursues its own pathway of differentiation: chick embryo retina, for example, expresses low levels of lens crystallin mRNA but as retina differentiation proceeds these levels decline and fall below detectable limits (Clayton, Thomson & de Pomerai, 1979). On the other hand, such a minor component may become a major one in growth conditions which permit the tissue to change its developmental pathway, as in the conversion of retina cells into lens cells ('transdifferentiation'), when crystallin mRNA and proteins increase steeply (discussed below).

We may define transdifferentiation as a change in the synthetic patterns of a group of cells so that they come to synthesise in quantity a product which is normally regarded as specific (i.e. superabundant), for a tissue of different developmental antecedents. This describes the situation in terminally differentiated lens fibre cells derived from neural retina, retina pigmented epithelium or iris, but it equally well describes the 'ectopic hormone syndrome', in which 'tissue-specific' substances may become abundant products in certain inappropriate tissues following malignant transformation (Table 1). The hypothesis that this latter phenomenon is due to migratory neural crest stem cells (Pearse & Blake, 1974) is no longer acceptable (Noden, 1978).

The application to cells or tissues of various chemical agencies including hormones, inducers or carcinogens may also change the abundance of a minority product in a cell. Investigations of the factors affecting crystallin synthesis in lens cells and in retina cells transdifferentiating into lens or differentiating along their own pathway may be helpful in considering the following general questions: is the spectrum of products in a cell at a low level related to developmental potential, and what are the mechanisms operating when a competent tissue enters into one of its possible alternative pathways of development? In the system to be discussed here, changes affecting the differentiation pathway, and changes affecting a modulation of the products of a cell without changing its overall character will both be described.

Molecular overlap between tissues and the possible relationship to development and differentiation

The concept that much tissue specificity is not exclusively due to the posession of some specific protein restricted to that tissue, but rather the product of a unique selection of the proteins genetically available, and their relative concentration: that is, that it is combinatorial and quantitative, was explicitly put forward for cell surface antigens (Clayton, 1964) and for the antigens of the different tissues in the eye (Clayton, Campbell & Truman, 1968; Clayton, 1970), and recent studies support this concept.

Estimates of the RNA complexity of different tissues of the same organism, show that each tissue expressed a large selection of the total mRNAs of the organism, that many mRNA species are widely distributed, but that they may not be in the same abundance class in all tissues. (Hastie & Bishop, 1976; Axel, Feigelson & Schutz, 1976; Galau *et al.*, 1976; Young, Birnie & Paul, 1976). A species of mRNA, abundant in a tissue characterised by the corresponding superabundant protein product, may also be present at low levels elsewhere. Axel *et al.* (1976) found that 85% of the mRNA sequences of chick oviduct and liver were common to both tissues, but in the case of oviduct, 70% of the mRNA represents only 10 abundant species, while the intermediate abundance class represented about 12. This makes it quite likely, as they point out, that some of the tissue-specific oviduct mRNAs were present at low levels in liver, and some 'liver-specific' mRNAs were also present in oviduct, but again at lower levels. There is some disagreement concerning trace levels of globin mRNA in other tissues such as brain and liver (Humphries, Windass & Williamson, 1976; Ono & Cutler, 1978; Stalder *et al.*, 1980). While it might be claimed that total exclusion of all traces of red cells from brain and liver preparations is difficult, in the lens-retina system, possible contamination of, for example, pigment epithelium by lens can be wholly excluded. Furthermore, in the $3\frac{1}{2}$ day embryo eye cup, no crystallin protein was detectable, although the level of crystallin mRNA was higher than that of 8 day embryo retina, at which time low levels of crystallin proteins were detectable (Clayton *et al.*, 1979).

Differences between tissues, especially in early embryogenesis, are far less striking than differences between developmental stages. In

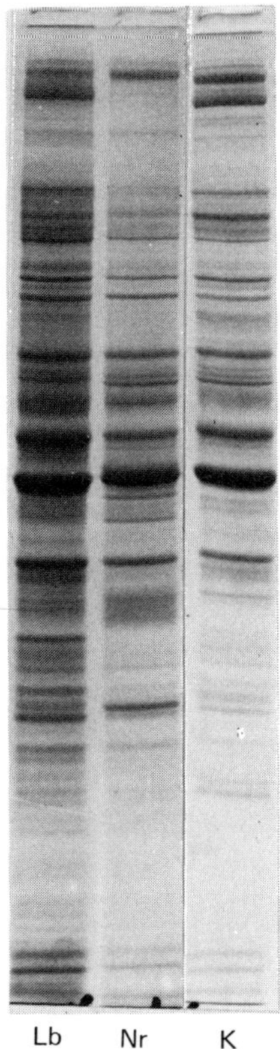

Fig. 1. Polyacrylamide gel SDS electrophoresis of proteins from cells of limb fibroblasts Lb, kidney fibroblasts K and neural retina Nr, all from 8 day chick embryos. I am indebted to Mr A. Zehir for this photograph.

Drosophila the imaginal discs show patterns in electrophoresis which although not identical to each other, share many components, discs being distinguished by their protein patterns rather than by any individual component. Stage-related changes are more striking, (Rodgers & Shearn, 1977; Seybold & Sullivan, 1978; Arking, 1978). Similar data are found for the mouse (Manes, 1974; Klose & von Wallenberg-Pachaly, 1976), and Bravo & Knowland (1979) also found stage-specific polypeptide components in *Xenopus* embryos, and that different tissues shared common components. Fig. 1 shows the considerable similarities which obtain between cells from different tissues of the 8 day chick embryo (A. Zehir, unpublished data).

Some proteins which characterise certain organs later in development may be detected in earlier stages, including the oocyte. Perlman, Ford & Rosbash (1977) detected both adult and tadpole globin mRNA in the oocyte. Both Bravo & Knowland and Perlman *et al.*, found some evidence for the presence in a tissue at an early stage of a gene product appropriate to a later stage of development and also of the loss from some tissues of a product which they may have contained earlier. During limb muscle development, some 50% of the mRNA species are lost (Anderson, Galau, Britten & Davidson, 1976).

These general findings for proteins and for mRNA species are in agreement with earlier immunological analyses of embryonic development. A study of the amphibian embryo from blastula to neural stages (Clayton, 1953) showed that antigens later characterising a specific embryonic region or structure were detectable at an early stage before that structure developed, and that different antigens each characterising distinct regions coexisted in their common precursor tissue. Common or widely distributed antigens were also found and some stage-specific components. The coexistence in a tissue (such as gastrula ectoderm) of antigens which are later segregated in their different descendent tissues (such as neural plate and neurula ectoderm) implies a metastable state capable of resolution in more than one way. This segregation of antigenic material occurred in association with the tissue separation following gastrulation and again, later, following induction of the neural plate. Ebert (1953) detected chick embryo heart antigens before heart formation at low levels in areas competent to form heart. As the levels rose in the developing heart area they fell elsewhere. Some cell surface markers segregate during development (Monroy & Rosati, 1979).

The similarities between *Drosophila* imaginal discs mentioned above may also be related to their developmental potential. Several chemical agencies can effect homeotic-type transformations *in vivo* (reviewed Postlethwaite & Schneiderman, 1974). Transdetermination (reviewed Ouweneel, 1976) is essentially a homeotic-like transposition obtained by serial passage *in vivo*. Transdetermination of *Drosophila* discs can now be obtained in tissue culture (Shearn, Davis & Hersperger, 1978), so that it should in principle be possible to assess whether the direction of transdetermination between discs is related to the degree of similarity of their composition, and to determine the immediate effects on this composition of those agencies which effect disc transformations.

Molecular overlap in tissues of the eye

The potential transformations amongst tissues of the ocular system (Fig. 2) suggest the possibility of underlying biochemical relationships, in addition to the evident requirement that they must acquire increas-

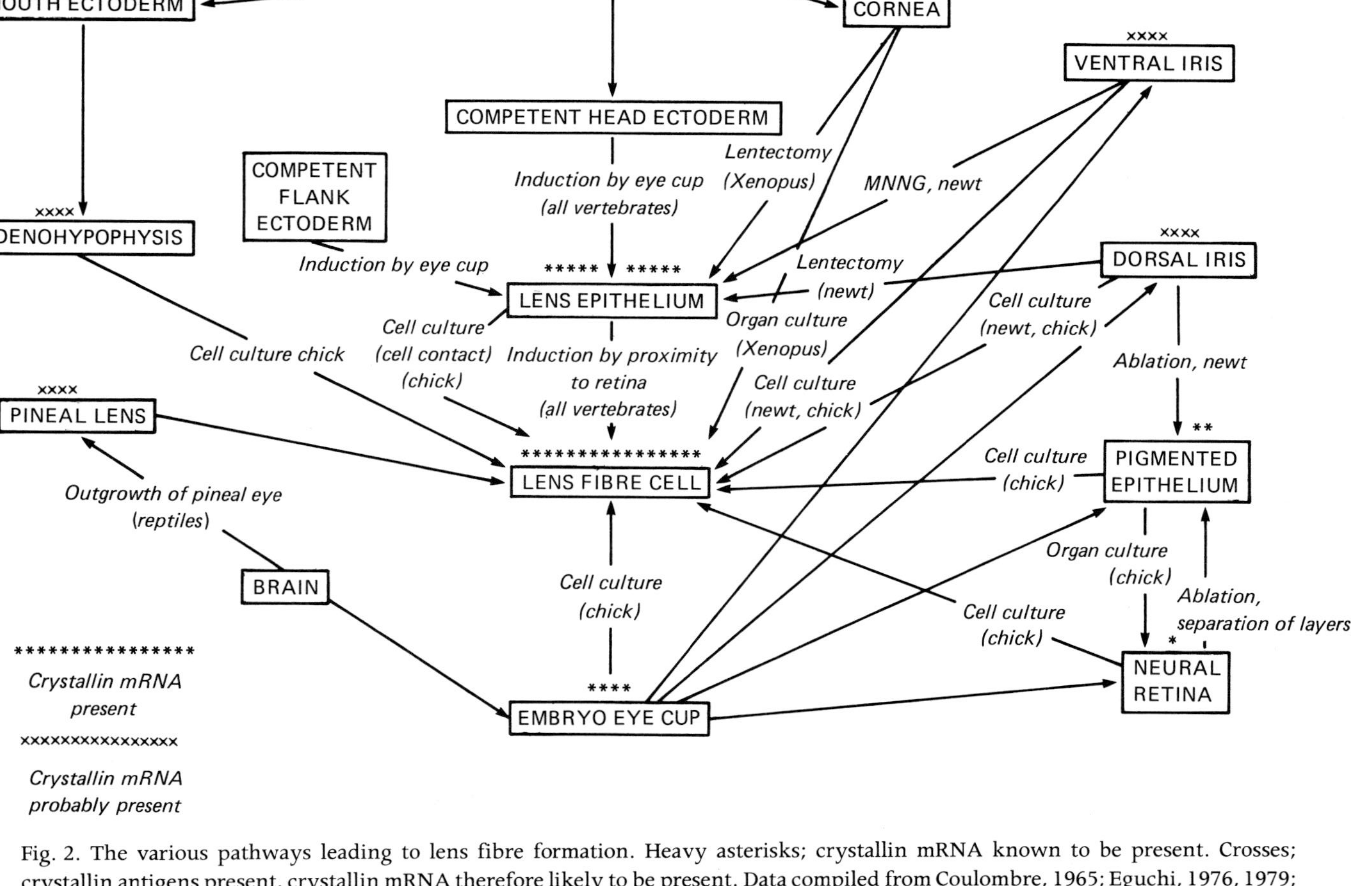

Fig. 2. The various pathways leading to lens fibre formation. Heavy asterisks; crystallin mRNA known to be present. Crosses; crystallin antigens present, crystallin mRNA therefore likely to be present. Data compiled from Coulombre, 1965; Eguchi, 1976, 1979; Okada, 1976, 1980; Reyer, 1977; Clayton, 1978, 1979b; Lopashov, 1978. Retina-like structures have been obtained from amphibian embryo pituitary *in vivo* (Sacerdote, 1971) and by malignant transformation of pineal in man (Stefanko and Maschot, 1979).

ing molecular divergences as they differentiate in structure and function.

There have long been discrepant reports in the literature concerning the organ specificity of lens antigens but these discrepancies can be accounted for in terms of differences in technological approach. The literature was reviewed by Clayton, Campbell & Truman (1968) who found several crystallins or crystallin determinants at low levels in other tissues including retina, iris and cornea. Some cross-reactivity was found at trace levels in some extra-ocular tissues. Similar results were found by Bours & van Doorenmaalen (1972).

It is possible that some cross-reactive antisera are detecting closely related but not identical molecules in other tissues. (An analogy would be the multiple forms of actin which are found in different tissues of the same organism; Garrels & Gibson, 1976; Vandekerckhove & Weber, 1979). Alternatively, widely distributed non-crystallins might aggregate with crystallins *in vitro* without affecting their charge or antigenicity. A large number of cytoarchitectural elements are found in lens cells (see for example, Longchampt *et al.*, 1976; Ireland, Maisel & Bradley, 1978; Ramaekers *et al.*, 1980), and some crystallins appear to be closely associated both with certain of those structures and with the cell membrane. However, the likelihood that at least in ocular tissues lens cross-reactions are due to the crystallins themselves comes from data on the transdifferentiation process. There is ample evidence (reviewed Clayton, 1978, 1979a; Eguchi, 1976, 1979; Okada 1976, 1980) that in all species, transdifferentiated lentoids contain authentic crystallins as defined by antigenic specificity, charge and molecular weight. Both retinal layers freshly isolated from the 8 day chick embryo contain RNA species in the intermediate abundance class which hybridise with cDNA to crystallin mRNA (Jackson *et al.*, 1978; Thomson, de Pomerai, Jackson & Clayton, 1979), and during the process of transdifferentiation into lens when crystallin proteins increase steadily, crystallin mRNA also increases until it attains one-tenth of the level in day old chick lens (Thomson *et al.*, 1979; Thomson, Yasuda, Okada & Clayton, in preparation; Yasuda, Thomson, de Pomerai, Clayton & Okada, 1979). The overlap in composition between lens, neural retina (NR), and pigmented epithelium (PE) must include other products, since cDNA to either of the two latter tissues hybridises with the intermediate as well as the low abundant class of lens mRNA (Jackson *et al.*, 1978), while some non-crystallin antigenic material shared by lens and retina has been noted (for example Mikhailov & Barabanov, 1975).

POSSIBLE LEVELS OF REGULATION OF LENS FIBRE FORMATION

The conditions which govern the formation of either normal or trans-differentiated lens fibre cells, *in vivo* or *in vitro*, do not, *per se*, determine the crystallin composition obtained, and it has been suggested that lens fibre formation and the eventual crystallin balance are regulated at several different levels (Clayton, 1979b, c): preconditions, permissive conditions, directive conditions, and modulatory conditions.

(1) *Preconditions.* The preconditions for differentiation into the lens fibre pathway are likely to be low levels of pre-existing crystallin mRNA; of the pathways shown in Fig. 2, pre-existing crystallin mRNA has been found in lens epithelium (LE) (Piatigorsky *et al.*, 1976; Thomson *et al.*, 1978), and lower levels have been found for neural retina (NR), pigment epithelium (PE) and eye cup (EC) (Jackson *et al.*, 1978; Thomson *et al.*, 1979; Clayton *et al.*, 1979). In the case of embryo adenohypophysis, which can also transdifferentiate into lens (Clayton, 1979b), crystallin mRNA has not been directly demonstrated, but the presence of δ crystallin, during the transitory period before early adenohypophysis histodifferentiation, has been shown by immuno-fluorescence (Barabanov, 1977) and confirmed by haemagglutination inhibition by Clayton (1979b). Crystallins have also been observed in the reptile pineal lens (McDevitt, 1972), and detected in iris extracts (Maisel & Harmison, 1963; Clayton *et al.*, 1968; Bours & van Doorenmaalen, 1972). In these tissues, therefore, it would be reasonable to assume the presence of crystallin mRNA. Evidence which might be interpreted to suggest the possible presence of crystallin mRNA in competent head ectoderm is outlined elsewhere (Clayton, 1979b). δ crystallin disappears from the adenohypophysis after 7 days of incubation (Barabanov, 1977), while crystallin mRNA also disappears from the retina some time between 8 days and hatching, which parallels the loss of ability to transdifferentiate (Clayton *et al.*, 1979). There is an apparent relationship between the initial level of crystallin mRNA, and the frequency and rate of transdifferentiation from NR (Clayton *et al.*, 1979; Table 2 and Fig. 3). However, other factors must be involved, since the rate of increase of crystallin mRNA in transdifferentiating cultures of NR follows a similar pattern when $3\frac{1}{2}$ day and 8 day NR cells are compared, yet 8 day PE which has initially a higher level of crystallin mRNA than 8 day NR (Jackson *et al.*, 1978) has a slower rate

Table 2. *Transdifferentiation into lens cells from chick embryo neural retina*

	Age of embryo (days)							Day old chick
	$3\frac{1}{2}$	6	8	12	15	17	19	
Time for lentoid formation, days	8–15	23	19–25	25	28–35	50	—	—
Percentage of lentoid clones formed	20		8	few	few	—		—
Multipotential clones	+	few	rare					
δ crystallin in lentoids, %	80	50	25	6	3	0	—	—
δ : α crystallin ratio	20 : 1	7 : 1	3 : 1	2 : 1	2 : 1	α, β only	—	—
Crystallin mRNA compared to day old lens	0.18%		0.025%					not detectable (<0.01%)
Final level of crystallin mRNA obtained in terminal culture compared to day old lens	17%		14%					
Time required	3 weeks		6 weeks					
Crystallin, as percentage of total soluble protein: haemagglutination inhibition assay. Limits of detection for antisera used: α 0.01%, β 0.01%, δ 0.005%	α not detectable β not detectable δ not detectable		0.15% 0.04% not detectable					0.04% 0.13% not detectable

After Clayton *et al.* (1979)

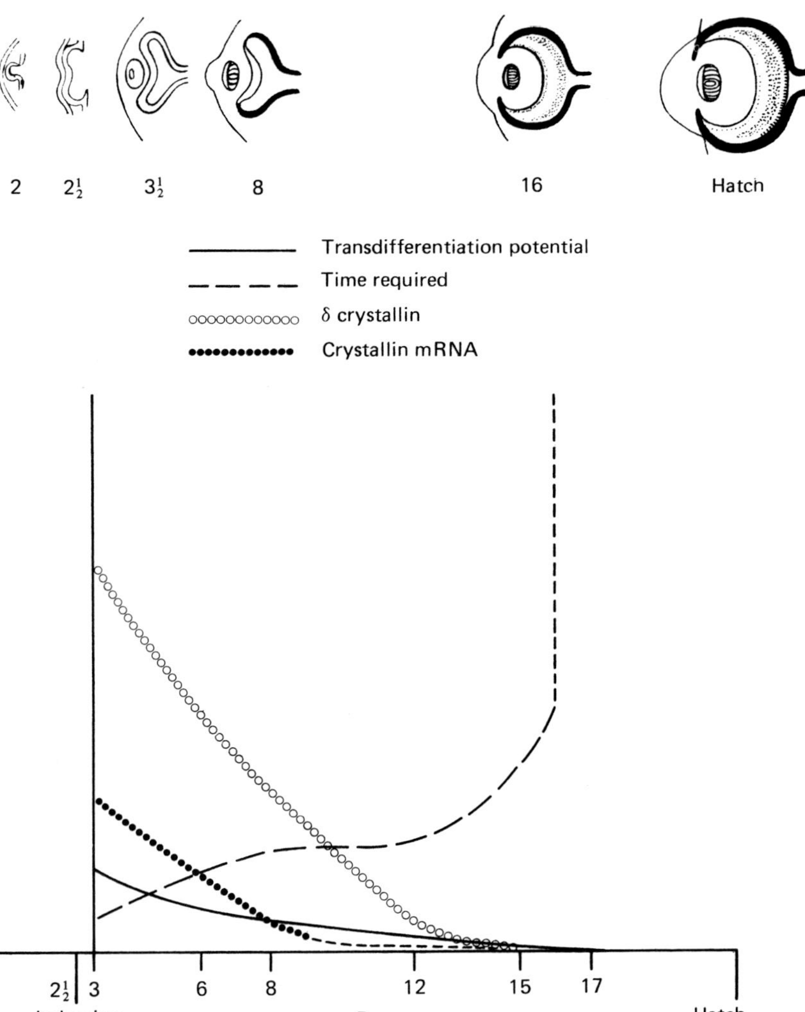

Fig. 3. The relationship between crystallin mRNA content of retina, frequency of transdifferentiation, rate of transdifferentiation and δ-crystallin content. The values on the vertical axis are proportional, but represent different scales for each curve.

of transdifferentiation (Okada, 1976), and the increase in crystallin mRNA follows a different pattern (Thomson *et al.*, 1979; and Fig. 4).

Another difference between lentoids derived from PE and NR was found when crystallin mRNAs from terminally transdifferentiated cultures were compared in a cell-free system. The relative proportions of α, β and δ crystallins are markedly different (Thomson *et al.*, 1979),

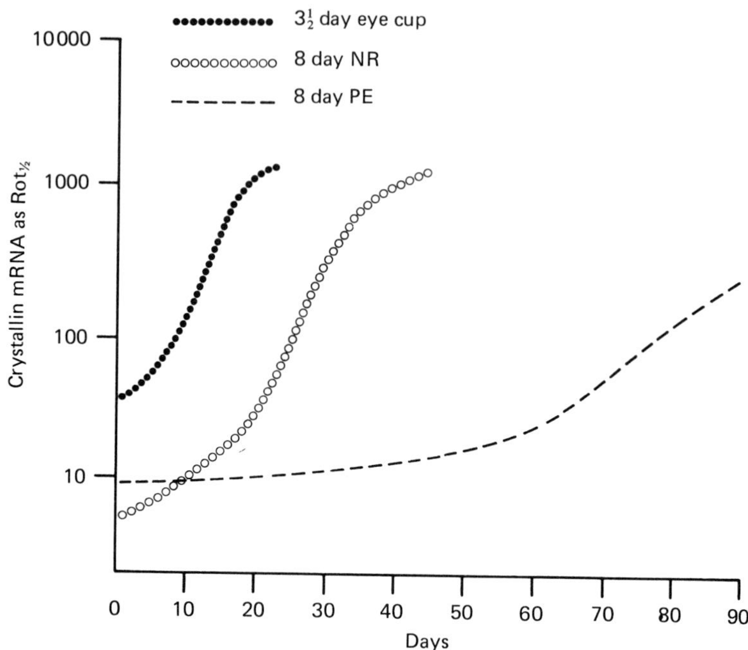

Fig. 4. The rate of increase of crystallin mRNA, plotted as $Rot_{1/2}$ values against time. Taken from Thomson *et al.*, 1979; Yasuda *et al.*, 1979; Thomson, Yasuda, de Pomerai, Okada & Clayton, 1981.

as is the proportion of a protein unidentified at the time, but now identified as actin (Wainwright, Zehir & Clayton, in preparation).

(2) *Permissive conditions.* The capacity of ocular tissues for trans-differentiation into lens is normally supressed *in vivo*, and the permissive conditions for transdifferentiation appear to be changes in cell – cell contacts. Lopashov (1977, 1978) has proposed that basement membranes act as stabilisers of differentiation, and Yamada & McDevitt (1974) obtained lentoid cells from epithelia of dorsal or ventral iris (DI and VI), stripped off the basement membrane and grown in organ culture. Yasuda (1979) found that PE cells fail to transdifferentiate if plated onto collagen. (Since cultured PE cells secrete their own collagen substratum the orientation of cells in contact may be significant.)

Lens regeneration from iris, or retina and iris transdifferentiations *in vivo* require changes in cell neighbours or contacts by excision, separation, or transplantation of retina tissues (reviewed Coulombre, 1965; Scheib, 1965; Reyer, 1977; Yamada, 1977). Lentectomy in the amphibian leads to a change in electrical coupling, but only in the region of dorsal iris (DI) which will transdifferentiate into the regenerated lens (Eguchi, 1979).

Lens fibre formation from lens epithelium (LE) or transdifferentiated from iris, whether *in vivo* or in organ culture, requires proximity to such tissues as retina or pituitary (reviewed Clayton, 1978). However, if LE cells are completely dissociated they give rise to fibre cells in the absence of other tissues (Okada, Eguchi & Takeichi, 1973) while PE, NR, DI or VI cells can transdifferentiate after dissociation without any admixture of other cell types (Eguchi & Okada, 1973; Eguchi, Abe & Watanabe, 1974; Okada, Itoh, Watanabe & Eguchi, 1975).

(3) *Directive conditions*. Cells will not inevitably transdifferentiate after dissociation and plating. The directive conditions include those that impede or facilitate transdifferentiation into one or more of the possible pathways and those that affect the time required or the apparent number of transdifferentiation events (Table 3).

The frequency and the rate of transdifferentiation are related to embryonic age (Araki & Okada, 1977; de Pomerai & Clayton, 1978; Nomura & Okada, 1979; Okada, Yasuda & Nomura, 1979a). Differentiation of NR expresses itself as a progressive restriction of possible pathways. $3\frac{1}{2}$ day NR contains some multipotential cells (able to give rise to both PE and lens cells (LC)), but their number is rapidly diminished during embryonic development, while the number of cells which can transdifferentiate into PE or LE only also declines with time (Okada, Yasuda & Nomura, 1979a; Table 2 and Fig. 3). The potential for LC transdifferentiation is lost before that for PE transdifferentiation (de Pomerai & Clayton, 1978). Since embryonic NR is heterogeneous for transdifferentiation potential, (Okada, Yasuda, Araki & Eguchi, 1979b), conditions affecting transdifferentiation frequency and direction could operate by cell selection. However some data suggest that mechanisms within the cell may also be important. For example, abundant neuronal cells appear in cultures of 8 day NR sown at initial densities of 10^7/dish, but only neuroepithelium appears, if sown at densities of 10^6/dish. PE cells appear early and before LC at the lower initial densities, later and after LC at the higher ones. LC appear precociously and in greater numbers in areas where cell − cell contact areas have been increased by folding of the cell sheet, (Clayton, de Pomerai & Pritchard, 1977). Inhibition of melanin synthesis by phenylthiourea accelerates LC transdifferentiation from PE (Eguchi, 1976).

Table 3. *Some effects of growth conditions on the formation of lens cells and pigment from neural retina*

	Increase	Decrease
Pigment (numbers of transdifferentia-tion events or intensity of pigmentation)	Prostaglandins, dibutyryl cAMP, triiodothyronine, Hams medium, bicar-bonate, low cell density	5 BUdR, 5 hydroxyindole, phenylthiourea, rapid growth, delayed effect of MNNG, *cdp* mutant, conditioned Hams medium, (no pigment)
Lens (numbers of transdifferentiation events)	Eagle's medium, con-ditioned Eagle's medium, high cell density, folding of cell sheet, ascorbic acid, insulin, *Hy*-1 and *Hy*-2 genotypes	Increasing age of donor, Ham's medium, delayed effect of MNNG, con-ditioned Hams medium (no lentoids), dialysed serum
crystallin	*Hy*-1 and *Hy*-2 geno-types, insulin	Increasing age of donor, dialysed serum

Compiled from Araki & Okada, 1978; Chader, Newsome, Bensinger & Fletcher, 1975; Clayton *et al.*, 1977; Eguchi, 1976; Itoh, 1976; Kondoh *et al.*, 1980; Nomura & Okada, 1979; de Pomerai & Clayton, 1978; Pritchard, Clayton & de Pomerai, 1978; Okada, 1976; Redfern *et al.*, 1976; Pollock, Kondoh, Zehir, Okada & Clayton (unpublished data).

The levels of bicarbonate in the medium affect the number of PE foci formed from NR cultures as well as their subsequent rate of growth, (Clayton *et al.*, 1977; Pritchard, Clayton & de Pomerai, 1978). Cell selection is not a likely mechanism for these effects, and selection within a cell for synthetic pathways simultaneously present is more plausible. If it transpires that in the early stages crystallin mRNA is widely distributed, albeit at heterogeneous levels, the concept of within-cell selection might be supported: if it transpires that this mRNA species is present at unvarying levels but in a small and decreasing proportion of the total cell population, the concept of selection between cells might be supported: however, these possibi-lities are not necessarily mutually exclusive.

Whether transdifferentiation is delayed or accelerated, the for-mation of lentoid or pigment cells occurs within a relatively narrow time band. Cells of an iris clone transform together, after 6 or 7 mitoses (Eguchi, 1979) and Yamada & Beauchamp (1978) suggested that a minimum number of mitoses were required before transdifferentiation could occur. This may not reflect the need for a 'quantal' mitotic change (Holtzer, 1978), since an increase in crystallins and in crystallin mRNA may be detected long before lens cells are detectable morphologically (de Pomerai, Pritchard & Clayton, 1977; Thomson *et al.*, 1979).

Only rapidly mitosing cells are likely to transdifferentiate (Yamada & Beauchamp, 1978) and the dilution of pigment in PE cells under conditions of rapid growth and the reappearance of pigment when cell growth is slowed down (Cahn & Cahn, 1966; Trinkaus, 1956) all suggest that uncoupling between the rate of mitosis and the rate of some synthetic pathways is possible. The effect of mitotic interval on the ratio of δ to α and β crystallins (see below) is compatible with this view, since the relative balance of crystallin syntheses changes over the duration of the cell cycle (Clayton *et al.*, 1980).

The concept of canalisation, that is, the relatively sudden entry of cells from a metastable state into a new stable state during embryonic development has been frequently discussed by Waddington (for example Waddington, 1956). The concept of canalisation is also compatible with epigenetic theories of tumour formation and is supported by numerous examples of reversion of tumour cells to a non-malignant differentiated state (see for example reviews by Simnet, 1974; Mintz, 1976; Uriel, 1979. Recent examples include Ilmensee & Mintz, 1976; Tomozawa & Sueoka, 1978; Imada & Sueoka, 1978; Aidells & Lee, 1979.) Retinoblastoma cells differentiated into neurones in cell culture, and later transdifferentiated to pigment cells and eventually into lens cells (Okada, 1978).

The corollary of the canalisation concept is that many external signals might disturb the metastable state, the effect depending in part on conditions in the system, as well as the level at which the signal may operate. Agencies affecting transdifferentiation of ocular tissues are shown in Table 3. Other systems are similarly capable of a switch in differentiation, if a suitable agency is used. Rous sarcoma virus activates globin gene expression in chick fibroblasts (Groudine & Weintraub, 1975) and globin mRNA appears following DMSO (or other inducer) treatment of Friend cells (Harrison *et al.*, 1976; Harrison, Conkie, Rutherford & Yeoh, 1978) (although it is not clear whether globin sequences are already present at low levels before induction: Affara & Daubas, 1979; Reeves & Cserjesi, 1979). A change from neuroblastoma cells to typical neurites can be brought about by a wide range of agencies including FudR, 5BUdR, prostaglandins, and dibutryl cAMP (reviewed Ross, Olmsted & Rosenbaum, 1975). Ross *et al.* found that this redifferentiation requires only protein synthesis. Many of these agencies are likely to effect disturbances in a range of cellular functions, their action on differentiation being therefore more akin to a bludgeon than a fine scalpel. The limited range of response must therefore reflect the innate capacities of the reacting systems.

The use of carcinogens might also be expected sometimes to lead to a change in differentiation or to illicit transdifferentiations, the effects obtained depending on the status of the treated cells. There is some evidence to this point. For example Breedis (1952) appears to have found only two cases of tumour formation but large numbers of accessory limbs following the injection into newts of various carcinogens, including methylcholanthrene, (and others rather poorly defined), and carcinogens may sometimes effect transdifferentiation among ocular cells.

Unlike amphibian DI, VI does not normally form lens *in vivo*, but will do so *in situ* and without lentectomy if exposed once to *N*-methyl-*N'*-nitro-*N*-nitrosoguanidine (MNNG), a potent carcinogen (Eguchi & Watanabe, 1973). The potential of the VI for transdifferentiation acquired by this treatment persisted for at least a year after the experiment yet no tumours were observed. This suggested to us that although the initial effect of MNNG may have included cell–cell uncoupling this would be transitory: some more permanent change in molecular composition of these cells was indicated. Since the effect was always to produce lentoid transformation, any mutations produced by the carcinogen, being random, would also be irrelevant.

On the assumption that a disturbance of the population of mRNAs available for translation might be more relevant to this phenomenon than the known mutagenic effects of most carcinogens, we have begun to treat lens epithelium (LE) and NR cultures with MNNG *in vitro* (Clayton *et al.*, 1980). Divergences between control and experimental cell cultures were found both immediately and during subsequent passages (Figs. 5 and 6).

Primary LE cultures were treated for one hour with MNNG at 6 days after plating. The immediate effect of exposure of LE cells to MNNG is a breakdown of polysomes, the release of mRNA into the cytoplasm and a deficit in newly synthesised mRNA but after the synthetic machinery of the cell has recovered, there is a marked quantitative change in the balance of syntheses, both *in vivo* and in a cell-free system (Clayton *et al.*, 1980 and unpublished data). Chakrabarty & Schneider (1978) found

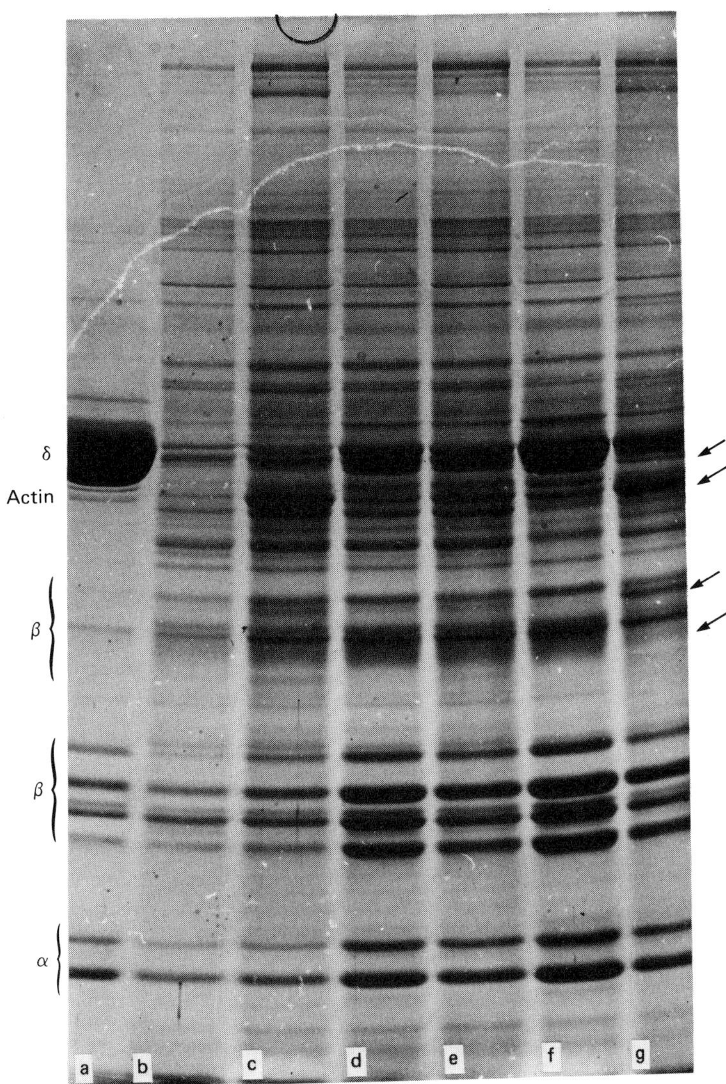

Fig. 5. Polyacrylamide gel SDS electrophoresis of proteins from LE cell cultures. (a) Day-old lens standard. (b, d, f) Controls. (c, e, g) MNNG-treated at 7 days of primary culture. (b, c) 9 days. (d, e) 13 days. (f, g) 21 days. Arrows indicate some affected components. I am indebted to Dr C. Patek for this photograph. Procedures as in Clayton *et al.*, 1980.

a similar series of changes in rat liver mRNA following the administration of thioacetamide. During serial passages, the ability to form lens fibre cells was gradually lost, both from control and MNNG treated cultures; and by the fifth passage crystallins were not detected, but

R. M. Clayton

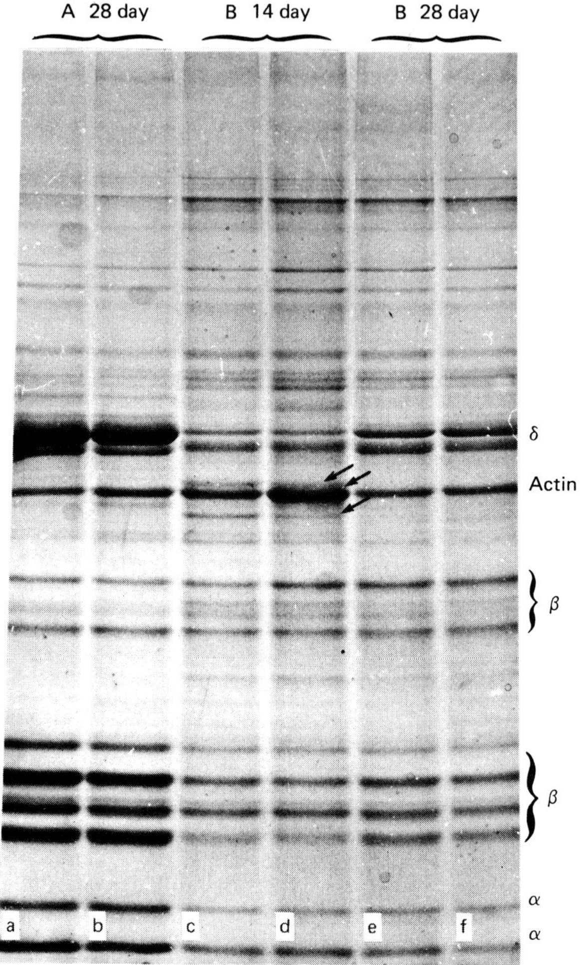

Fig. 6. Primary and secondary cultures of LE. (a, c, e) Controls. (b, d, f) Treated with MNNG at 7 days of primary culture. (a, b) Day 28 of primary culture. (c, d) Day 14 of secondary culture. (e, g) Day 28 of secondary culture. The differences are less noticeable in late cultures when most of the protein comes from lentoids. I am indebted to Dr C. Patek for these photographs. Procedures as in Clayton *et al.*, 1980.

until this point the balance of crystallins remained different between control and treated cultures. A fusiform morphology characterised MNNG treated cells from the secondary passage onwards, and in all passages these cells contain a higher level of actin than controls. Rarely, neurone-like cells and 'black' cells were observed in secondary cultures of LE which were treated with MNNG as primary cultures (Clayton *et al.*, 1980). Abundant NR and abundant PE mRNAs are normally represented in the intermediate abundance class of lens (Jackson *et al.*, 1978), but speculation that these mRNAs may occasionally be favoured after MNNG must await confirmation of the nature of the neurone-like and the 'black' cells. Interestingly, a neural retina antigen was also found in transformed rat LE cell lines (Hamada, Watanabe, Aoyama & Okada, 1979).

Control 8 day embryo NR cultures produce LE lentoids by about day 25 and pigment cells from about day 30; but cultures treated with MNNG at 9 and 11 days after plating had still formed no pigment cells by day 45, and lentoids were delayed by 7–10 days.

Pre-pigment cells were not seen until about day 36 in NR cultures treated with MNNG on days 19 and 21, but only about a fifth of these develop melanin (whereas all pre-pigment cells do so in controls), and about 7–10 days later, these delayed pigment cells began to degenerate (Fig. 7). Lentoid formation is delayed by a few days as compared to controls, but whereas the amount of α crystallin is unaffected following MNNG treatment on days 9 and 11, it is severely diminished after treatment.

Neuronal cells normally disappear over the third week of culture of 8 day chick embryo. Choline acetyl transferase (CAT) is a neurotransmitter synthesised in chick embryo NR cultures, and its levels fall dramatically as δ crystallin levels rise (Nomura, Takagi & Okada, 1980). In MNNG-treated cultures, we observed small numbers of neurone-like cells had reappeared by 30–50 days of culture (Fig. 7). A pilot experiment showed that levels of CAT had risen over this period in treated cultures (Clayton *et al.*, 1980). Further tests are needed to assess whether this is another possible type of transdifferentiation.

These delayed effects. demonstrate that the time of treatment is relevant to the effect, and that there are molecular events leading to transdifferentiation which take place several cell divisions before morphological change is seen. Both immediate and long-term effects following exposure to a carcinogen *in vitro* have been reported in other systems, (e.g. Yoshida & Cravioto, 1978; Haugen & Laerum, 1978) and evidently obtain for LE and NR cultures (Clayton *et al.*, 1980). Changes in cell differentiation following MNNG treatment have been reported for gut cells and for neuroblastoma cells (Fukumachi & Takayama, 1979; Yoda & Fujimura, 1979).

There are also several reports that the outcome of exposure to a carcinogen is affected by the stage of development and differentiation and the stage of the cell cycle at the time of exposure (e.g. Claisse *et al.*, 1978; Jurgelski, Hudson & Falk, 1979; Peterson, Bertram & Heidelber-

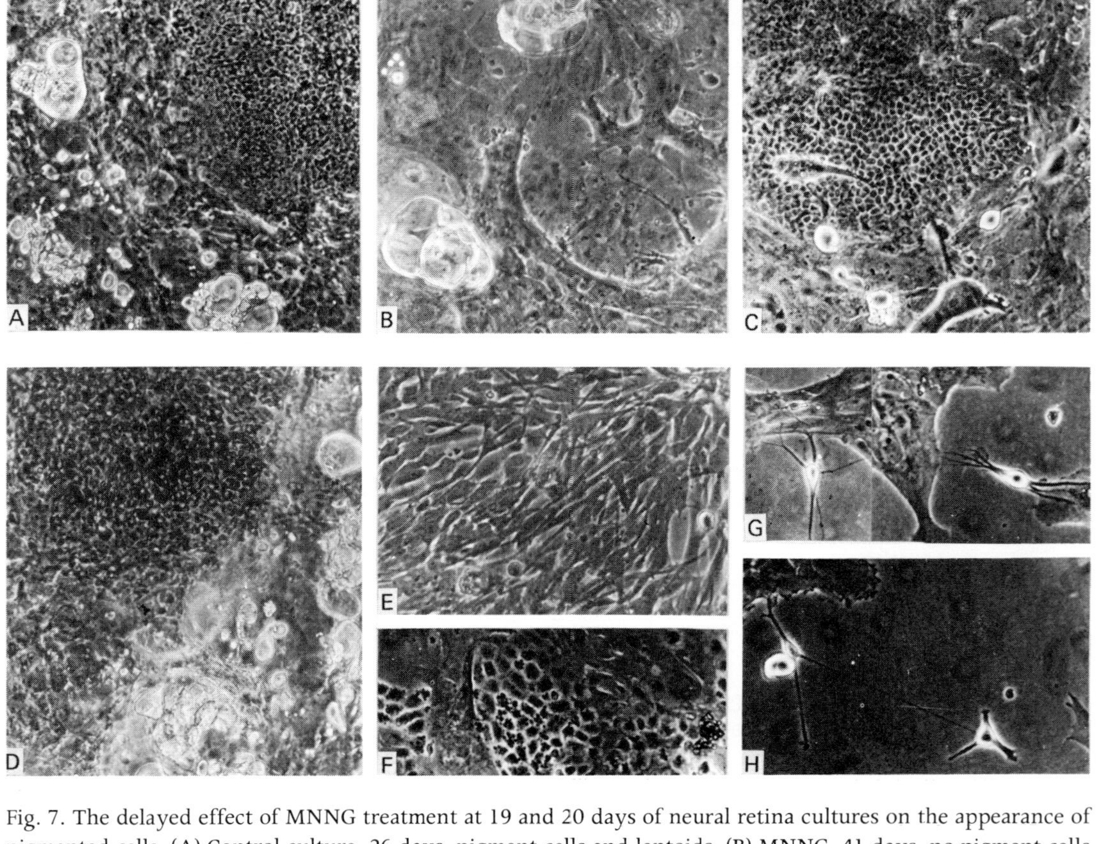

Fig. 7. The delayed effect of MNNG treatment at 19 and 20 days of neural retina cultures on the appearance of pigmented cells. (A) Control culture, 26 days, pigment cells and lentoids. (B) MNNG, 41 days, no pigment cells seen, lentoids present. (C) MNNG, 44 days, pigment cells have appeared but show abnormality. (D) 49 days, pigment cells rounded up. (E) 49 days, fusiform cells. (F) MNNG, 52 days, breakdown of pigment cells. (G) 46 days, and (H) 37 days: cells of neurone-like appearance in MNNG treated NR cultures. I am indebted to Dr C. Patek for these photographs.

Table 4. *The effect of insulin in cell culture on crystallin synthesis in a cell-free system. mRNA prepared from 15 day cultures, treated for 11 days with insulin at 10 µg/ml. Figures with one or two asterisks indicate those polypeptides showing a significant difference in synthesis between control and insulin-treated cells. The italic figure has the higher value. A difference in response is observed between polypeptides but the pattern of response also differs between genotypes, (Db and Hy-2 strain). See Clayton* et al. *(1980).*

crystallin	Db strain		Hy-2 strain	
	Cont.	Ins.	Cont.	Ins.
$\alpha 1$	5.0	5.0	*13.0****	7.7
$\alpha 2$	3.9	*14.4****	12.8	*15.7**
$\beta 1$	6.6	8.5*	5.3	4.6
$\beta 2$	1.7	*3.8**	2.7	2.9
$\beta 3$	*6.7**	5.7	5.4	6.1
$\beta 4$	*21.7****	12.8	11.7	11.0
$\beta 5$	14.8	14.4	13.8	14.2
$\beta 6$	*13.7**	10.6	9.4	9.1
$\delta 1$	7.5	*8.4**	5.5	5.8
$\delta 2$	*4.9**	3.6	2.0	*3.2**
	3.5	*4.8**	3.3	*4.5**

From Zehir, Wainwright & Clayton (unpublished data).

ger, 1974). Factors such as these may also account for the differences in response to MNNG shown by NR or LE cells of the different genotypes used (Clayton *et al.*, 1980), which are distinguished by cell cycle characteristics and by crystallin representation.

Reversible tumour differentiation, the ectopic hormone syndrome and transdifferentiation are similar in that normal gene products are expressed in cells where they would not be expected.

(4) *Modulatory conditions.* Given that transdifferentiation is directed along the pathway to LC formation, the crystallin composition may vary considerably. The source (LE, NR, PE) affects the crystallin balance, (de Pomerai *et al.*, 1977; Thomson *et al.*, 1979; Clayton, 1978). The age of the animal is also a significant modulatory factor (de Pomerai, Clayton & Pritchard, 1978; de Pomerai & Clayton, 1978; Araki & Okada, 1978; Nomura & Okada, 1979; Clayton, 1979a, Clayton *et al.*, 1979): in general the younger the animal the higher the δ content,

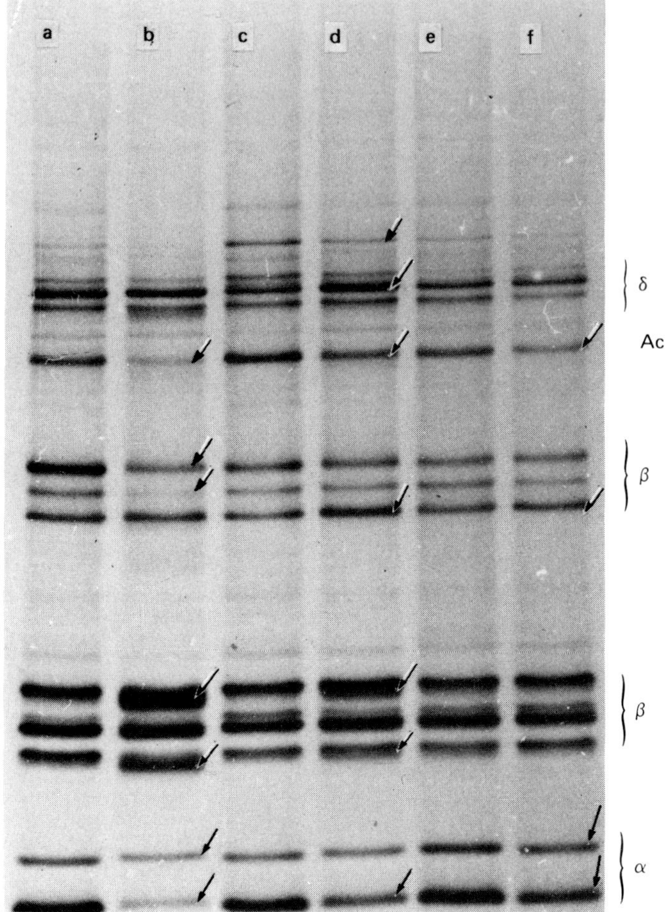

Fig. 8. Fluorography of polyacrylamide SDS gel electrophoresis of products translated in a cell-free system primed by mRNA from control and insulin-treated cell cultures as shown in Fig. 4. (a, c, e) Insulin-treated. (b, d, f) Controls (a, b) mRNA at 10 µg/ml. (c, d, e, f) mRNA at 5µg/ml. (a, b, c, d) Db genotype. (e, f) Hy-1 genotype. 5 µg/ml is the optimum for translation in our system. At 10 µg/ml there is a reduction in efficiency but some competitive effects may be noted. Products labelled with ^{35}S methionine. Procedures as in Clayton *et al.*, 1980. I am indebted to Dr N. Wainwright and Mr A. Zehir for these photographs.

whatever the tissue of origin. Since high levels of δ crystallin characterise earlier stages of development when the mitotic rate is high, we suggested that the mitotic interval might act as a regulator of the level of δ crystallin synthesis (Clayton, 1979a,c; de Pomerai & Clayton, 1978; de Pomerai, Clayton & Pritchard, 1978).

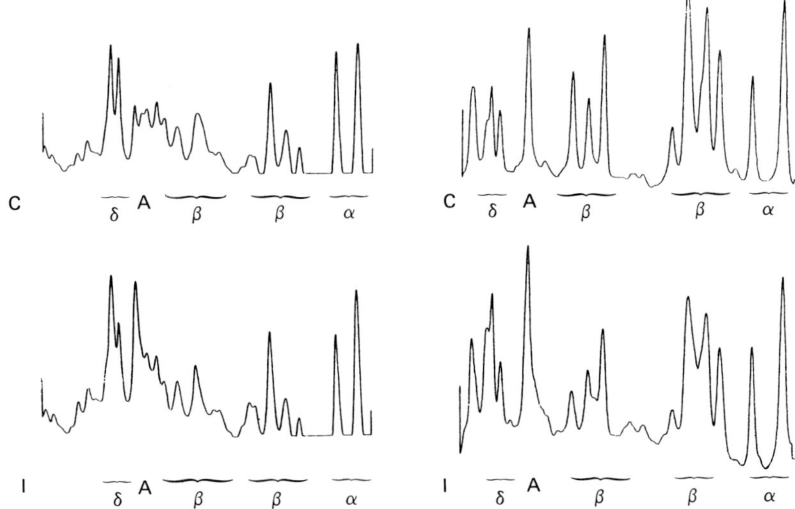

Fig. 9. Densitometer traces of fluorographs of control and insulin treated cultures. Cultures harvested at 12 days, cells grown with 10 µg/ml insulin for 8 days before harvesting. (c) Control. (d) Insulin. (a) Synthesis in cell culture. ^{3}H amino acid mixture label. (b) Synthesis in cell-free system. ^{35}S methionine label. Procedures as in Clayton *et al.*, 1980.

Insulin stimulates the synthesis of hybridisable δ crystallin mRNA and of δ crystallin preferentially in organ cultures of embryonic LE (Milstone, Zelenka & Piatigorsky, 1976; Milstone & Piatigorsky, 1977), and modifies the overall balance of crystallin synthesis in short-term organ cultures of day old LE (Clayton *et al.*, 1976b). The effect of insulin on LE cell cultures is to affect, non-coordinately, the relative proportions of the crystallin classes and of the subunits of these classes synthesised, both in these cells in culture, and also in a cell-free system (Table 4, Figs. 8 and 9). Differences in the proportions of polypeptides in these two conditions imply translational regulation (Clayton *et al.*, 1980; Wainwright, Zehir & Clayton, in preparation; Zehir, Wainwright & Clayton, in preparation). There is also a genotypic-specific difference in response to insulin (Table 4, Fig. 8).

Mitosis is depressed in medium containing dialysed serum and accelerated by the addition of ascorbic acid (Itoh, 1976). Depressed growth rate of transdifferentiating NR in dialysed serum caused a preferential decline in the proportion of δ crystallin. Ascorbic acid only partly restored this, increasing the levels of α and β crystallins relatively more than δ, while the addition of insulin causes a considerable preferential increase of δ crystallin (de Pomerai & Clayton, 1980). The effect of insulin on NR depends on the stage of differentiation of the culture at which the insulin was added: crystallin proportions are not entirely pre-set from the beginning of the culture period, but are also affected by conditions over the period of maximum crystallin synthesis. Retina extract (RE) (Arruti & Courtois, 1978) also affects the rate of mitosis, cell morphology, and ultrastructural elements (Barritault, Arruti, Whalen & Courtois, 1979). The relative proportions of crystallins are changed but there are differences between the effects of RE and insulin (Clayton *et al.*, 1980). Factors other than the rate of mitosis must therefore also act as regulators; probably insulin and RE operate in a different receptor effector system.

Piatigorsky and colleagues have studied the effects of Na^+/K^+ ratios on crystallin synthesis. Changes in Na^+/K^+ ratios modify the relative rates of synthesis of the two δ crystallin polypeptides in the chick lens both *in vitro*, and in a cell-free system. The normal ratio *in vitro* of the 50 k and 48 k polypeptide chains is 1 : 3, but an elevated Na^+/K^+ ratio changes this to 3 : 1 (Piatigorsky, 1980).

Effects on particular crystallins of colchicine, dexamethasone, dibutyryl cAMP, colecemid and cytochalasin and of various metabolic inhibitors have been reported by several authors and are reviewed by Clayton (1979a, 1980).

Cell – cell contacts are regulatory for crystallin synthesis in culture (Clayton *et al.*, 1976a, b; Creighton, Mousa & Trevithick, 1976; Clayton *et al.*, 1977). Cell substrate and cell position affect the relative proportions of bovine crystallin polypeptides (Vermorken, Hilderink, van de Ven & Bloemendal, 1978).

Cell genotype may also affect crystallin balance (Clayton, 1979a; Piatigorsky, 1980). Two genotypes (*Hy*-1 and *Hy*-2), with modified cell membranes and cell-contact behaviour respond to all signals to which normal epithelial cells respond but the quantitative pattern of different crystallin synthesis is modulated by these genotypes (reviewed Clayton, 1979a).

Although the relative balance of synthesis of the different crystallin polypeptides is affected by numerous factors, the response of the polypeptides is non-coordinate, and different for each agency. It is evident that there must be several levels at which the synthesis of each crystallin is regulated, and that the cell membrane, known to differ markedly in some of the genotypes examined, might be able to modulate the effect of the various agencies.

Apparent loss of crystallins

Transformed cell lines have been obtained from bovine LE (Courtois *et al.*, 1978), rat LE (Miller *et al.*, 1979; Hamada *et al.*, 1979; Rink & Vornhagen, 1979), and mouse (Clayton, Wylie & Jones, unpublished). Crystallin synthesis is at least partially lost in such lines, but Courtois *et al.*, found capsular glycoprotein and α crystallin still detectable in solid tumours derived from the transformed cells, and we have found a β crystallin in such a tumour. β crystallins and a β-like component common to neural retina were found in continuous rat cell lines by Hamada *et al.*, while Miller *et al.*, found that γ was lost immediately, and α and β crystallins after several passages. In all these cases, there is no evidence of reversibility, but also no evidence as to whether the loss is at the level of the genome, transcription, or translation.

POST-TRANSCRIPTIONAL REGULATION

Evidence for post-transcriptional regulation in other systems comes from comparisons of polysomal and nuclear mRNAs. mRNA species from polysomes of one tissue may be found in another tissue or developmental stage in the nuclear fraction (Shepherd & Flickinger, 1979; Wold *et al.*, 1978). Traces of a species of mRNA may be found before development switches into its production, and sensitivity to DNAase-1 suggests that a locus may be partially recognised before it is expressed (Stalder *et al.*, 1980; Chapman & Tobin, 1979; Perlman *et al.*, 1977). There is also evidence from other systems that the types of regulatory events outlined here for crystallin regulation (cell – cell and cell – substratum contacts, induction, hormonal signals, and abnormal events such as viral infection or carcinogen treatment) can affect mRNA synthesis, processing, turnover and translation (Cox, 1977; Benecke, Ben Ze'ev & Penman, 1978; Kurtz, Chan & Feigelson, 1978; Harrison *et al.*, 1978; Guyette, Matusik & Rosen, 1979; Hynes *et al.*, 1979; Lowenhaupt & Lingrel, 1978; Farmer *et al.*, 1978.)

The availability, stability and translatability of crystallin mRNAs have been shown to vary and to be modifiable. There are numerous reports of increase in crystallin mRNA stability during ontogeny, and also that different crystallin mRNAs differ both in their stability and in the period of change in their stability (reviewed Clayton, 1979a). Lavers (1978), Asselbergs, Peters, van Venrooij & Bloemendal (1978) and Asselbergs, van Venrooij & Bloemendal (1979) have demonstrated differential responses of crystallin mRNA in cell-free systems to conditions affecting mRNA translatability. The translational efficiency of mRNAs may vary amongst crystallins (Clayton, Truman & Hannah 1974) or change during development (Beebe & Piatigorsky, 1981).

The immediate response to insulin was a transitory increase in translational efficiency of δ crystallin mRNA followed by an increase in transcription (Milstone & Piatigorsky, 1977). A comparison of the rate of increase of crystallin proteins with the increase of translatable and hybridisable crystallin mRNAs during transdifferentiation (de Pomerai *et al.*, 1977; Thomson *et al.*, 1979) suggests a similar sequence of molecular events. There is evidence for sequestration of β crystallin mRNAs in the post-polysomal RNA, and a deficit of δ in the poly-A fractions (Thomson *et al.*, 1978), while crystallin mRNA in the $3\frac{1}{2}$ day chick embryo eye cup is apparently not translated (Clayton *et al.*, 1979).

This article has attempted to assemble evidence suggesting that changes, in response to various signals, of the availability and relative abundance of mRNA species already present in a cell may be the prerequisite for a change in differentiation; whether this be to exhibit an embryonic competence to redifferentiate, or to synthesise an unexpected product, as in transdifferentiation of ocular tissues, and perhaps in the ectopic hormone syndrome.

The exceptional plasticity of ocular tissues in experimental conditions should make this system particularly advantageous in the exploration of molecular and cellular regulatory mechanisms.

Acknowledgements

It is a pleasure to acknowledge the enjoyable and profitable discussions, on problems of transdifferentiation, over a period of many years, with G. Eguchi, T. S. Okada, D. de Pomerai, I. Thomson and D. E. S. Truman.

It is a pleasure also to thank those colleagues and visiting scientists who have shared in the more recent and the unpublished experiments described here: J. Bower, J. Brodie, L. Errington, J. Jackson, H. Kondoh, C. Patek, D. de Pomerai, B. Pollock, C. Randall, F. Randall, I. Thomson, N. Wainwright, K. Yasuda and A. Zehir. In particular I am grateful to C. Patek, N. Wainwright and A. Zehir for making available the photographs and traces from their unpublished experiments which are reproduced here.

I am glad to thank S. Herrick, J. Jack, C. Sime, D. Simmons and S. Williamson for invaluable technical assistance, and I am greatly indebted to M. Alexander and L. Dobbie for their indispensable and patient help in the preparation of the manuscript.

I thank the MRC and the CRC for their continued support and also thank Sterling Poultry Products, Ltd, Newbridge, D. B. Marshall, Newbridge, Pfizer H & N Inc., Dunbar and the Poultry Research Centre, Edinburgh, for chicks and fertile eggs of various genotypes.

REFERENCES

AFFARA, N. & DAUBAS, P. (1979). Regulation of a group of abundant mRNA sequences during Friend cell differentiation. *Developmental Biology*, **72**, 110–25.

AIDELLS, B. D. & LEE, A. E. (1979). Transplanted cultured cells from pregnancy-dependent mammary tumors have a heterogeneous developmental potential. *International Journal of Cancer*, **23**, 718–21.

ANDERSON, D. M., GALAU, C. A., BRITTEN, R. J. & DAVIDSON, E.H. (1976). Sequence complexity of the RNA accumulated in oocytes of *Arbacia punctulata*. *Developmental Biology*, **51**, 138–45.

ARAKI, M. & OKADA, T. S. (1977). Differentiation of lens and pigment cells in cultures of neural retinal cells of early chick embryos. *Developmental Biology*, **60**, 278–86.

ARAKI, M. & OKADA, T. S. (1978). Effects of culture media on "foreign" differentiation of lens and pigment-cells from neural retina *in vitro*. *Development, Growth and Differentiation*, **20**, 71–8.

ARKING, R. (1978). Tissue-, age-, and stage-specific patterns of protein synthesis during the development of *Drosophila melanogaster*. *Developmental Biology*, **63**, 118–27.

ARRUTI, C. & COURTOIS, Y. (1978). Morphological changes and growth stimulation of bovine epithelial lens cells by a retinal extract *in vitro*. *Experimental Cell Research*, **117**, 283–92.

ASSELBERGS, F. A., PETERS, W. H., VAN VENROOIJ, W. J. & BLOEMENDAL, H. (1978). Inhibition of translation of lens mRNAs in a messenger dependent reticulocyte lysate by cap analogues. *Biochimica et Biophysica Acta*, **520**, 577–87.

ASSELBERGS, F. A., VAN VENROOIJ, W. J. & BLOEMENDAL, H. (1979). Messenger-RNA competition in living *Xenopus* oocytes. *European Journal of Biochemistry*, **94**, 249–54.

AXEL, R., FEIGELSON, P. & SCHUTZ, G. (1976). Analysis of the complexity and diversity of mRNA from chicken liver and oviduct. *Cell*, **7**, 247–54.

BARABANOV, V.M. (1977). Expression of δ-crystallin in the adenohypophysis of chick embryos. *Doklady Akademii Nauk SSSR*, **234**, 195–8.

BARCLAY, A. N., LETARTE-MUIRHEAD, M., WILLIAMS, A. F. & FAULKES, R. A. (1976). Chemical characterisation of the Thy-1 glycoproteins from the membranes of rat thymocytes and brain. *Nature*, **263**, 563–7.

BARRITAULT, D., ARRUTI, C., WHALEN, R. G. & COURTOIS, Y. (1979). Adult bovine epithelial lens cell in culture. Electrophoretic pattern of total protein on longterm cultures and morphological changes induced by retinal extract. *Ophthalmic Research*, **11**, 316–21.

BEEBE, D. C. & PIATIGORSKY, J. (1981). Translational regulation of δ-crystallin synthesis during lens development in the chicken embryo. *Developmental Biology,* (in press).

BENECKE, B. J., BEN ZE'EV, A. & PENMAN, S. (1978). The control of mRNA production, translation and turnover in suspended and reattached anchorage-dependent fibroblasts. *Cell,* 14, 931–9.

BERTAGNA, X. Y., NICHOLSON W. E., TANAKA, K., MOUNT, C. D., SORENSON, G. D., PERRENGILL, O. S. & ORTH, D. N. (1979). Ectopic production of ACTH, lipotropin and β-endorphin by human cancer cells. Structurally related tumour markers. *Recent Results in Cancer Research,* 67, 16–25.

BOURS, J. & VAN DOORENMAALEN, W. J. (1972). The presence of lens antigens in the intra-ocular tissues of the chick eye. *Experimental Eye Research,* 13, 236–47.

BRADLEY, R. H., IRELAND, M. & MAISEL, H. (1979). The cytoskeleton of chick lens cells. *Experimental Eye Research,* 28, 441–53.

BRAUNSTEIN, G. D., KAMDAR, V., RASOR, J., SWAMINATHAN, N. & WADE, M. (1979). Widespread distribution of a chorionic gonadotropin-like substance in normal human tissues. *Journal of Clinical Endocrinology and Metabolism,* 49, 917–25.

BRAVO, R. & KNOWLAND, J. (1979). Classes of proteins synthesized in oocytes, eggs, embryos, and differentiated tissues of *Xenopus laevis. Differentiation,* 13, 101–8.

BREEDIS, C. (1952). Induction of accessory limbs and of sarcoma in the newt (*Triturus viridescens*), with carcinogenic substances. *Cancer Research,* 12, 861–6.

CAHN, R. D. & CAHN, M. B. (1966). Heritability of cellular differentiation: clonal growth and expression of differentiation in retinal pigment cells *in vitro. Proceedings of the National Academy of Sciences, USA,* 55, 106–14.

CHADER, G. J., NEWSOME, D. A., BESINGER, R. E. & FLETCHER, R. T. (1975). Studies on the differentiation of retinal pigmented epithelium cells in culture. *Investigative Ophthalmology,* 14, 108–13.

CHAKRABARTY, P. R. & SCHNEIDER, W. C. (1978). Increased activity of rat liver messenger RNA and of albumin messenger RNA modulated by thioacetamide. *Cancer Research,* 38, 2043–7.

CHAN, L., MEANS, A. R. & O'MALLEY, B. W. (1978). Steroid hormone regulation of specific gene expression. *Vitamins and Hormones. Advances in Research and Application,* 36, 259–95.

CHAPMAN, B. S. & TOBIN, A. J. (1979). Distribution of developmentally regulated hemoglobins in embryonic erythroid populations. *Developmental Biology,* 69, 375–87.

CLAISSE, P. J., LANTOS, P. L. & ROSCOE, J. P. (1978). Analysis of N-ethyl-N-nitrosourea-induced brain carcinogenesis by sequential culturing during the latent period. II. Morphology of the tumour induced by cell cultures. *Journal of the National Cancer Institute,* 61, 381–90.

CLAYTON, R. M. (1953). Distribution of antigens in the developing newt embryo. *Journal of Embryology and Experimental Morphology,* 1, 25–42.

CLAYTON, R. M. (1964). Differentiation. In *Penguin Science Survey, Series B,* ed. A McLaren & A. Bennet, pp. 68–88. Penguin Publications, London.

CLAYTON, R. M. (1970). Problems of differentiation in the vertebrate lens. In *Current Topics in Developmental Biology,* vol. 5, ed. A. Moscona & A. Monroy, pp. 115–80.Academic Press, London.

CLAYTON, R. M. (1974). Comparative aspects of lens proteins. In *The Eye*, vol. 5, ed. H. Davson, pp. 394–94. Academic Press, London.

CLAYTON, R. M. (1978). Divergence and convergence in lens cell differentiation: Regulation of the formation and specific content of lens fibre cells. In *Stem Cells and Tissue Homeostasis*, ed. B. I. Lord, C. S. Potten & R. J. Cole, pp. 115–38. Cambridge University Press.

CLAYTON, R. M. (1979a). Genetic regulation in the vertebrate lens cell. In *Mechanisms of Cell Change*, ed. J. D. Ebert & T. S. Okada, pp. 129–67. Wiley, New York.

CLAYTON, R. M. (1979b). Regulatory factors for lens fibre formation in cell culture. I. Possible requirement for pre-existing levels of crystallin mRNA. *Ophthalmic Research*, 11, 324–8.

CLAYTON, R. M. (1979c). Regulatory factors for lens fibre formation in cell culture. II. The role of growth conditions and factors affecting cell cycle duration. *Ophthalmic Research*, 11, 329–34.

CLAYTON, R. M. (1980). An in vitro system for teratogenicity testing. In *Alternatives in Drug Research*, ed. A. Rowan & C. Stratman, pp. 153–73. Macmillan, London.

CLAYTON, R. M., BOWER, D. J., CLAYTON, P. R., PATEK, C. E., RANDALL, F. E., SIME, C., WAINWRIGHT, N. R. & ZEHIR, A. (1980). Cell culture in the investigation of normal and abnormal differentiation of eye tissues. In *Tissue Culture Methods in Medicine*, ed. R. Richards & K. Rajan, pp. 185–94. Pergamon Press, Oxford and New York.

CLAYTON, R. M. CAMPBELL, J. C. & TRUMAN, D. E. S. (1968). A re-examination of organ specificity of lens antigens. *Experimental Eye Research*, 7, 11–29.

CLAYTON, R. M., EGUCHI, G., TRUMAN, D. E. S., PERRY, M. M., JACOB, J. & FLINT, O. P. (1976a). Abnormalities in the differentiation and cellular properties of hyperplastic lens epithelium from strains of chickens selected for high growth rate. *Journal of Embryology and Experimental Morphology*, 35, 1–23.

CLAYTON, R. M., ODEIGAH, P. G., DE, POMERAI, D. I., PRITCHARD, D. J., THOMSON, I. & TRUMAN, D. E. S. (1976b). Experimental modification of the quantitative pattern of crystallin synthesis in normal and hyperplastic lens epithelium. In *Biology of the Epithelial Lens Cells in Relation to Development, Ageing and Cataract*, ed. Y. Courtois & F. Regnault. Les Colloques de L'Institut National de la Sante et de la Recherche Medicale, vol. 60, pp. 123–36.

CLAYTON, R. M. DE, POMERAI, D. I. & PRITCHARD, D. J. (1977). Experimental manipulation of alternative pathways of differentiation in cultures of embryonic chick neural retina. *Development, Growth and Differentiation*, 19, 165–76.

CLAYTON, R. M., THOMSON, I. & DE POMERAI, D. I. (1979). Relationship between crystallin mRNA expression in retina cells and their capacity to redifferentiate into lens cells. *Nature*, 282, 628–9.

CLAYTON, R. M., TRUMAN, D. E. S. & HANNAH, A. I. (1974). RNA turnover and translational regulation of specific crystallin synthesis. *Cell Differentiation*, 3, 135–45.

COULOMBRE, A. J. (1965). The Eye. In *Organogenesis*, ed. R. L. de Haan & H. Ursprung, pp. 219–51. Holt, Rinehart and Winston, New York.

COURTOIS, Y., SIMONNEAU, L., TASSIN, J., LAURENT, M. V. & MALAISE, E. (1978). Spontaneous transformation of bovine lens epithelial cells: kinetic analysis and differentiation in monolayers and in nude mice. *Differentiation*, 10, 23–30.

Cox, R. P. (1977). Estrogen withdrawal in chick oviduct. Selective loss of high abundance classes of polyadenylated messenger RNA. *Biochemistry* **16**, 3433–43.

Cox, R. P. & Ghosh, N. K. (1978). Current concepts in the ectopic production of fetal proteins and hormones by neoplastic cells. *American Journal of the Medical Sciences*, **275**, 232–48.

Creighton, M. O., Mousa, G. Y. & Trevithick, J. R. (1976). Differentiation of rat lens epithelial cells in tissue culture. I. Effects of cell density, medium and embryonic age of initial culture. *Differentiation*, **6**, 155–67.

Ebert, J. D. (1953). An analysis of the synthesis and distribution of the contractile protein, myosin, in the development of the heart. *Proceedings of the National Academy of Sciences, USA*, **39**, 333–44.

Eguchi, G. (1976). Transdifferentiation of vertebrate cells in cell culture. In *Ciba Foundation Symposium 40: Embryogenesis in Mammals*, pp. 241–58. Elsevier, Amsterdam.

Eguchi, G. (1979). Transdifferentiation in pigment epithelial cells of the vertebrate eye *in vitro*. In *Mechanisms of Cell Change*, ed. J. D. Ebert & T. S. Okada, pp. 273–91. Wiley, New York.

Eguchi, G., Abe, S. I. & Watanabe, K. (1974). Differentiation of lens-like structures from newt iris epithelial cells *in vitro*. *Proceedings of the National Academy of Sciences, USA*, **71**, 5052–6.

Eguchi, G. & Okada, T. S. (1973). Differentiation of lens tissue from the progeny of chick retinal pigment cells cultured *in vitro*: a demonstration of a switch of cell types in clonal cell culture. *Proceedings of the National Academy of Sciences, USA*, **70**, 1495–9.

Eguchi, G. & Watanabe, K. (1973). Elicitation of lens formation from the 'ventral iris' epithelium of the newt by a carcinogen, N-methyl-N-nitro-N-nitroso-guanidine. *Journal of Embryology and Experimental Morphology*, **30**, 63–71.

Farmer, S. R., Ben Ze'ev, A., Benecke, B. J. & Penman, S. (1978). Altered trans-latability of messenger RNA from suspended anchorage-dependent fibroblasts: reversal upon cell attachment to a surface. *Cell*, **15**, 627–37.

Galau, G. A., Klein, W. H., Danis, M. M., Wold, B. J., Britten, R. J. & Davidson, E. H. (1976). Structural gene sets active in embryos and adult tissues of the sea urchin. *Cell*, **7**, 487–505.

Garrels, J. I. & Gibson, W. (1976). Identification and characterization of multiple forms of actin. *Cell*, **9**, 793–805.

Gilmour, R. S., Harrison, P. R., Windass, J. W., Affara, N. A. & Paul, J. (1974). Globin messenger RNA synthesis and processing during haemoglobin induction in Friend cells. I. Evidence for transcriptional control in clone M2. *Cell Differentiation*, **3**, 9–22.

Groudine, M. & Weintraub, H. (1975). Rous sarcoma virus activates embryonic globin genes in chicken fibroblasts. *Proceedings of the National Academy of Sciences, USA*, **72**, 4464–8.

Guyette, W. A., Matusik, R. J. & Rosen, J. M. (1979). Prolactin-mediated transcrip-tional and post transcriptional control of casein gene expression. *Cell*, **17**, 1013–23.

Hamada, Y., Watanabe, K., Aoyama, H. & Okada, T. S. (1979). Differentiation and dedifferentiation of rat lens epithelial cells in short- and long- term cultures. *Development, Growth and Differentiation*, **21**, 205–20.

HARRISON, P. R., AFFARA, N., CONKIE, D., RUTHERFORD, T., SOMERVILLE, J. & PAUL, J. (1976). Regulation of erythroid differentiation in Friend erythroleukemic cells. In *Progress in Differentiation Research*, ed. W. Muller-Berat, pp. 135–46. Elsevier, Amsterdam.

HARRISON, P. R., CONKIE, D., RUTHERFORD, T. & YEOH, G. (1978). Molecular aspects of erythroid cell regulation. In *Stem Cells and Tissue Homeostasis*, ed. B. I. Lord, C. S. Potten & R. J. Cole, pp. 241–58. Cambridge University Press.

HASTIE, N. D. & BISHOP, J. O. (1976). The expression of three abundance classes of messenger RNA in mouse tissues. *Cell*, 9, 761–74.

HAUGEN, A. & LAERUM, O. D. (1978). Surface structure of fetal rat brain cells during neoplastic transformation in cell culture. *Journal of the National Cancer Institute*, 61, 1415–22.

HOLTZER, H. (1978). Cell lineages, stem cells of the 'quantal' cell cycle concept. In *Stem Cells and Tissue Homeostasis*, ed. B. I. Lord, C. S. Potten & R. J. Cole, pp. 1–27. Cambridge University Press.

HUMPHRIES, S., WINDASS, J. WILLIAMSON, R. (1976). Mouse globin gene expression in erythroid and non-erythroid tissues. *Cell*, 7, 267–77.

HYNES, N. E., GROVER, B., SIPPEL, A. E., JEEP, S., WURTZ, T., NGUYEN-HUU, M. C., GRESEKE, K. & SCHUTZ, G. (1979). Control of cellular content of chicken egg white protein specific RNA during estrogen administration and withdrawal. *Biochemistry*, 18, 616–24.

ILMENSEE, K. & MINTZ, B. (1976). Totipotency and normal differentiation of single teratocarcinoma cells cloned by injection into blastocysts. *Proceedings of the National Academy of Sciences, USA*, 73, 549–53.

IMADA, M. & SUEOKA, N. (1978). Clonal sublines of rat neurotumour RT4 and cell differentiation II. A conversion coupling of tumorigenicity and a glial cell property. *Developmental Biology*, 66, 109–16.

IRELAND, M., MAISEL, H. & BRADLEY, R. H. (1978). The rabbit lens cytoskeleton: An ultrastructural analysis. *Ophthalmic Research*, 10, 231–6.

ITOH, Y. (1976). Enhancement of differentiation of lens and pigment cells by ascorbic acid in cultures of neural retinal cells of chick embryos. *Developmental Biology*, 54, 157–62.

JACKSON, J. F., CLAYTON, R. M., WILLIAMSON, R., THOMSON, I., TRUMAN, D. E. S. & DE POMERAI, D. I. (1978). Sequence complexity and tissue distribution of chick lens crystallin mRNAs. *Developmental Biology*, 65, 383–95.

JEFFCOATE, W. J. & REES, L. H. (1978). Adrenocorticotropin and related peptides in non endocrine tumors. *Current Topics in Experimental Endocrinology*, 3, 57–74.

JURGELSKI, W., HUDSON, D. & FALK, H. L. (1979). Tissue differentiation and susceptibility to embryonal tumor induction by ethylnitrosourea in the oppossum. *National Cancer Institute Monographs*, 51, 123–58.

KAWAMURA, J., MACHIDA, S., YOSHIDA, O., OSEKO, F., IMURA, H. & HATTORI, M. (1978). Bladder carcinoma associated with ectopic production of gonadotropin. *Cancer*, 42, 2773–80.

KLOSE, J. & VON WALLENBERG-PACHALY, H. (1976). Changes of soluble protein populations during organogenesis of mouse embryos as revealed by protein mapping. *Developmental Biology*, 51, 324–31.

KONDOH, H., OKADA, T. S., RANDALL, C., BRODIE, J., ZEHIR, A. & CLAYTON, R. M. (1980). Intrinsic programming of neural retina degeneration in a mutant chick.

Abstracts of the Annual Meeting of the Japanese Society of Developmental Biologists,
Japan.

KURTZ, D. T., CHAN, K-M. & FEIGELSON, P. (1978). Translational control of hepatic
α2U globulin synthesis by growth hormone. *Cell,* **15**, 743–50.

LAVERS, G. G. (1978). Differential synthesis of lens proteins in the presence of
M76(5′) pppG and cleavage product m7GMP in an embryonic chick lens cell
lysate. *Molecular Biology Reports,* **4**, 233–6.

LENNON, V. A., UNGER, M. & DULBECCO, R. (1978). Thy-1: A differentiation marker
of potential mammary myoepithelial cells *in vitro. Proceedings of the National
Academy of Sciences, USA,* **75**, 6093–7.

LESLEY, J. F. & LENNON, V. A. (1977). Transitory expression of Thy-1 antigen in
skeletal muscle development. *Nature,* **268**, 163–5.

LONGCHAMPT, M. O., LAURENT, M., COURTOIS, Y., TRENCHEV, P. & HUGHES, R. C.
(1976). Microtubules and microfilaments of bovine lens epithelial cells. Electron
microscopy and immunofluorescence staining with specific antibodies. *Experimental Eye Research,* **23**, 505–18.

LOPASHOV, G. V. (1977). Levels in stabilisation of cell differentiation and its
experimental transformation. *Differentiation,* **9**, 131–7.

LOPASHOV, G. V. (1978). Developmental approaches to organ restoration. *Medical
Biology,* **56**, 344–8.

LOWENHAUPT, K. & LINGREL, J. B. (1978). A change in the stability of globin mRNA
during the induction of murine erythroleukemia cells. *Cell,* **14**, 337–44.

McDEVITT, D. S. (1972). Presence of lateral eye lens crystallins in the median eye of
the American chameleon. *Science,* **175**, 763–4.

MAISEL, H. & HARMISON, C. (1963). An immunoembryological study of the chick
iris. *Journal of Embryology and Experimental Morphology,* **11**, 483–91.

MANES, C. (1974). Phasing of gene products during development. *Cancer Research,*
34, 2044–52.

MIKHAILOV, A. T. & BARABANOV, V. M. (1975). Immunochemical analysis of water-
soluble antigens of chick retina in the course of embryogenesis. *Journal of
Embryology and Experimental Morphology,* **34**, 531–57.

MILLER, G. G., BLAIR, D. G., HUNTER, E., MOUSA, G. Y. & TREVITHICK, J. R. (1979).
Differentiation of rat lens epithelial cells in tissue culture. III. Functions *in vitro* of
a transformed rat lens epithelial cell line. *Development, Growth and Differentiation,*
21, 19–27.

MILSTONE, L. M. & PIATIGORSKY, J. (1977). Delta crystallin gene expression in
embryonic chick lens epithelia cultured in the presence of insulin. *Experimental
Cell Research,* **105**, 9–14.

MILSTONE, L. M., ZELENKA, P. & PIATIGORSKY, J. (1976). Delta-crystallin mRNA in
chick lens cells: mRNA accumulates during differential stimulation of delta-
crystallin synthesis in cultured cells. *Developmental Biology,* **48**, 197–204.

MINTZ, B. (1976). Gene expression in neoplasia and differentiation. *Harvey Lectures,*
71, 193–245.

NODEN, D. M. (1978). Interactions directing the migration and cytodifferentiation of
avian neural crest cells. In *Specificity of Embryological Interactions,* ed. Dr Garrod,
pp. 3–49. Chapman & Hall, London.

NOMURA, K. & OKADA, T. S. (1979). Age dependent change in the transdifferentiation ability of chick neural retina in cell-culture. *Development, Growth and Differentiation*, **21**, 161–8.

NOMURA, K., TAKAGI, S. & OKADA, T. S. (1980). Expression of neuronal specificities in 'transdifferentiating' cultures of neural retina. *Differentiation*, **16**, 141–7.

OKADA, T. S. (1976). 'Transdifferentiation' of cells of specialised eye tissues in cell culture. In *Tests of Teratogenicity in Vitro*, ed. J. D. Ebert & M. Marois, pp. 91–105. Elsevier, Amsterdam.

OKADA, T. S. (1978). Redifferentiation and transdifferentiation of retinoblastoma cells. In *Proceedings of the VIth International Congress on Eye Research, Osaka, Japan*.

OKADA, T. S., EGUCHI, G. & TAKEICHI, M. (1973). The retention of differentiated properties of lens epithelial cells in clonal cell culture. *Developmental Biology*, **34**, 321–33.

OKADA, T. S., ITOH, Y., WATANABE, K. & EGUCHI, G. (1975). Differentiation of lens in cultures of neural retinal cells of chick embryos. *Developmental Biology*, **45**, 318–29.

OKADA, T. S., YASUDA, K. & NOMURA, K. (1979a). Presence of multipotential progenitor cells in embryonic neural retina as revealed by clonal cell culture. In *Cell Lineage, Stem Cells and Cell Determination*, ed. N. le Douarin, *INSERM Symposium Series, 10*, pp. 335–46. Elsevier, Amsterdam.

OKADA, T. S., YASUDA, K., ARAKI, M. & EGUCHI, G. (1979b). Possible demonstration of multipotential nature of embryonic neural retina by clonal cell culture. *Developmental Biology*, **68**, 600–17.

ONO, T. & CUTLLER, R. G. (1978). Age dependent relaxation of gene expression: Increase of endogenous murine leukemia virus-related and globin-related RNA in brain and liver of mice. *Proceedings of the National Academy of Sciences, USA*, **75**, 4431–5.

OUWENEEL, W. J. (1976). Developmental genetics of homoeosis. *Advances in Genetics*, **18**, 179–248.

PEARSE, A. G. E. & BLAKE, J. M. (1974). Endocrine tumours of neural crest origin. Neurolophomas, apudomas and the APuD concept. *Medical Biology*, **52**, 3–18.

PENMAN, S. (1979). Metabolic-regulation in suspended and reattatched anchorage-dependent fibroblasts. *Journal of Supramolecular Structure*, 1979 (S3), 174.

PERLMAN, S. M., FORD, P. J. & ROSBASH, M. M. (1977). Presence of tadpole and adult globin RNA sequences in oocytes of *Xenopus laevis*. *Proceedings of the National Academy of Sciences, USA*, **74**, 3835–9.

PETERSON, A. R., BERTRAM, J. S. & HEIDELBERGER, C. (1974). Cell cycle dependency of DNA damage and repair in transformable mouse fibroblasts treated with N-methyl-N′-nitro-N-nitrosoguanidine. *Cancer Research*, **34**, 1600–7.

PIATIGORSKY, J. (1980). Intracellular ions, protein metabolism and cataract formation. In *Current Topics in Eye Research*, vol 3, pp. 1–39. Academic Press, New York.

PIATIGORSKY, J., BEEBE, D. C., ZELENKA, P., MILSTONE, L. M. & SHINOHARA, T. (1976). Regulation of delta crystallin gene expression during development of the embryonic chick lens. In *Biology of the Epithelial Lens Cells*, ed. Y. Courtois & F. Regnault, *Les Colloques de l'Institut National de la Sante et de la Recherche Medicale*, *60*, pp. 85–112. INSERM, Paris.

DE POMERAI, D. I. & CLAYTON, R. M. (1978). Influence of embryonic stage on the transdifferentiation of chick neural retina cells in culture. *Journal of Embryology and Experimental Morphology*, **47**, 179–93.

DE POMERAI, D. I. & CLAYTON, R. M. (1980). The influence of growth-inhibiting and growth-promoting medium conditions on crystallin accumulation in trans-differentiating cultures of embryonic chick neural retina. *Development, Growth and Differentiation*, **22**, 49–60.

DE POMERAI, D. I., CLAYTON, R. M. & PRITCHARD, D. J. (1978). Delta crystallin accumulation in chick lens epithelial cultures: Dependence on age and genotype. *Experimental Eye Research*, **27**, 365–75.

DE POMERAI, D. I., PRITCHARD, D. J. & CLAYTON, R. M. (1977). Biochemical and immunological studies of lentoid formation in cultures of embryonic chick neural retina and day-old chick lens epithelium. *Developmental Biology*, **60**, 416–27.

POSTLETHWAITE, J. H. & SCHNEIDERMAN, H. A. (1974). Developmental genetics of *Drosophila* imaginal discs. *Annual Review of Genetics*, **7**, 381–433.

PRITCHARD, D. J., CLAYTON, R. M. & DE POMERAI, D. I. (1978). Transdifferentiation of chicken neural retina into lens and pigment epithelium in culture: controlling influences. *Journal of Embryology and Experimental Morphology*, **48**, 1–21.

RAMAEKERS, F. C. S., OSBORN, M., SCHMID, E., WEBER, K., BLOEMENDAL, H. & FRANKE, W. W. (1980). Identification of the cytoskeletal proteins in lens-forming cells. A special epithelioid cell type. *Experimental Cell Research*, **127**, 309–27.

REDFERN, N., ISRAEL, P., BERGSMA, D., ROBINSON, W, G., Jnr, WHIKEHART, D. & CHADER, G. (1976). Neural retinal and pigment epithelial cells in culture: patterns of differentiation and effects of prostaglandins and cyclic AMP on pigmentation. *Experimental Eye Research*, **22**, 559–68.

REEVES, R. & CSERJESI, P. (1979). Constitutive synthesis of globins within most of the members of an uninduced, proliferating population of Friend erythroleukemic cells. *Developmental Biology*, **69**, 576–88.

REYER, R. W. (1977). The amphibian eye: development and regeneration. In *The Visual System in Vertebrates*, ed. F. Crescitelli, pp. 309–90. Springer-Verlag, Berlin.

RINK, H. & VORNHAGEN, R. (1979). Crystallins of lens epithelial cells during aging and differentiation. *Ophthalmic Research*, **11**, 355–9.

RODGERS, M. E. & SHEARN, A. (1977). Patterns of protein synthesis in imaginal discs of *Drosophila melanogaster. Cell*, **12** 915–21.

ROSEN, S. W., KAMINSKA, J., CALVERT, I. S. & AARONSON, S. A. (1979). Human fibroblasts produce 'pregnancy-specific' Beta-1 glycoprotein in vitro. *American Journal of Obstetrics and Gynecology*, **134**, 734–8.

ROSENZWEIG, J. L., HAVRANKOVA, J., LESNIAK, M. A., BROWNSTEIN, M. & ROTH, J. (1980). Insulin is ubiquitous in extrapancreatic tissues of rats and humans. *Proceedings of the National Academy of Sciences, USA*, **77**, 572–6.

ROSS, J., OLMSTED, J. B. & ROSENBAUM, J. L. (1975). The ultrastructure of mouse neuroblastoma cells in tissue culture. *Tissue and Cell*, **7**, 107–35.

SCHEIB, D. (1965). Recherches recentes sur la regeneration du cristallin chez les vertebres. Evolution du probleme entre 1931 et 1963. *Ergebnisse der Anatomie und Entwicklungsgeschichte*, **38**, 47–114.

SEYBOLD, W. D. & SULLIVAN, D. T. (1978). Protein synthetic patterns during differentiation of imaginal discs *in vitro*. *Developmental Biology*, **65**, 69–80.

SHEARN, A., DAVIS, K. T. & HERSPERGER, F. (1978). Transdetermination of *Drosophila* imaginal discs cultured *in vitro. Developmental Biology*, **65**, 536–40.

SHEPHERD, G. W. & FLICKINGER, R. (1979). Post transcriptional control of messenger RNA diversity in frog embryos. *Biochimica et Biophysica Acta*, **563**, 413–21.

SIMNET, J. D. (1974). Nuclear differentiation in the development of normal and neoplastic tissues. *In Neoplasia and Cell Differentiation*, ed. G. V. Sherbet, pp. 1–26. Karger, Basel.

STALDER, J., LARSEN, A., ENGEL, J. D., DOLAN, M., GROUDINE, M. & WEINTRAUB, H. (1980). Tissue specific DNA cleavages in the globin chromatin domain induced by DNAase 1. *Cell*, **20**, 451–60.

THOMSON, I., DE POMERAI, D. I., JACKSON, J. F. & CLAYTON, R. M. (1979). Lens specific mRNA in transdifferentiating cultures of embryonic chick neural retina and pigmented epithelium. *Experimental Cell Research*, **122**, 73–81.

THOMSON, I., WILKINSON, C. E., JACKSON, J. F., DE POMERAI, D. I., CLAYTON, R. M., TRUMAN, D. E. S. & WILLIAMSON, R. (1978). Isolation and cell-free translation of chick lens crystallin mRNA during normal development, and transdifferentiation of neural retina. *Developmental Biology*, **65**, 372–82.

TOMOZAWA, Y. & SUEOKA, N. (1978). In vitro segregation of different cell lines with neuronal and glial properties from a stem cell line of rat neurotumour RT4. *Proceedings of the National Academy of Sciences, USA*, **75**, 6305–9.

TRINKAUS, J. P. (1956). The differentiation of tissue cells. *American Naturalist*, **90**, 273–89.

TSAI, S. Y., BAI, M-J., LIN, C-T. & O'MALLEY, B. W. (1979). Effect of estrogen on ovalbumin gene-expression in differentiated non-target tissues. *Biochemistry*, **18**, 5726–31.

URIEL, J. (1979). Retrodifferentiation and the fetal patterns of gene expression in cancer. *Advances in Cancer Research*, **29**, 127–74.

VANDEKERCKHOVE, J. & WEBER, K. (1979). The complete amino acid sequence of actins from bovine aorta, bovine heart, bovine fast skeletal muscle, and rabbit slow skeletal muscle. *Differentiation*, 14, 123–33.

VERMORKEN, A. J. M., HILDERINK, J. M., VAN DE VEN, W. J. & BLOEMENDAL, H. (1978). Lens differentiation. Crystallin synthesis in isolated epithelia from calf lenses. *Journal of Cell Biology*, **76**, 175–83.

WADDINGTON, C. H. (1956). *Principles of Embryology*. Allen & Unwin, London.

WEKERLE, H., PATERSON, B., KETELSON, U-P. & FELDMAN, M. (1975). Striated muscle fibres differentiate in monolayer cultures of adult thymus reticulum. *Nature*, **256**, 493–4.

WINCHESTER, R. J., ROSS, G. D., JAROWSKI, C. I., WANG, C. Y., HALPER, J. & BROXMEYER, H. E. (1977). Expression of Ia-like antigen molecules on human granulocytes during early phases of differentiation. *Proceedings of the National Academy of Sciences, USA*, **74**, 4012–16.

WINCHESTER, R. J., WANG, C-Y., GIBOFSKY, A., KUNKEL, H. G., LLOYD, K. O. & OLD, L. J. (1978). Expression of Ia-like antigens on cultured human malignant melanoma cell lines. *Proceedings of the National Academy of Sciences, USA*, **75**, 6235–9.

WOLD, B. J., KLEIN, W. H., HOUGH-EVANS, B. R., BRITTEN, R. J. & DAVIDSON, E. H. (1978). Sea urchin embryo mRNA sequences expressed in the nuclear RNA of adult tissues. *Cell*, **14**, 941–50.

YAMADA, T. (1977). Control mechanisms in cell-type conversion in newt lens regeneration. *Monographs in Developmental Biology*, **13**, 1–126.

YAMADA, T. & BEAUCHAMP, J. J. (1978). The cell cycle of cultured iris epithelial cells: its possible role in cell-type conversion. *Developmental Biology*, **66**, 275–8.

YAMADA, T. & McDEVITT, D. S. (1974). Direct evidence for transformation of differentiated iris epithelial cells into lens cells. *Developmental Biology*, **38**, 104–18.

YASUDA, K. (1979). Transdifferentiation of 'lentoid' structures in cultures derived from pigmented epithelium inhibited by collagen. *Developmental Biology*, **68**, 618–23.

YASUDA, K., THOMSON, I., DE, POMERAI, D. I., CLAYTON, R. M. & OKADA, T. S. (1979). Abstracts of the Annual Meeting of the Japanese Society of Developmental Biologists.

YOSHIDA, J. & CRAVIOTO, H. (1978). Nitrosourea-induced brain tumors: An in vivo and in vitro tumor model system. *Journal of the National Cancer Institute*, **61**, 365–70.

YOSHIMOTO, Y., WOLFSEN, A. R., HIROSE, F. & ODELL, W. D. (1979). Human chorionic gonadotropin-like material: presence in normal human tissues. *American Journal of Obstetrics and Gynecology*, **134**, 729–33.

YOSHIMOTO, Y., WOLFSEN, A. R. & ODELL, W. D. (1977). Human chorionic gonadotropin-like substance in nonendocrine tissues of normal subjects. *Science*, **197**, 575–7.

YOUNG, B. D., BIRNIE, G. D. & PAUL, J. (1976). Complexity and specificity of polysomal poly (A |) RNA in mouse tissues. *Biochemistry*, **15**, 2823–9.

References added at proof stage

FUKUMACHI, H. & TAKAYAMA, S. (1979). Promotion of epithelial keratinization by N-methyl-N′-nitro-N-nitrosoguanidine in rat forestomach in organ culture. *Experientia*, **35**, 666–8.

MONROY, A. & ROSATI, F. (1979). Cell surface differentiations during early embryonic development. *Current Topics in Developmental Biology*, **13**, 45–69.

OKADA, P. S. (1980). Cellular metaplasia or transdifferentiation as a model for retina cell differentiation. In *Current Topics in Developmental Biology*, vol. 16(2), ed. A. Moscona & A. Monroy, pp. 349–80. Academic Press, London.

SACERDOTE, M. (1971). Differentiation of ectopic retinal structures in the hypothalamus-hypophysial area in the adult newt bearing a permanent hypothalamic lesion. *Zeitschrift für Anatomie und Entwicklungsgeschichte*, **134**, 49–60.

STEFANKO, S. Z. & MASCHOT, W. A. (1979). Pinealoblastoma with retinoblastoma differentiation. *Brain*, **102**, 321–32.

THOMSON, I., YASUDA, K., DE POMERAI, D. I., CLAYTON, R. M. & OKADA, T. S. (1981). The accumulation of lens-specific protein and mRNA in cultures of neural retina from $3\frac{1}{2}$ day chick embryos. *Experimental Cell Research* (in press).

YODA, K. & FUJIMURA, S. (1979). Induction of differentiation in cultured mouse neuroblastoma cells by N-methyl-N′-nitro-N-nitrosoguanidine. *Biochemical and Biophysical Research Communications*, **87**, 128–34.

Factors affecting gene expression and differentiation during myogenesis in tissue culture

HUW A. JOHN

Department of Genetics, West Mains Road, University of Edinburgh EH9 3JN

INTRODUCTION

Gene expression and differentiation during myogenesis may be influenced by many intrinsic and extrinsic factors.

Intrinsic factors

(1) Factors involved in the inheritance of the epigenetic state, that is, how the myogenic cell can divide repeatedly and still apparently 'remember' its particular state of differentiation (Carlsson, Ringertz & Savage, 1974).

(2) A critical final mitosis during which reprogramming of the genomic DNA may occur (Bischoff & Holtzer, 1969; Gurdon & Woodland, 1968).

(3) Prostaglandin synthesis to generate a transient increase in intracellular cAMP concentration which brings about the cellular changes necessary for fusion to occur (Zalin, 1977).

(4) The act of cell fusion which may trigger off muscle gene expression.

(5) Storage of mRNA in ribonucleoprotein particles and therefore in an unexpressed state (Heywood, Kennedy & Bester, 1975; Bag & Sarkar, 1975, 1976) which may be due to the action of a translational control RNA (Bester, Kennedy & Heywood, 1975).

(6) Tissue specific initiation factors of protein synthesis (Bester *et al.*, 1975).

(7) Production of creatine, a unique product of muscle contraction, which stimulates synthesis of contractile proteins and so may act as a positive feedback effector (Ingwall, Morales & Stockdale, 1972).

Extrinsic factors

(1) Neighbouring cells. The differentiation of a myogenic cell is almost certainly influenced by neighbouring myogenic cells and fibroblasts. Konigsberg (1971) has suggested that the progressive increase in the myogenic cell population may have two effects. At the

higher cell densities cell-to-cell contacts might occur more frequently thus increasing the probability of encounters between cells competent to fuse or alternatively increased cell density would accelerate any alterations of the medium generated by the metabolic activities of the cells. In this way high cell density might facilitate the establishment of a microenvironment in some way more favourable for differentiation. Neighbouring fibroblasts are responsible for laying down the muscle connective tissue system which involves synthesis of large amounts of collagen which may have an important influence acting as a substratum for the muscle cells.

(2) Factors circulated in the blood including the serum proteins, nutrients and hormones which may have a role in regulating muscle cell differentiation.

(3) Nerve cells exert an extended and continuous influence on muscle development and maintenance of the differentiated state by patterns of impulses (Salmons & Sreter, 1976) and neurotrophic factors (Markelonis & Oh, 1978).

MYOGENESIS IN TISSUE CULTURE

When myogenic cells are disaggregated from muscle by trypsinization and allowed to grow in tissue culture they divide for a few days and then the cells fuse rapidly to form multinucleate syncytia which can further differentiate into myotubes showing contractile twitches (Fig. 1). Cell fusion is not an artefact of tissue culture. The occurrence of myoblast fusion *in vivo* was rigorously demonstrated in experiments by Mintz & Baker (1967). Cleavage-stage mouse embryos were produced by aggregating blastomeres of two genotypes. Sets of blastomeres were combined, each of which contained the genome for production of only one of two different genetically determined forms of the dimeric enzyme isocitric dehydrogenase. During the development of these artifical embryos, fusion of myoblasts with inclusion of nuclei of the two genotypes in one muscle myotube was expected to lead to the appearance of subunits of both enzyme types in the common sarcoplasm, hybridization of the different subunits, and appearance of a third enzyme form. Since such an extra enzyme form could be electrophoretically identified in muscle but not in other tissues the assumption of myotube formation in the intact embryo by myoblast fusion seemed fully supported.

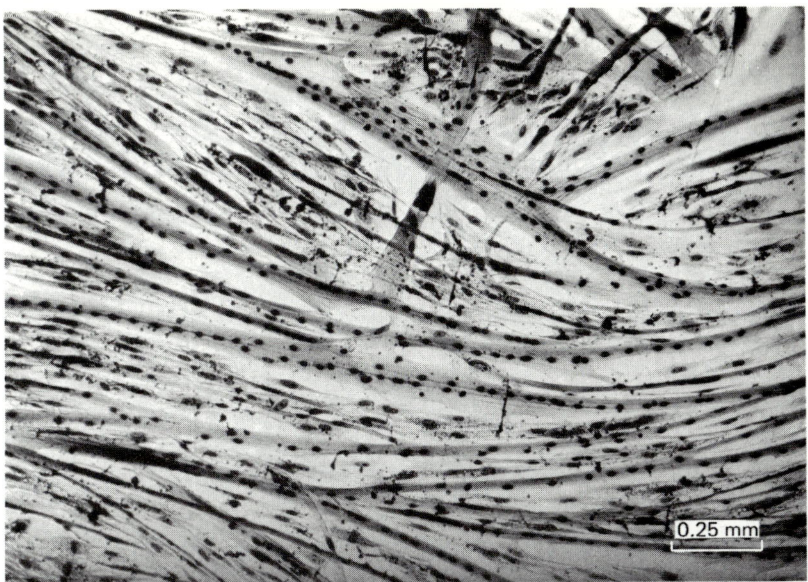

Fig. 1. Myotubes in tissue culture. Myogenic cells were disaggregated by trypsinization from the thigh muscle of 11 day chick embryos and 2×10^6 cells were cultured on gelatinized 9 cm dishes using Eagle's MEM (Gibco powder medium containing glutamine) with 10% horse serum (Flow) and 2% chick embryo extract (50% w/v). The cultures were fed fresh medium on the second day. After 5 days, cultures were fixed with 3:1 ethanol acetic acid, dehydrated through an alcohol series, air dried and stained with Giemsa, pH 6.8.

FACTORS AFFECTING GENE EXPRESSION AND DIFFERENTIATION DURING MYOGENESIS IN TISSUE CULTURE

The tissue culture system has been extensively used to investigate the effect of various factors on gene expression and differentiation of muscle cells. In this chapter investigations to determine the importance of two extrinsic factors (the collagen substratum and blood serum) and two intrinsic factors (final mitosis and fusion) will be described.

COLLAGEN SUBSTRATA

A substratum of collagen influences gene expression and differentiation of muscle cells in tissue culture. When culture surfaces were coated with rat tail type I collagen, growth and clonal differentiation of

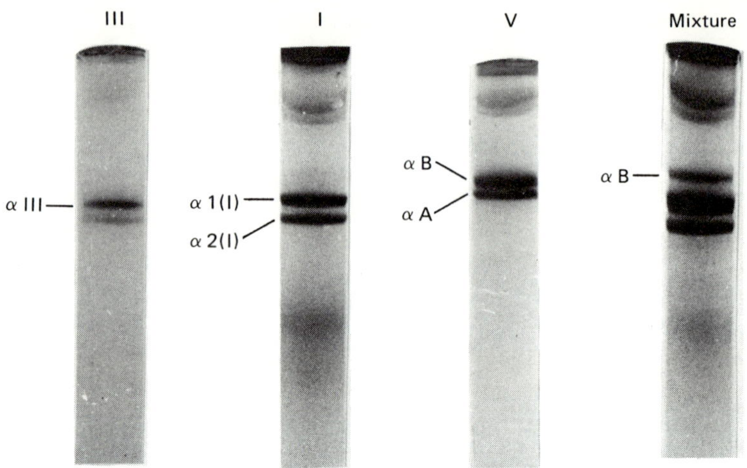

Fig. 2. SDS gel electrophoresis of collagen types. Pepsin-solubilized collagens type I, III and V were prepared from the total leg and breast musculature of 16 week hens and analysed by SDS gel electrophoresis as previously described (John & Lawson, 1980).

chick embryo myoblasts were stimulated (Hauschka & Konigsberg, 1966).

In vivo, muscle tissue contains a connective tissue network consisting mainly of collagen. The muscle cells come to lie in bundles enclosed by layers of connective membranes: the endomysium or basement lamina surrounding each individual myotube, the perimysium enclosing a bundle of myotubes, and the epimysium which surrounds a whole series of bundles making up the anatomical muscle. Recent biochemical and immunofluorescence localization studies (Bailey & Sims, 1977; Duance *et al.*, 1977; Bailey *et al.*, 1979) have demonstrated that the different membranes contain distinct polymorphic forms of collagen: type I mainly in the epi- and peri-mysium, type III in the perimysium and type V in the endomysium.

To investigate the possibility that there may be a direct relationship between the type of collagen laid down and muscle cell differentiation, John & Lawson (1980) cultured chick muscle cells on culture-dishes coated with different polymorphic forms of collagen prepared from adult chicken muscle (Fig. 2). When muscle type I collagen was dried down on dishes according to the procedure of De La Haba, Kamali & Tiede (1975) prior to culturing muscle cells, there were marked differences in cell growth and differentiation in different regions of the dish. The collagen solution in dishes began to dry in the centre within about 1 hour at room temperature (Fig. 3*a* – marked (x)) and in this region cells were sparse resembling cultures on untreated dishes. In the region marked (y) (Fig. 3*a*) muscle cells fused to form myotubes aligned parallel to the drying edge (Fig. 3*b*). The parallel myotubes were usually unbranched (Fig. 3*c*) and single myotubes could extend for several centimetres. Even at an early stage when the cultures were sparse and consisted of mononucleate cells many cells in zone y were aligned in parallel (Fig. 3*f*). In early and later cultures fibroblast-like cells were non-aligned. In the zone within 1 cm of the sidewall of the dish which was the last to dry (Fig. 3*a* (z)) the cells were denser and the myotubes non-aligned (Fig. 3*d*) resembling those in dishes coated by the other techniques.

The capacity of other muscle collagen types to cause alignment of myotubes was compared using the drying down procedure. Parallel myotubes were occasionally found on type III collagen substrates but these were usually heavily contaminated with type I collagen. Cells grew evenly over the dishes coated with type V collagen but these cultures were characterized by myotubes in which the cytoplasm was not extended and the nuclei were bunched together (Fig. 3*e*). No parallel alignment of cells was found in these cultures.

The alignment of cells suggested that the underlying cell recognition sites on collagen molecules were organized in such a way as to produce this effect, for example, by the molecules being aligned in parallel. Hauschka & White (1972) identified the sequences in rat skin type I collagen which are active in promoting cell attachment and clonal differentiation of chick myogenic cells as the α1 (I) polypeptide chain cyanogen bromide peptides α1 CB7 (273 amino acid residues) and α1 CB8 (263 amino acid residues). More recently these same sites have been shown to bind the cell surface glycoprotein fibronectin which mediates the attachment of some cells to collagen (Kleinemann, McGoodwin & Klebe, 1976; Dessau, Adelmann, Timpl & Martin, 1978). In type II collagen (cartilage-type) the fibronectin binding sites are located on the CB10 peptide of the α1 (II) polypeptide chain (Kleinemann *et al.*, 1976). So at the level of native collagen molecules there are different densities or lattice spacings of fibronectin binding sites for which some cells may have an affinity. For example, for each type I collagen molecule approximately 300 nm long there are four fibronectin binding sites on the two α1 (I) polypeptide chains which may have a rectangular spacing approximately 40–200 × 1.4 nm. In contrast each type II collagen molecule has three fibronectin binding

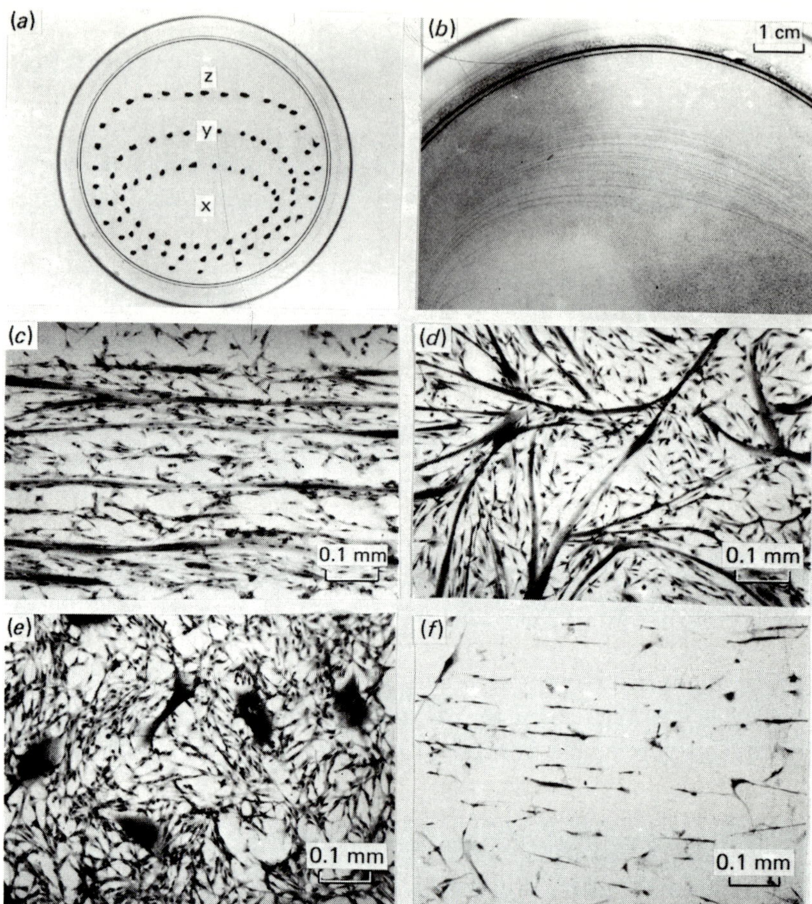

Fig. 3. The effect of different collagen types used as substrata on myogenesis in tissue culture. Tissue culture dishes (Falcon plastic, 100 × 20 mm) were coated with collagen types using the method of De La Haba *et al.* (1975). 4 ml 0.004% collagen in 0.15 N acetic acid (160 μg collagen) was spread on the dish and allowed to dry and sterilize overnight under an ultraviolet light (254 μm). Dishes were rinsed briefly with Dulbecco physiological saline before use. (*a*) Drying of type I collagen solution on a culture dish (100 × 10 mm). The collagen solution begins to dry in the centre (zone x), forms parallel drying streaks in zone y, and dries last in zone z. (*b*) Myotubes (5 days in culture) aligned parallel to the drying edge in zone y of a type I collagen substratum. (*c*) *ibid.* (*d*) Non-aligned myotubes (5 days in culture) in zone z of a type I collagen substratum. (*e*) Non-extended myotubes (5 days in culture) on a type V collagen substratum. (*f*) Mononucleate cells (2 days in culture) aligned parallel to the drying edge in zone y of a type I collagen substratum.

sites on the three $\alpha 1$ (II) polypeptide chains approximately 1.4 nm apart.

Further characteristics of a lattice pattern would depend on the longitudinal and lateral relationship to neighbouring collagen molecules. Type I collagen molecules can aggregate to form fibres with a periodicity of approximately 65 nm which would put all fibronectin binding sites along the long axis of the fibre and this may lead to alignment of the cells in the same direction. However, when coated dishes were stained for collagen (Bermann *et al.*, 1978) there was no evidence of fibres. Instead, a series of faint parallel streaks parallel to the drying edge were seen suggesting that the collagen dries unevenly resulting in concentric zones of relatively high concentration. Thus the presence of a high concentration of collagen in such a zone may be all that is required for cells to grow densely and therefore fuse in this region. However, a number of observations suggest that the explanation is not this simple. First, the streaks of collagen were considerably wider (1000 μm) than individual cells or myotubes. Second, within a short time of settling, many of the myoblast-like mononucleate cells become aligned in parallel along the streaks. Third, fibroblast-like cells which are not incorporated into myotubes appear to be orientated at random.

It is to be expected that long molecules such as collagen would dry down in the plane of the film and have a tendency to align parallel to one another. Furthermore the asymmetry of the surface forces at the drying edge of the film could well induce alignment in the direction of the drying edge. However there was no evidence which would indicate alignment of the molecules when similar collagen layers were examined by polarizing microscopy. This does not rule out the possibility that alignment of the molecules could occur at the surface of the layer only, without birefringence being detected.

Different cell types may have their collagen recognition sites located in different lattice spacings. Some cells, for example myoblasts and myotubes, may have patterns that result in a particular affinity for groups of four fibronectin binding sites on type I collagen molecules arranged in parallel and could therefore result in the cells being aligned in the same direction. Other cells, for example, fibroblasts may have a pattern that is amenable to adhering in any orientation on any collagen substratum.

There is previous evidence that functionally diverse cell types show differences in their interaction with collagen types. Murray *et al.*

(1979) compared the attachment of fibroblasts and epidermal cells to different collagen substrates and their dependence on fibronectin for the mediation of attachment. Epidermal cells attached and differentiated preferentially on substrates of type IV collagen compared to those of types I–III. In contrast, fibroblasts attached equally well to all four collagen types. Fibronectin enhanced the attachment of fibroblasts but not of epidermal cells to collagen.

SERUM PROTEINS

In vivo, muscle cells are exposed to factors circulating in the blood. Many of these are probably represented in tissue culture media by the serum and embryo extract components. These non-defined supplements may contain nutrients, hormones and other substances which have a role in regulating muscle differentiation *in vivo*.

The type of serum used in tissue culture of muscle cells can affect both growth and differentiation. Foetal calf serum is considered to favour division of myogenic cells in culture, while horse serum promotes differentiation (Yaffe, 1971; Morris & Cole, 1972).

Fig. 4*a* shows that when muscle cells are grown in medium containing 10% horse serum (HS medium) the myotubes have well defined cross striations indicating substantial synthesis and organization of the myofibrillar proteins. In medium containing 10% foetal calf serum (FCS medium) there is either little fusion and continuing cell division leads to overgrowth, or, if myotubes are formed, they lack cross striations. (Fig. 4*b*). Protein components of medium recovered at different stages of myogenesis were analysed by SDS gel electrophoresis (Figs. 5, 6). Analysis of both HS and FCS medium showed no detectable change between the early stages of culture during which cells divide (1–3 days) and the period during and after fusion (days 3–5). However, certain proteins (indicated in Figs. 5, 6) whose functions are known (reviewed by Allen, Hill & Stokes, 1977) were increased in HS medium:
(1) Serum albumin which may have a role similar to that *in vivo* of transport, sequestration and binding of physiologically important molecules.
(2) Biologically active glycoproteins such as α_2 macroglobulin which binds insulin, α_1 acid glycoprotein which is increased in situations of cell proliferation and β_2 glycoprotein.
(3) Transferrin which transports iron to cells (see also, Imbenotte & Verger, 1980).

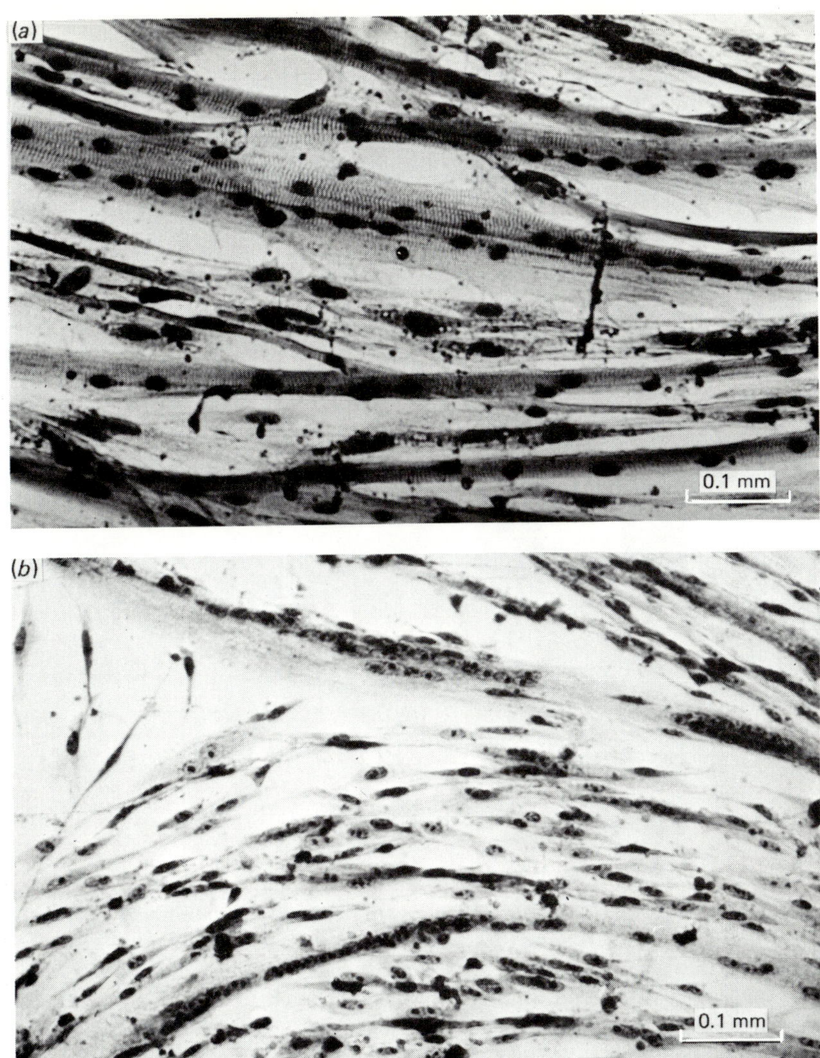

Fig. 4. Comparison of myotubes in culture grown in Eagle's MEM containing 2% chick embryo extract and either (*a*) 10% horse serum, or (*b*) 10% foetal calf serum.

(4) Haemopexin which is a haem-scavenging protein.
(5) Ceruloplasmin, a copper-binding protein.
(6) Immunoglobulins.
(7) Total protein judged by the criterion of staining intensity of the protein bands.

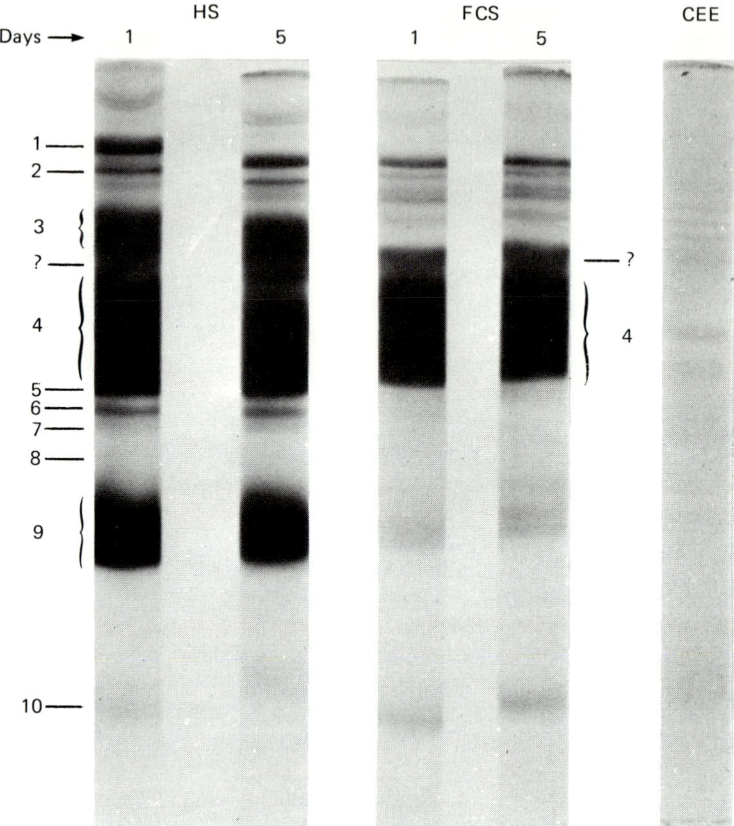

Fig. 5. SDS gel electrophoresis of tissue culture medium. Cultures were set up as described in the legend to Fig. 1 using Eagle's MEM with 2% chick embryo extract and either 10% horse serum (HS) or 10% foetal calf serum (FCS). Medium was recovered after 1 and 5 days, centrifuged to remove floating cells and debris and stored frozen at −20 °C. Samples (25–50 μl) were lyophilized to dryness and analysed by SDS gel electrophoresis in tube gels (Fig. 5) or slab gels (Fig. 6). A sample of 2% chick embryo extract (CEE) was also analysed (Fig. 5). 1, macroglobulin; 2, ceruloplasmin; 3, transferrin; 4, serum albumin; 5, haemopexin; 6, α_2 HS glycoprotein and antitrypsin; 7, α_1 acid glycoprotein; 8, β glycoprotein; 9, immunoglobulins; 10, lipoproteins.

The amount of these proteins in medium may affect myogenesis by promotion or inhibition of either cell division or cell differentiation.

Earlier investigations (see Paul, 1970) indicated that the active components of serum were α_1 acid glycoprotein and α_2 macroglobulin. Both disappeared from the medium during cultivation of cells. Each alone had a small effect in stimulating growth; together they had a marked

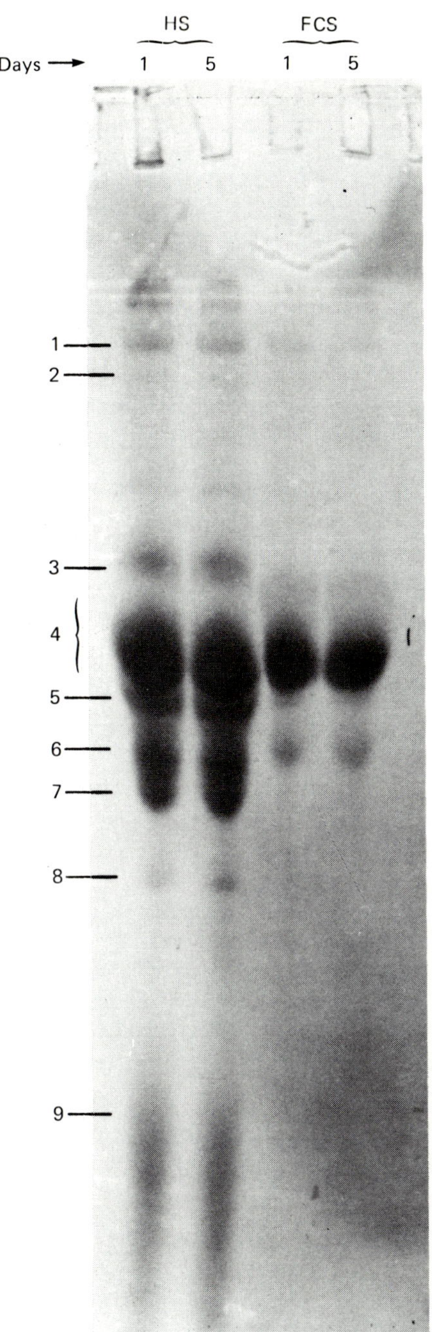

Fig. 6. See legend to Fig. 5; slab gels used instead of tube gels.

effect and could replace serum. The macroglobulin could be replaced by non-protein molecules but the acid glycoprotein could not be so replaced. These findings suggest that glycoproteins associated with Colin's fraction V are the important components in serum.

Kohama & Ozawa (1978a) reported that low concentrations of chicken serum stimulated the proliferation of cultures of 12 day embryonic chick muscle cells. The activity of serum varied during the course of development. The activity of serum from the embryo was high from day 11 to day 13 of incubation and then decreased, reaching a minimum on day 16. This decrease correlated with the period of decreased cell division and increased fusion of myotubes *in vivo* (Herrmann, Heywood & Marchok, 1970). Subsequently, the activity of the serum increased, reaching a peak around the day of hatching. After hatching, the activity of the serum began to decrease on day 4, reaching a minimum at day 18.

Ozawa & Kohama (1978) tested the ability of hormones or extracts from adult chicken tissues to replace or influence the activity of the serum factor. Pituitary gland extracts did not replace the serum factor in a range equivalent to normal serum concentrations of somatotropin, although high concentrations of such extract showed a significant ability to mimic the serum activity. Insulin did not show any ability to mimic serum. Both pituitary extract and insulin showed a potentiating effect on the serum factor when added to the medium. However, the concentrations necessary for this effect were too high to be considered physiological. Since Kohama & Ozawa (1978a) considered that chick embryo extract contained the same active factor as serum they proposed that the results obtained by De La Haba, Cooper & Etling (1966) and Powers & Florini (1975) might be due to the hormones having a potentiating effect on this factor activity. However, testosterone did not potentiate the serum factor when tested in tissue culture (Ozawa & Kohama 1978) but when testosterone was injected into chickens the activity of the serum then recovered was increased, suggesting that testosterone may play a physiological role in regulating the level of serum factor (Kohama & Ozawa, 1978b).

Chick embryo extract (CEE) at 2% is used to supplement medium for myogenic cultures. When this is omitted from FCS medium, myotube formation can still occur. However, when it is left out of HS medium, cultures are sparse, with no myotubes and a great deal of cell detachment. The amount of protein contributed to medium by 2% CEE is small compared with 10% serum (Fig. 5) but clearly some components are of crucial importance.

Yaffe (1971) found that reduced CEE concentration caused proliferating cells to cease dividing and to undergo fusion suggesting that high concentration may be a non-specific stimulator of cell proliferation.

De La Haba *et al.* (1975) separated CEE into two fractions, a non-filterable fraction (including molecules of molecular weight greater than 10 000) which promotes both the fusion of myoblasts to form syncytia and the further differentiation of these syncytia into myotubes, and a filterable fraction which stimulated myoblast fusion to form syncytia only. The filterable fraction promoted myotube formation if the cells were grown on a collagen substratum. The non-filterable fraction already contained a collagen-like molecule.

It seems possible that there may be active components found in CEE which are also present in serum used as a medium supplement. Kohama & Ozawa (1978a) reported that both CEE and chicken serum stimulated proliferation of chick muscle culture to about the same degree.

DETECTION OF mRNA IN CYTOLOGICAL PREPARATIONS BY *IN SITU* HYBRIDIZATION: THE ROLE OF CELL FUSION AND THE FINAL MITOSIS IN SWITCHING ON MUSCLE SPECIFIC GENES

In fully differentiated muscle the myofibrillar proteins comprise about 60% of the total protein and the appearance of these, as well as other muscle proteins, has often been taken as indicative of differentiation in experiments designed to illuminate the role of various factors in switching on muscle specific genes.

These studies have been complicated by two observations. First, that myofibrillar proteins occur in most cell types (myogenic and non-myogenic cells) and, in particular, occur in myogenic cells during the cell division phase as well as immediately before and after fusion (Chi, Fellini & Holtzer, 1975a; Chi, Rubinstein, Stahs & Holtzer, 1975b). Therefore reliable methods to distinguish between the muscle-specific (myogenic) proteins and the non-myogenic proteins are obviously necessary. The second observation is that substantial amounts of myosin and actin mRNA occur in cytoplasmic particles (not in poly-somes) in what has been described as stored mRNA (Heywood *et al.*, 1975; Bag & Sarkar, 1975, 1976). Therefore, in muscle cells it is possible that the transcription of a gene may be widely separated in time from the translation of the mRNA into protein. Clearly methods for detecting muscle specific mRNAs would be of considerable advantage.

An initial approach to detection and measurement of mRNA was by hybridization with a radioactive complementary DNA sequence (cDNA) prepared from purified mRNA by reverse transcription. More recently, radioactive nick translated copies of cloned double-stranded cDNA or cloned sequences from genomic DNA libraries have become available. Radioactive probes of this type can be used to detect mRNA in cytological preparations by *in situ* hybridization and this technique is particularly appropriate for investigations of cells differentiating in tissue culture. After hybridization of the radioactive probe to the mRNA in intact cells the distribution in the cytoplasm and nucleus can be visualized by autoradiography. The technique enables individual cells in mixed populations such as primary myogenic cultures to be identified by their mRNA content and not by morphological criteria which can be variable (Konigsberg, 1963). Because mRNA is detected, translational products need not be present and the cell can be studied at the earliest developmental stage.

Ideally, in the technique for detection of mRNA in intact cells by *in situ* hybridization the cells are:

(1) Grown on glass slides and histologically fixed so as to maximize preservation of endogenous mRNA.
(2) Dehydrated so that the radioactive probe in an appropriate solution can enter the cells efficiently.
(3) Sealed with the probe solution under a coverslip.
(4) Incubated under conditions which allow the most efficient and stringent hybridization reaction between the radioactive sequences and mRNA compatible with maximum preservation of cell morphology and minimum loss of endogenous mRNA.
(5) Treated to remove unreacted or non-specifically bound radioactive probe prior to autoradiography.

John, Morss, Curtis & Jones (in preparation) have compared the published procedures for *in situ* hybridization. These techniques show variation in fixation, hybridization conditions and post-hybridization treatment. Chick myogenic cultures at the myotube stage were hybridized with two types of radioactive probe, either with cDNA prepared from the mRNA purified from large polysomes isolated from chick embryonic leg skeletal muscle (Patrinou-Georgoulas & John, 1977) or with a cloned double-stranded cDNA (ds cDNA) from the same source (Morss & Jones, unpublished; Jones, 1980). The method which

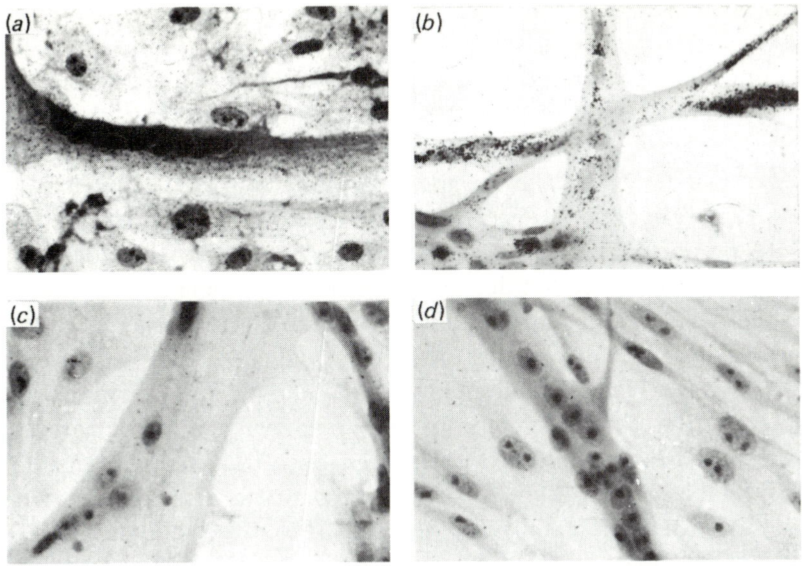

Fig. 7. *In situ* hybridization to myotubes. Myogenic cells grown on gelatinized glass slides were fixed in 3 : 1 methanol : acetic acid at 4 °C for 20 minutes, rinsed in 70% ethanol followed by 1 : 1 absolute ethanol : 0.2 M sodium cacodylate pH 7.2, and post fixed in 0.1% glutaraldehyde, 0.1 M sodium cacodylate pH 7.2, at 4 °C for 20 minutes. The slides were dehydrated through an alcohol series: 50% (in 0.1 M cacodylate), 70%, 90%, 100% (2) and dried *in vacuo*. 3–5μl of probe solution (^{3}H cDNA – 20 000 c.p.m., ^{3}H cloned ds cDNA – 90 000 c.p.m.) in 2 × SSC, 24% formamide were spotted on to the glass slides, covered with an acid washed, siliconized and heat sterilized 18 × 18 mm glass coverslip and sealed with a 1 : 1 mixture of cowgum and diethyl ether. The slides were incubated at 45 °C for 10 minutes (cDNA) or 16 hours (cloned ds cDNA) in a humidified atmosphere. The cowgum was peeled off and the slides were placed in cold 2 × SSC in which the coverslips could be loosened. After rinsing in 2 × SSC the glass slides were incubated with S_1 nuclease (1 unit per μl of S_1 buffer – 0.3 M NaCl, 2 mM $ZnSO_4$, 0.3 M sodium acetate pH 4.5) at 42 °C for 1 hour, rinsed with 2 × SSC at 45 °C for 2 hours and rinsed with a large excess of 2 × SSC at 4 °C overnight. Slides were dehydrated through an ethanol series dried in air and dipped for autoradiography in a 1 : 1 mixture of Ilford K2 photographic emulsion : water at 45 °C. Exposure was for 7 weeks. (*a*) cDNA. (*b*) cloned ds cDNA without S_1 nuclease treatment. (*c*) cloned ds cDNA with S_1 nuclease treatment. (*d*) pMB9.

allowed most efficient hybridization and preservation of cell morphology is outlined in the legend to Fig. 7 and is a modification of that described by Maitland & Jones (in preparation).

The cDNA preparation used was enriched in sequences complementary to the mRNA for the myosin heavy chain (Patrinou-Georgoulas & John, 1977). Previous estimates of approximately 4000 molecules per myotube nucleus of this mRNA (Emerson & Beckner, 1975; John, Patrinou-Georgoulas & Jones, 1977) indicated that the hybridization reaction should be complete in a very short time (5 min). That this was probably the case is indicated by Fig. 7*a* in which cells show considerable hybridized cDNA, resistant to nuclease S_1 treatment, after 10 minutes incubation. The number of grains was reduced with prolonged hybridization time (16 hours – not shown). This particular cDNA preparation hybridized to both myogenic cells and fibroblasts.

When cloned ds cDNA derived from heavy polysomes was used as a probe, there was substantial hybridization to both myogenic cells and fibroblasts prior to S_1 nuclease treatment (Fig. 7*b*). This was substantially reduced after S_1 treatment (Fig. 7*c*) suggesting that radioactive copies of the cloning vector (pMB9) present as a single stranded covalently linked 'tail' to the inserted ds cDNA sequence made a considerable contribution to the initial radioactivity bound. This is obviously useful in 'amplifying' the true hybridized sequences. When radioactive copies of pMB9 without inserted sequences were used as probes there was no hybridization before or after S_1 nuclease treatment (Fig. 7*d*). One interpretation of these results is that only a very small inserted sequence was present in this particular cloned ds cDNA.

To ascertain to what extent RNA was lost during this *in situ* hybridization procedure, myotubes cultured on glass slides were incubated with ^{3}H uridine to label the endogenous RNA and then processed through the stages of the procedure except that radioactive probes were omitted from hybridization solutions. Fig. 8 shows that RNA content judged by the criterion of autoradiographic grains was slightly reduced in the cells treated with the hybridizing solutions under the coverslip. Some procedures showed a marked reduction in the RNA content of cells under the coverslip.

The usefulness of the *in situ* hybridization technique in investigating differentiation in tissue culture is illustrated from data from an earlier study (John *et al.*, 1977). Highly purified mRNA for chicken skeletal muscle myosin heavy chain was used to prepare cDNA which hybridized to mRNA in myotubes but not in fibroblasts (Fig. 9*a*) although these contained almost as much myosin heavy chain protein. The muscle-cell-specific cDNA was used to detect mRNA in experiments designed to test the importance of cell fusion and the final mitosis in triggering off the expression of the muscle-specific myosin heavy chain gene.

When cell fusion in tissue culture was blocked by EGTA, *in situ*

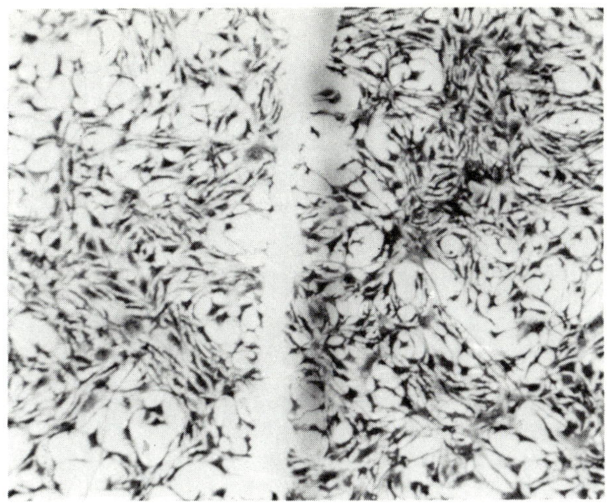

Fig. 8. Stability of endogenous RNA during *in situ* hybridization. Myotubes were incubated with 10 µCi/ml of ^{3}H uridine for 24 hours to label endogenous RNA and then processed through all the stages of the *in situ* hybridization procedure described in the legend to Fig. 7 except that radioactive probes were left out of the hybridization solutions and treatment with S_1 nuclease was omitted. The cells on the left were incubated for 16 hours at 45 °C in hybridization solution (2 × SSC, 24% formamide) under the coverslip. The cells on the right remained dry at this stage. The subsequent autoradiographs were not stained so that the photographs here are of 'grain images' of the cells.

hybridization indicated the presence of substantial amounts of the mRNA (Fig. 9*b*) indicating that fusion was not important in switching on this gene.

Bischoff & Holtzer (1969) suggested that the expression of muscle-specific genes occurs only after a final critical mitosis and is therefore restricted to post-mitotic cells irreversibly committed to differentiation. However, John *et al.* (1977) detected skeletal muscle myosin mRNA in dividing cells (Fig. 9*c*) suggesting that this is not the case. This observation suggests that, either, the mRNA is synthesized prior to, and remains in the cytoplasm during mitosis, or, that cells that have been committed to differentiation can reverse the process, re-enter the cell cycle and yet still retain the muscle specific myosin mRNA. Investigations by Konigsberg (see this book) suggest that a myogenic cell can re-enter the cell cycle and still show expression of skeletal muscle myosin.

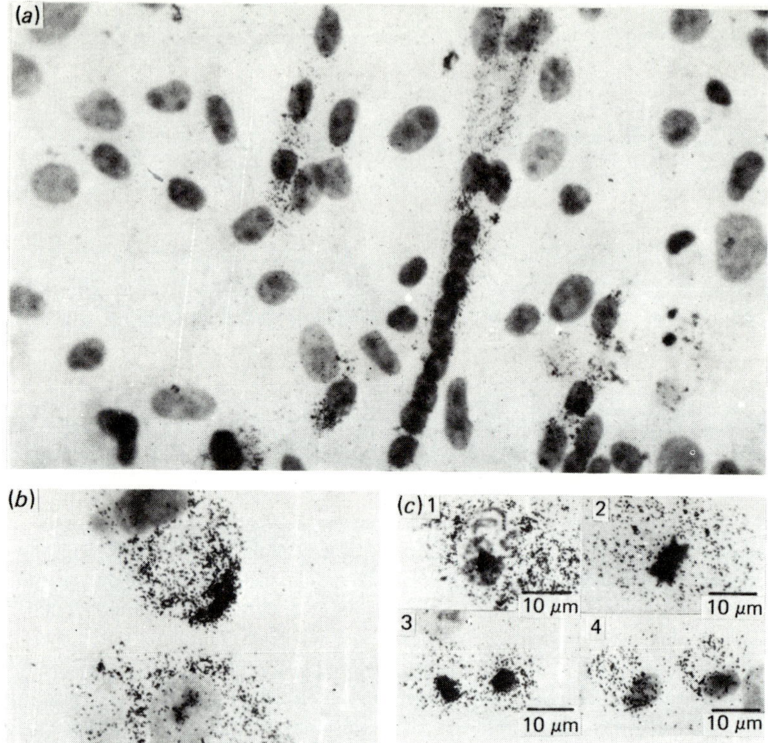

Fig. 9. Detection of myosin heavy chain mRNA by *in situ* hybridization of myosin heavy chain cDNA to (*a*) myotube culture (*b*) EGTA treated, fusion blocked myoblasts (*c*) cells in different stages of mitosis in dividing myoblast cultures (1) prophase, (2) metaphase, (3) telophase, (4) daughter cells.

REFERENCES

ALLEN, P. C., HILL, E. A. & STOKES, A. M. (1977). *Plasma Proteins. Analytical and Preparative Techniques.* Oxford: Blackwell Scientific Publications.

BAG, J. & SARKAR, S. (1975). Cytoplasmic nonpolysomal messenger ribonucleoprotein containing actin messenger RNA in chicken embryonic muscles. *Biochemistry*, **14**, 3800–7.

BAG, J. & SARKAR, S. (1976). Studies on a nonpolysomal ribonucleoprotein coding for myosin heavy chains from chick embryonic muscles. *Journal of Biological Chemistry*, **251**, 7600–9.

BAILEY, A. J., SHELLSWELL, G. B. & DUANCE, V. C. (1979). Identification and change of collagen types in differentiating myoblasts and developing chick muscle. *Nature*, **278**, 67–9.

BAILEY, A. J. & SIMS, T. J. (1977). Meat tenderness: distribution of molecular species of collagen in bovine muscle. *Journal of the Science of Food and Agriculture*, **28**, 565–70.

BERMANN, J., STONER, G., DAWE, C., RICE, J. & KINGSBURY, E. (1978). Histochemical demonstration of collagen fibres in ascorbic-acid-fed cell cultures. *In Vitro*, **14**, 673–85.

BESTER, A. J., KENNEDY, D. S. & HEYWOOD, S. M. (1975). Two classes of translational control RNA: their role in the regulation of protein synthesis. *Proceedings of the National Academy of Sciences, USA*, **72**, 1523–7.

BISCHOFF, R. & HOLTZER, H. (1969). Mitosis and the process of differentiation of myogenic cells in vitro. *Journal of Cell Biology*, **41**, 188–200.

CARLSSON, S-A., RINGERTZ, N. R. & SAVAGE, R. E. (1974). Intracellular antigen migration in interspecific myoblast heterokaryons. *Experimental Cell Research*, **84**, 255–60.

CHI, J. C., FELLINI, S. A. & HOLTZER, H. (1975a). Differences among myosins synthesized in non myogenic cells, presumptive myoblasts and myoblasts. *Proceedings of the National Academy of Sciences, USA*, **72**, 4999–5003.

CHI, J. C. H., RUBINSTEIN, N., STAHS, K. & HOLTZER, H. (1975b). Synthesis of myosin heavy and light chains in muscle cultures. *Journal of Cell Biology*, **67**, 523–37.

DE LA HABA, G., COOPER, G. W. & ETLING, V. (1966). Hormonal requirements for myogenesis of striated muscle in vitro: insulin and somatotropin. *Proceedings of the National Academy of Sciences, USA*, **56**, 1719–23.

DE LA HABA, G., KAMALI, H. M. & TIEDE, D. M. (1975). Myogenesis of avian striated muscle in vitro: role of collagen in myofiber formation. *Proceedings of the National Academy of Sciences, USA*, **72**, 2729–32.

DESSAU, W., ADELMANN, B. C., TIMPL, R. & MARTIN, G. R. (1978). Identification of the sites in collagen α-chains that bind serum antigelatin factor (c-lg). *Biochemical Journal*, **169**, 55–9.

DUANCE, V. C., RESTALL, D. J., BEARD, H., BOURNE, F. J. & BAILEY, A. J. (1977). The location of three collagen types in skeletal muscle. *FEBS Letters*, **79**, 248–252.

EMERSON, C. P. & BECKNER, S. K. (1975). Activation of myosin synthesis in fusing and mononucleated myoblasts. *Journal of Molecular Biology*, **93**, 431–47.

GURDON, J. B. & WOODLAND, H. R. (1968). The cytoplasmic control of nuclear activity in animal development. *Biological Reviews*, **43**, 233–67.

HAUSCHKA, S. D. & KONIGSBERG, I. R. (1966). The influence of collagen on the development of muscle clones. *Proceedings of the National Academy of Sciences, USA*, **55**, 119–26.

HAUSCHKA, S. D. & WHITE, N. K. (1972). Studies of myogenesis *in vitro*. I Temporal changes in the proportion of muscle-colony-forming cells during the early stages of limb development. II Myoblast-collagen interaction: molecular specificity required. In *Research Concepts in Muscle Development and the Muscle Spindle*, ed. B. Banker, R. Preybylski, J. Van der Meulen and M. Victor, pp. 53–71. Amsterdam: Excerpta Medica.

HERRMANN, H., HEYWOOD, S. M. & MARCHOK, A. C. (1970). Reconstruction of muscle development as a sequence of macromolecular synthesis. *Current Topics in Developmental Biology*, **5**, 181–234.

HEYWOOD, S. M., KENNEDY, D. S. & BESTER, A. J. (1975). Stored myosin messenger in embryonic chick muscle. *FEBS Letters*, **53**, 69–74.

IMBENOTTE, J. & VERGER, C. (1980). Nature of the iron requirement for chick embryo cells cultured in the presence of horse serum. *Cell Biology International Reports*, **4**, 447–52.

INGWALL, J. S., MORALES, M. F. & STOCKDALE, F. E. (1972). Creatine and the control of myosin synthesis in differentiating skeletal muscle. *Proceedings of the National Academy of Sciences, USA*, **69**, 2250–3.

JOHN, H. A. & LAWSON, H. (1980). The effect of different collagen types used as substrata on myogenesis in tissue culture. *Cell Biology International Reports*, **4**, 841–50.

JOHN, H. A., PATRINOU-GEORGOULAS, M. & JONES, K. W. (1977). Detection of myosin heavy chain mRNA during myogenesis in tissue culture by *in vitro* and *in situ* hybridization. *Cell*, **12**, 501–8.

JONES, K. W. (1980). Muscle cell differentiation and the prospects for genetic engineering. *British Medical Bulletin*, **36**, 173–80.

KLEINEMANN, H. K., MCGOODWIN, E. B. & KLEBE, R. J. (1976). Localization of the cell attachment region in type I and II collagen. *Biochemical and Biophysical Research Communications*, **72**, 426–32.

KOHAMA, K. & OZAWA, E. (1978a). Muscle trophic factor: II Ontogenic development of activity of a muscle trophic factor in chicken serum. *Muscle and Nerve*, **1**, 236–41.

KOHAMA, K. & OZAWA, E. (1978b). Muscle trophic factor: IV Testosterone-induced increase in muscle trophic factor in chicken serum. *Muscle and Nerve*, **1**, 320–1.

KONIGSBERG, I. R. (1963). Clonal analysis of myogenesis. *Science*, **140**, 1213–24.

KONIGSBERG, I. R. (1971). Diffusion-mediated control of myoblast fusion. *Developmental Biology*, **26**, 133–52.

MARKELONIS, G. J. & OH, T. H. (1978). A protein fraction from peripheral nerve having neurotrophic effects on skeletal muscle cells in culture. *Experimental Neurology*, **58**, 285–95.

MINTZ, B. & BAKER, W. W. (1967). Normal muscle differentiation and gene control of isocitrate dehydrogenase synthesis. *Proceedings of the National Academy of Sciences, USA*, **58**, 592–8.

MORRIS, G. E. & COLE, R. J. (1972). Cell fusion and differentiation in cultured chick muscle cells. *Experimental Cell Research*, **75**, 191–9.

MURRAY, J. C., STINGL, G., KLEINMAN, H. K., MARTIN, G. R. & KATZ, S. I. (1979). Epidermal cells adhere preferentially to type IV (basement membrane) collagen. *Journal of Cell Biology*, **80**, 197–202.

OZAWA, E. & KOHAMA, K. (1978). Muscle trophic factor: III Effect of hormones and tissue extracts on muscle trophic-factor activity. *Muscle and Nerve*, **1**, 314–19.

PATRINOU-GEORGOULAS, M. & JOHN, H. A. (1977). The genes and mRNA coding for the heavy chains of chick embryonic skeletal myosin. *Cell*, **12**, 491–9.

PAUL, J. (1970). *Cell and Tissue Culture*. Edinburgh and London: Livingstone.

POWERS, M. L. & FLORINI, J. R. (1975). A direct effect of testosterone on muscle cells in tissue culture. *Endocrinology*, **97**, 1043–7.

SALMONS, S. & SRETER, F. A. (1976). Significance of impulse activity in the transformation of skeletal muscle type. *Nature*, **263**, 30–5.

YAFFE, D. (1971). Developmental changes preceding cell fusion during muscle differentiation in vitro. *Experimental Cell Research*, **66**, 33–48.

ZALIN, R. J. (1977). Prostaglandins and myoblast fusion. *Developmental Biology*, **59**, 241–8.

Progression through G_1 and myogenic differentiation *in vitro*

IRWIN R. KONIGSBERG

Department of Biology, The University of Virginia, Charlottesville, Virginia, USA

INTRODUCTION

The term cellular differentiation has been used, sometimes loosely, to describe the processes of diversification and specialization of the cell lineages descended from the developing zygote. These processes consist, in fact, of two separate, operationally defined phenomena. The first of these, the process of cell determination, restricts genic expression in any given cell lineage to just that segment of the genome which specifies the distinct structural and functional phenotype which uniquely characterizes the cell type. The time interval between determination and the actual measurable expression of this restricted subset of gene products varies from cell type to cell type.

In most vertebrate tissues a group of reserve cells can be identified which, by the various operational criteria usually applied, is determined. The neurons of the central nervous system represent a notable exception, in as much as no reserve cells are set aside. Such stem cell populations may be actively proliferating as seen in those tissues which are continuously being replaced (viz. digestive epithelium, keratinized epithelium, haematopoietic system and so forth) or may remain dormant until recruited by an activating stimulus (trauma) to proliferate and participate in the subsequent repair or regenerative process (as in skeletal muscle, skeletal and hepatic tissues). These determined stem cells whether dormant or proliferating are devoid of detectable levels of cell-type specific proteins or organelles. Certain ocular tissues (see the chapter by R. M. Clayton, this volume) are exceptional in that not only do differentiated cells re-enter the cell cycle but re-differentiate into cells of different function within the ocular system.

Consideration of what mechanisms might control and regulate transit from the determined precursor cell pool to the differentiated cell compartment has long been an intriguing problem. As frequently happens in problems of long standing, old theories re-cycle and must be dealt with in each succeeding generation.

                    I. R. Konigsberg

Rolshoven (1951) nearly 30 years ago suggested that all cell renewal systems might require a differential stem cell division to simultaneously provide separate proliferative and differentiating populations. He proposed that a unique cell division confers properties on the daughter stem cells different from the parental stem cell. These daughters then divide giving rise to pairs of unlike cells, one of which remains a stem cell, the other subsequently differentiating.

Nature often proves to be less complicated than the mind of man, however. Using a simple but decisive test, C. P. LeBlond (1964) examined the same theory in the stratum germinativum of the squamous epithelium of the rat oesophagous. Following short exposures to tritiated thymidine (^{3}H TdR), LeBlond could recognize both products of a single cell division by the close proximity of the nuclear-labelled daughters. He found probabilites that both daughters remained in the stem cell layer, that both were found in the differentiating spinous layer or that one daughter was in each of the different layers to be identical. His conclusion, quite logically, was that crowding within the basal layer forces some cells up into the spinous layer where they initiate differentiation in response to the new environment and that each daughter has an equal probability of remaining in the basal layer or being forced up into the spinous layer.

Some years later, unaware of LeBlond's studies, an elaboration of the differential division hypothesis was offered (Holtzer, 1972) suggesting that two classes, not just one, of stem cells exist; one giving rise only to other stem cells, the other class of stem cells, after undergoing a 'quantal' (differential) mitosis produces daughter cells which differentiate in the distal keratinizing layers. *Entia non sunt multiplicanda praeter necessitatum* (Occam, 1300–1349).

Unfortunately the geometry of few tissues is as ideally suited to an investigation of the differentiative transition *in vivo* as are the stratified epithelia. For such less well organized tissues, cell culture offers a way out. Skeletal muscle tissue provides a case in point. Embryonic hind limb and pectoralis, the most accessible and generous sources of developing muscle, are extremely heterogenous as to cell type. The two major cell types of the soft tissue, fibroblasts and myoblasts are, furthermore, closely interspersed.

It is difficult to pin-point the precise time when the differentiation of the muscle fibres – the contractile elements – is initiated, but once initiated it proceeds very asynchronously. At any one time during myogenic differentiation (between 9 days of incubation and hatching, in the chick embryo) the tissue is a mélange of varying percentages of fibroblasts, myoblasts, nascent multinucleated fibres and differentiated muscle fibres.

When the investigations reported here were first begun, the prevailing opinion was that the newer (for that period) techniques of dispersed cell culture were inimical to the attainment of the differentiated state. Differentiation in culture, at that time, had been attained only by the application of organ culture techniques and it was assumed (on the basis of negative evidence) that the rapid proliferation of cells not organized into some tissue fabric would prove antagonistic to the processes of differentiation. As a matter of fact, a paper appeared which documented the fact that the authors (Holtzer, Abbott, Lash & Holtzer, 1960) were unable to obtain differentiation in pelleted vertebral chondroblasts which had been permitted to undergo more than five cell divisions in dispersed cell culture. Fortunately we were far enough along with our own studies to know that if this limitation was, as suggested, a general phenomenon, it did not apply to embryonic myoblasts. We were subsequently reassured to learn that there were other exceptions including the cell type used by the above cited authors, that is, vertebral chondrocytes (Coon, 1966).

We published our first studies of myogenic differentiation in mass cultures of embryonic chick myoblasts in 1960 (Konigsberg, 1960; Konigsberg, McElvain, Tootle, & Herrmann, 1960) and the following year demonstrated that single myoblasts, plated at clonal density, could form macroscopic colonies containing multinucleated muscle fibres (Konigsberg, 1961a). Subsequently the conditions for obtaining clonal differentiation were examined in more detail, improved and standardized (Konigsberg, 1963).

These procedures have been modified and refined through the years in both our laboratory and others (reviewed in Konigsberg, 1979). One can state with confidence that the proliferation and subsequent differentiation of embryonic myoblasts in cell culture reproduce with remarkable fidelity the sequence of developmental events which have been reconstructed from timed cytological examination of embryonic skeletal muscle tissue (Boyd, 1960).

As an experimental system, muscle cell culture offers distinct advantages that are unmatched in the intact embryo. For example, one can, by exploiting the differential adhesive properties of myoblasts, selectively detach large numbers of these myogenic cells from primary cultures relatively free of other contaminating cell types. One can, with care, prepare suspensions of quail myoblasts in which 96% of the cells are myogenic by the criterion of the percentage of nuclei which become incorporated into muscle fibres (Buckley & Konigsberg, 1974a). Moreover, the use of cell culture allows the investigator to rigidly control the environment and to employ a spectrum of experimental interventions difficult if not impossible to achieve in the organism. Finally, if one appreciates the inverse relationship between rates of cell proliferation and the initiation of fusion (see below), one can manipulate culture conditions to impose a high degree of synchrony of differentiation and thereby an extremely reproducible pattern of development.

CHARACTERISTICS OF THE MUSCLE CELL CULTURE SYSTEM

Our earlier studies employed the leg musculature of the chick embryo as the tissue source (Konigsberg, 1960, 1961a, b, 1967, 1970; Konigsberg *et al*, 1960; Cooper & Konigsberg, 1961a, b; Reporter, Konigsberg & Strehler, 1963; Strehler, Konigsberg & Kelley, 1963; Konigsberg & Hauschka, 1966; Hauschka & Konigsberg, 1966). The following studies, however, employ myoblasts obtained from the breast muscle of embryonic Japanese quail. One of the advantages of this species is that, using selective techniques, we can obtain suspensions of muscle cells which are virtually free of non-myogenic contaminants. In Fig. 1 the percentage of the total number of nuclei which are present in multinucleated fibres is plotted as a function of time in culture. Since by the 8th day of culture, 96% of the nuclei are in syncytia certainly no more than 4% (and probably less) of the population at this stage could be non-myogenic. Microscopic examination of these cultures at earlier times also suggests that they are relatively free of fibroblastic cells. In Fig. 2, for example, the vast majority of the as yet unfused cells are obviously fusiform in shape. Re-suspending these mononucleated cells (sieving out the myotubes) from cultures as late as 5 days (see Fig. 2) and re-plating them at clonal density, the percentage of muscle colonies obtained ranged from 90 to 99%.

In our experience the bipolar morphology of embryonic myoblasts, particularly quail myoblasts, has proven to be a reliable criterion of cell type (Konigsberg, 1963) in cultures prepared and maintained under rigorously controlled conditions, although it can be obliterated by inadequate conditions (see Lipton, 1977). In selecting early myoblast

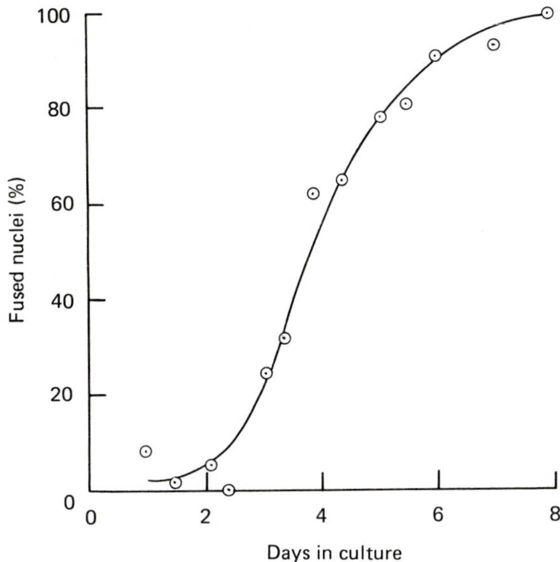

Fig. 1. The increase in the percentage of nuclei in multinuclear cells ('fused nuclei') with time in culture. Slices were prepared from mass cultures fixed and stained on successive days. The number of mononucleated myoblasts and the number of nuclei within multinucleated myotubes were scored in random fields. To preclude a bias towards fused nuclei, especially during the later periods of culture time when the number of fused nuclei predominates, at least 200 mononucleated cells were counted in scoring the total nuclei represented by each point. Beginning on day 3, there is a rapid increase in the percentage of nuclei within multinucleated myotubes. The rate of appearance of fused nuclei remains relatively high for the following 2 days (increasing to a level of approx. 90% on day 5) and begins to decrease and approach a plateau level early on day 6. (After Buckley & Konigsberg, 1974a).

clones for time-lapse analysis, for example, (see below) 114 clones were selected on the basis of bipolar morphology, *all* of which gave rise to colonies of differentiated muscle fibres.

Finally, for reasons which are still unclear, quail myoblasts are apparently hardier than chick. They tolerate well, for example, treatments (such as continuous circulation of the medium) which we found earlier could not be successfully performed using chick cells.

Using standardized conditions in mass cultures established at *moderate* cell density, one regularly observes that culture development occurs in two distinct phases. The first of these is marked simply by rapid proliferation with no indication at the cellular or molecular level of any further differentiation (Fig. 2*a*). Then, at a predictable time, an abrupt and rapid transition occurs to a population of cells in which the

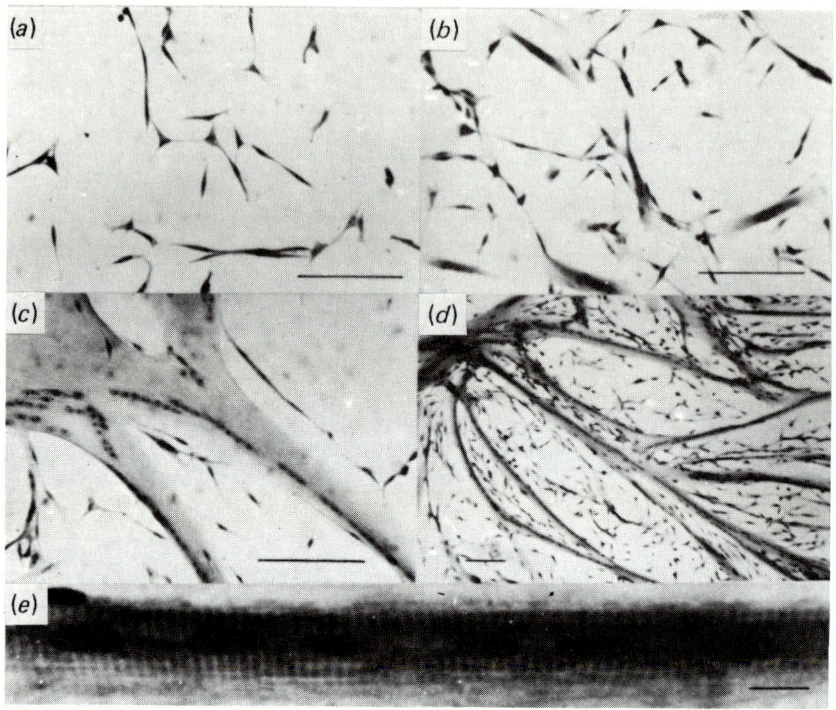

Fig. 2. Photomicrographs of quail muscle cells in culture. All cultures were grown in 5-cm Falcon tissue culture plates and were inoculated with 31 250 cells. The cells were plated in 3 ml of growth medium which was replaced only once (at 24 h after plating). On representative days, cultures were fixed, stained, and photographed. (a) Day 2 culture consisting of mononucleated myoblasts, recognizable by their characteristic bipolar shape. Bar = 0.1 mm. (b) Day 3 culture in which fusion is first observed as the formation of short, 'stubby' multinucleated myotubes. Bar = 0.1 mm. (c) Day 5 culture demonstrating the progressive increase in the number of nuclei within multinucleated myotubes. Bar = 0.1 mm. (d) Lower magnification photograph of a day 5 culture demonstrating the extensiveness of fusion. Bar = 0.1 mm. (e) Higher magnification photograph of a multinucleated myotube demonstrating the characteristic cross striations. Bar = 0.01 mm. (After Buckley & Konigsberg, 1974a.)

predominant activities are now cell fusion (see Fig. 1) and the synthesis of those macromolecules characteristic of the differentiated muscle fibre.

CELL DENSITY, MEDIUM VOLUME AND THE INITIATION OF FUSION

The rapidity, regularity, and degree of synchrony which can be imposed upon this differentiative transition (O'Neill & Stockdale, 1972a; Emerson & Beckner, 1975) make it an attractive system in which to study the mechanisms which control the timing of the differentiative shift and the transcriptional and translational concomitants of myogenic fusion.

Two general theories of the timing of the differentiative event have been considered. The first of these is that some unique programme, intrinsic to the cell, limits the number of mitotic divisions and that following the last of these divisions the cell is programmed to fuse and to initiate cell-type specific synthesis (Holtzer, 1972). Although the epigenetic programme, encoded in the myoblast, restricts its biosynthetic capabilities to those activities required for the maintenance, proliferation, and subsequent differentiation of this cell type, we see no compelling evidence that suggests that included in the programme are the instructions dictating when, *in time*, the switch to the synthesis of cell-type specific macromolecules will occur.

Quite the contrary. During the past nine years, evidence has accumulated which demonstrates that the time of initiation of differentiation in muscle cell culture is not rigidly determined but can be easily and predictably shifted, in time, by simply manipulating such factors as cell density and the concentration of growth-promoting components of the medium (Konigsberg, 1971; O'Neill & Stockdale, 1972a, b; Buckley & Konigsberg, 1974a, b; Doering & Fischman, 1974; Emerson & Beckner, 1975; Emerson, 1977; Nadal-Ginard, 1978; Linkhart, Clegg & Hauschka, 1980).

The interest in resolving these alternatives is not simply an exercise in polemics. If an intrinsic timing mechanism were operating, one would want to elucidate the molecular mechanisms of the cellular clock. If, on the other hand, the time of initiation of fusion and cell-type specific synthesis can be regulated by environmental cues, one would like to know the nature of these cues and how they operate.

                           I. R. Konigsberg

Our own inquiries into these questions were prompted by observations made quite early during the development of the muscle cell culture system which suggested that the initiation of fusion was a function of the density of the cells in the monolayer. The first of these was that there is a gradual increase in cell density preceding fusion and second, as the inoculum size used to set up a series of cultures is increased, those cultures initiated with larger cell numbers fuse earlier. Two alternative explanations were offered, at the time, to explain these observations: (1) that higher cell density increases the probability that competent cells contact one another and fuse or, alternatively, (2) that increased cell density accelerates metabolic alterations of the medium which, in turn, provide a microenvironment more favourable for the initiation of fusion.

In order to discriminate between these alternatives we assumed that if higher cell density promoted fusion by altering the medium, earlier fusion should occur when we either increased the number of cells inoculated into a petri plate or, keeping the cell number constant, decreased the volume of medium bathing the cells. If, however, intercellular distance is the critical parameter, decreasing the volume of medium should be without effect. The results of such experiments indicated quite clearly, that the correct hypothesis is that the myoblasts modify the medium which in turn promotes the initiation of fusion (Konigsberg, 1971). If this were the case, then constant perturbation of the medium should delay fusion by preventing the local accumulation of such changes.

In experiments designed to test this prediction we found a well-developed network of multinucleated cells in static control cultures, while cultures in the same volume of medium, continuously circulated, consisted of dense populations of unfused, mononucleated cells. If the experimental period is extended, however, fusion is eventually initiated and progresses despite the continuous perturbation of the medium.

If medium withdrawn from fusing mass cultures is used (rather than fresh medium), fusion is not delayed despite continuous perturbation. Similarly one can override the delay by simply increasing the initial inoculum size. For example, if a number of high-density cultures are included in the same petri plate with a lower density 'test' culture it too will fuse in the common medium circulating over these 'parabiotic' cultures (Konigsberg, 1971). It is clear from these results that the time of initiation of fusion of myoblasts in cell culture is controlled by some metabolic processing of the medium which is diffusion-mediated and dependent upon population size.

NATURE OF THE MEDIUM ALTERATION

These experiments do not, however, discriminate between the depletion of media constituents and, alternatively, the addition of cell-type specific components to the medium. Since we subsequently found (see below) that medium collected from fusing cultures promotes, not only early fusion, but the cell cycle as well, the more reasonable hypothesis is that the effects are due to the depletion of growth-stimulatory components of the medium. The observation that the 'fusion-promoting' component of the conditioned medium resided in the molecular weight fraction greater than 300 000 MW (Konigsberg, 1971) suggested, in turn, that the putative growth factors were contributed by either the serum or the embryo extract. If so, then the addition of such putative growth factors to fusion-permissive medium should nullify the effect on fusion. When the greater than 300 000 MW fraction of embryo extract is added to fusion-permissive medium, that is precisely what happens. Conversely, by simply decreasing the concentration of serum and embryo extract (EE) in our growth medium, fusion is again initiated at an earlier time (Emerson & Beckner, 1975).

These observations, although suggestive, do not rule out the accumulation with time of a cell-specific component (possibly a surface component) which specifically promotes fusion. To test this possibility we asked whether the initiation of fusion depends on the accumulation of any cell product whatsoever in the medium. To examine this question we cultured myoblast clones in perfusion chambers containing micro spin bars. These were perfused with complete growth medium at a rate of 1.5 chamber volumes per hour. When perfused for 42 hours, large colonies of very closely spaced cells form, with little, if any, evidence of fusion. Clones grown under static conditions for the same period of time, however, form a discrete network of multinuclear cells. This experiment, in itself proves nothing, since not only are any cell products which might be released continuously flushed out but, at the same time, fresh complete growth medium is constantly added. If, however, perfusion is continued for an additional 24 hours, changing the perfusion medium to a maintenance medium lacking both serum and embryo extract, fusion is observed as early as 12 hours after the medium shift and at the end of the additional 24 hours an extensive

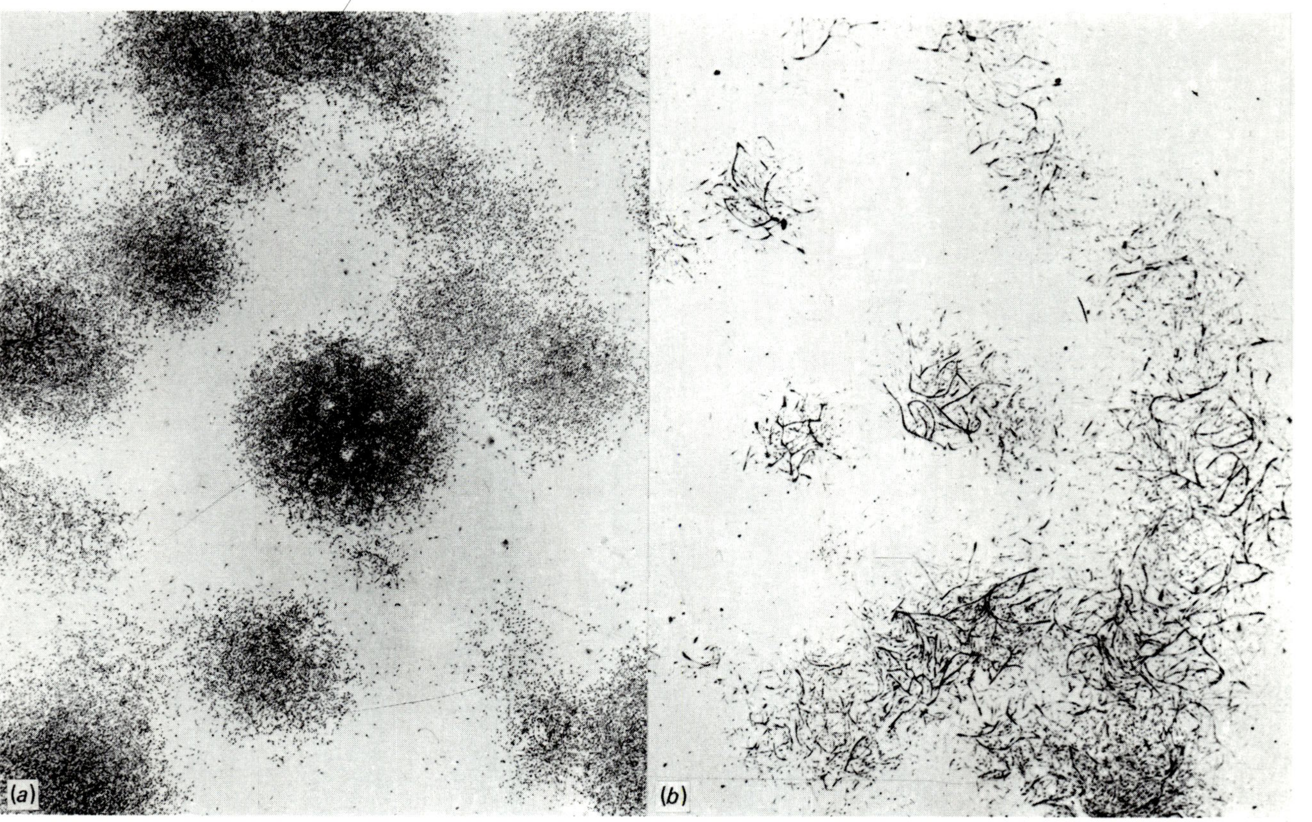
(a)
(b)

network of myotubes is formed. Control clones, in which perfusion with complete growth medium is continued for the additional 24 hours, are more crowded but are still essentially unfused (Fig. 3 and 4).

We conclude that the initiation of fusion does not require the accumulation of any cell product or products in the medium. In addition, since fusion is not initiated when complete growth medium is continuously replaced but does occur when a completely defined *maintenance* medium is perfused, it seems logical to assume that the necessary condition for fusion to occur is the absence (or depletion) of growth factors normally supplied by serum and embryo extract.

One can, in fact, control the time of initiation of fusion by switching cells from high-growth potentiating to a low-growth medium or by establishing cultures in media containing varying concentrations of embryo extract (Emerson,1977). In Fig. 5 the percentage nuclei in multinucleated fibres as a function of time is plotted for three media of different composition. Although percentage of syncytial nuclei reaches the same maximum in each curve and the rates of increase of percentage syncytial nuclei are quite similar, the slope of the sharp rise is progressively displaced on the time axis. The greatest delay in the appearance of syncytial cells is observed in cultures seeded in high-growth medium (10% EE; 15% serum)(closed squares). A significantly shorter delay in the onset of fusion occurs in our low-growth medium (1% EE; 10% serum; closed circles). In the cultures exhibiting the shortest time lag between seeding and the beginning of the rapid fusion stage (closed triangles) the latter growth medium was again employed; however, this medium had been conditioned before use by prior exposure for 3 days to a growing population of myoblasts (Sutherland, 1977).

These data illustrate that fusion occurs sooner when lower levels of

Fig. 3. Clones cultured under conditions of constant perfusion of medium (see text). Fifty cells were seeded in 1 ml of complete growth medium restricted to 1.6 mm diameter on the 2.4 mm circular cover slip (collagen-coated). Twenty-four hours after seeding, the cover slips were set up in perfusion chambers and all cultures were perfused with complete growth medium for 42 h. At 42 h the flow, through the experimental group, was switched to maintenance medium containing neither embryo extract nor serum. (*a*) Low magnification of control culture in which perfusion was continued with complete medium for the additional 24 h. No obvious fusion is apparent at this magnification but note the striking difference in cell density in the clones as compared to *b*. This is even more apparent at higher magnification (see Fig. 13). (*b*) Area of experimental culture, at low magnification, showing clones containing multinuclear fibres at the end of 24 h of perfusion with maintenance medium.

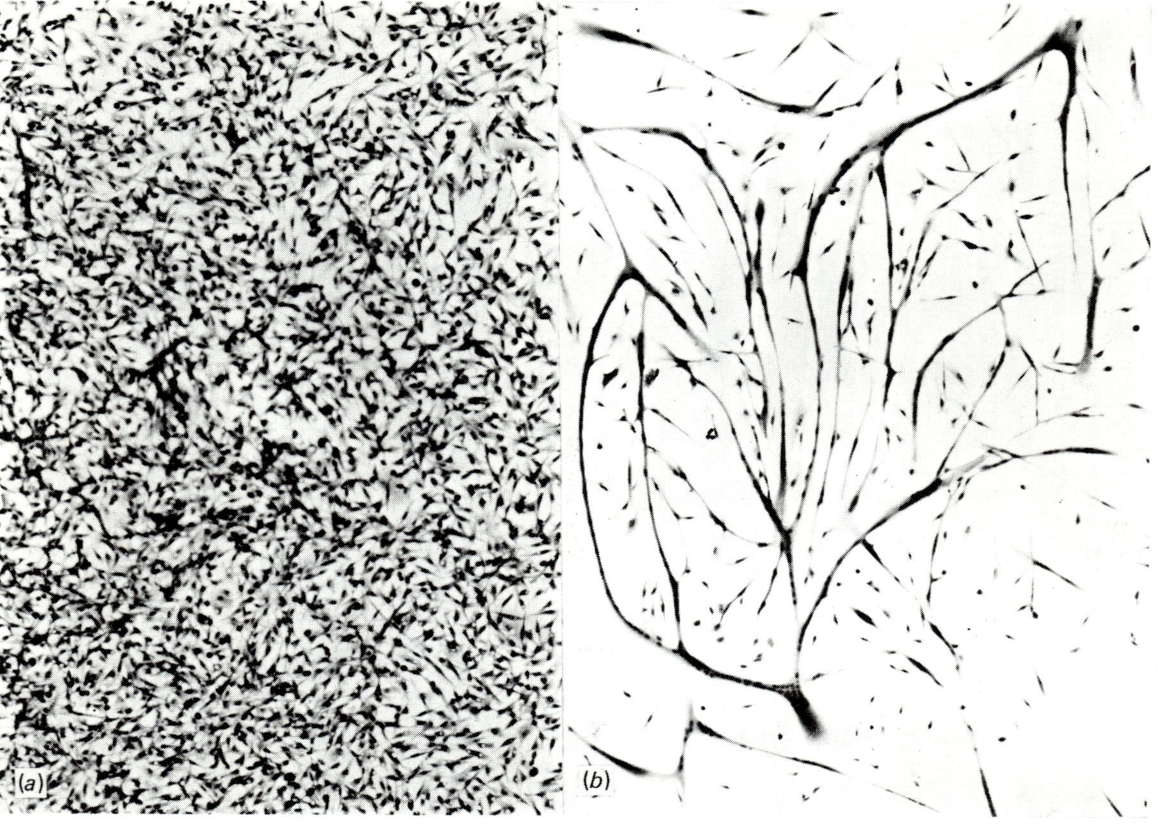

Fig. 4. Clones cultured under conditions of constant perfusion, as described in the figure legend of Fig. 3 but photographed at higher magnification. (*a*) Central portion of a clone in a culture perfused, throughout the experiment, with growth medium. (*b*) Single clone (virtually complete) from the experimental group perfused with maintenance medium during the last 24 h of the experiment. Note the network of multinucleated cells.

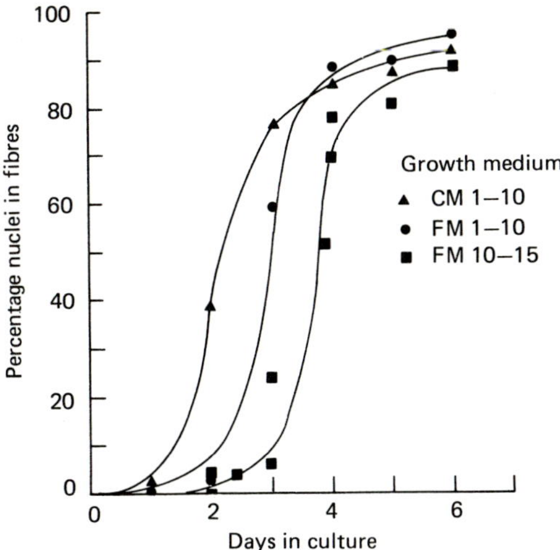

Fig. 5. The increase in percentage of nuclei in nascent multinuclear muscle fibres as a function of time in culture, using media of differing growth-supporting properties. Each point represents the average of two cultures each inoculated with 10^4 cells of a secondary suspension of quail myoblasts. Fusion is initiated earliest in conditioned low-growth medium, CM 1–10; then in freshly prepared low-growth medium, FM 1–10; and finally in high-growth medium, FM 10–15. (From Wm. M. Sutherland, Doctoral Thesis, University of Virginia, 1977.)

growth-stimulating factors are provided initially as well as when growth factors are depleted by prior conditioning. The reciprocal relationship between cell proliferation and the time of initiation of fusion was also demonstrated in this study. Measurements of the accumulation of micrograms DNA per culture as a function of culture time were made and are plotted in Fig. 6. Population doubling is inversely related to the time of initiation of fusion. Cells cultured in conditioned, low-growth medium barely double (1.70 ×) before reaching the stationary phase. Employing the same medium, used prior to conditioning, the population increases by a factor of 4.8 × while in high-growth medium, in which the onset of fusion is most delayed, DNA per culture increases 23.8 ×.

The rates of DNA accumulation during the second day of culture in either high- or low-growth medium are virtually identical being 22.4 ng/h/culture (high growth) and 21.7 ng/h/culture (low growth). The curves differ only in the time points at which the rates of DNA accumulation break. This break occurs approximately 24 hours earlier in low- than in high-growth medium, when the levels of mitogen in the medium presumably are depleted to some threshold concentration.

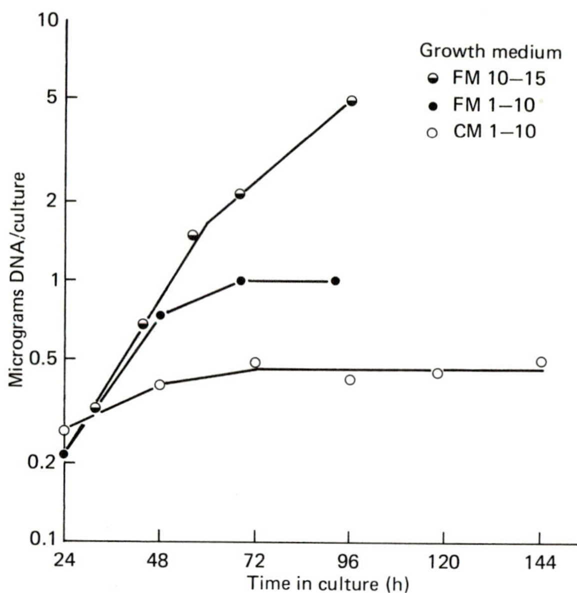

Fig. 6. The accumulation of DNA per culture under the same conditions as in Fig. 5. DNA accumulates most rapidly in high-growth medium, FM 10–15, and low-growth medium, FM 1–10. A marked break in the rate of DNA accumulation is observed in both media occurring later in high-growth medium in which a longer delay in the initiation of fusion is observed (Fig. 5). DNA accumulation is minimal in conditioned low-growth medium, CM 1–10 in which fusion is initiated earliest (see Fig. 5). (From Wm. M. Sutherland, Doctoral Thesis, University of Virginia, 1977.)

The break coincides with the time of initiation of the fusion phase, as previously observed (Konigsberg, 1971) when the media volume was varied to manipulate the time of initiation of fusion (see Fig. 7). Despite the differences in the volume of medium, both groups again grew at the same rate for the first 24 hours. The rate of DNA accumulation changed abruptly in both groups, with the break occurring 12 hours earlier in those cultures fed the smaller volume of medium; the time of initiation of fusion corresponding to the time of the break in rate of DNA accumulation. Thus when fusion is delayed either by manipulating the volume of medium or by restricting the concentration of mitogens, cell proliferation continues for a longer period of time. In this experiment, a two-fold difference in DNA per culture is generated between the two groups suggesting that the delay permitted an additional round of cell divisions which correlates well with the average generation time of these cells which is 10.2 hours (Buckley & Konigsberg, 1974a).

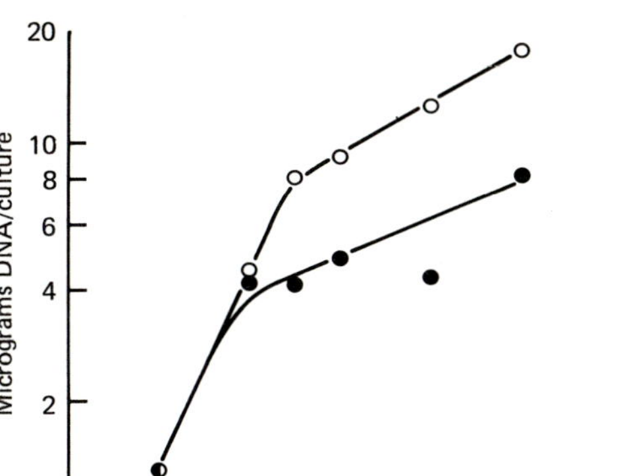

Fig. 7. DNA per culture as a function of time in cultures growing in different volumes of medium. All cultures were set up with 31 250 cells per 5 cm diameter Falcon TC petri dish (collagen-treated) in 3 ml of medium. After 24 h (day 1 on abscissa) medium was replaced with either 1 ml (closed circles) or 3 ml (open circles). (After Konigsberg, 1971.)

These results suggest the operation of an environmental cue which affects cell proliferation and fusion in an inverse fashion rather than some rigid intrinsic programme encoded in the proliferating myoblast.

EFFECT OF CELL DENSITY ON THE CELL CYCLE

The decreased growth rate which coincides with the initiation of fusion could simply reflect the withdrawal, by fusion, of cells from the proliferative pool. However, a similar break occurs at high cell density in all cultured cells, not only muscle cells. This abrupt change in growth rate as cells enter the stationary phase is accompanied by a change in generation time. To determine whether this occurs in pro-liferating myoblasts as well, the cell cycle was measured using the pulse-chase method of Quastler & Sherman (1959).

Two time intervals were examined: the 24 hour period immediately preceding the initiation of fusion (day 2) and a time well beyond the break in growth rate (day 5) (see Fig. 7). Cultures were pulsed for 30 min at both stages; the older, day 5 cultures were chased, however, with medium withdrawn from sister cultures of the same age since at this stage a burst of mitosis occurs in response to re-feeding with fresh medium (Buckley & Konigsberg, 1974b). The percentage labelled mi-totic figures were plotted as a function of time after the pulse for both

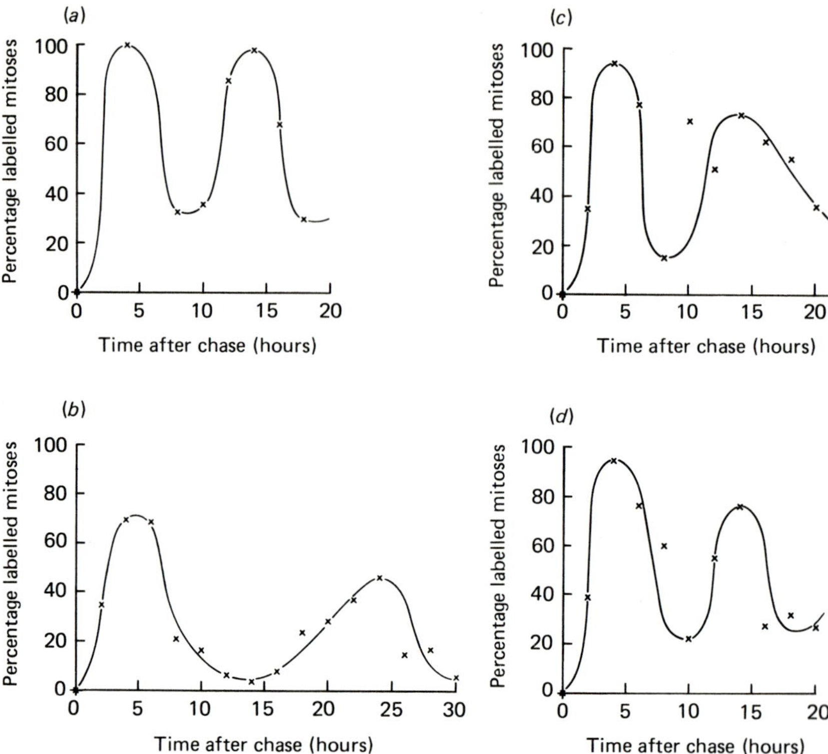

Fig. 8. The progression through mitosis of cohorts of cells pulse-labelled with tritiated thymidine. At zero minus 30 minutes a small volume of tritiated thymidine (s.a. = 6.7 curies/mM) was added to the culture and mixed with the medium to give a final concentration of 0.5 microcuries per ml. At zero time the medium was withdrawn and replaced with chase medium supplemented with 'cold' thymidine to a final concentration of 10^{-7} M. Sample cultures were fixed immediately after the pulse and at subsequent 2 hour intervals. Following autoradiography and staining, the percentage of labelled mitoses was scored. Each point represents 100–200 scored mitoses. (a) Day 2 cultures pulsed and chased using fresh medium. The curve exhibits two homologous and symmetrical peaks which reach a maximum of 100%. (b) Day 5 cultures pulsed and chased in conditioned medium pooled from equivalent day 5 cultures. At this stage of culture development, the two peaks are no longer homologous. The second peak appears later than the second peak in the day 2 curve (8a) and is also much broader than the first peak and consequently considerably dampened. (c). Day 2 cultures pulsed and then chased using conditioned medium collected from day 5 cultures. The homology between the two peaks is lost (compare to 8a). The second peak is broader and dampened. The curve appears more like the day 5 curve (8b). (d) Day 5 cultures pulsed and then chased with fresh medium such as was used for day 2 cultures. The two peaks exhibit the symmetry observed in day 2 cultures chased with fresh medium (8a) rather than the lack of homology exhibited on day 5 (8b). (After Buckley & Konigsberg, 1974a.)

Table 1. *Mathematical analysis of cell cycle curves; the curves illustrated in Fig. 8 were analysed according to the method indicated*

Mean cell cycle times (Half maximum method of Quastler & Sherman, 1959)

	day 2, FM†	day 5, CM
$G_2 + M$*	2.0 h	2.0 h
S	4.8 h	5.2 h
G_1	3.2 h	12.0 h
G_T	10.0 h	19.2 h

* The average duration of M, measured by time-lapse cinematography, was calculated to be 20 minutes. Due to its relatively short duration, it may be considered together with the G_2 phase.

† FM (fresh medium) was found in preliminary studies to be equivalent to medium collected from day 2 cultures. CM (conditioned medium) refers to medium collected from sister cultures of comparable age (day 5 in this study).

sets of data (Buckley & Konigsberg, 1974a). One can see that the shape of the plots of percentage labelled mitotic figures for day 2 and day 5 are identical for the first mitosis but differ markedly with respect to the second wave of mitosis (see Fig. 8). The first peak represents the cohort which was in S during the pulse and progressed to mitosis during the chase while the second peak represents the daughter cells of that cohort which have traversed a *complete* cell cycle (including G_1) and were in mitosis when the culture was fixed. Thus the differences in the plots of labelled mitotic figures indicate that the G_1 phase is longer and highly variable in fusing cultures and accounts, in fact, wholly for the increase in generation time at this stage (see Table 1).

It is clear from these data that the break in the rate of DNA accumulation is due, in part, to a change in the length of the G_1 phase of the cell cycle and is *not* simply due to a reduction of the proliferating cell pool due to fusion. Furthermore, this protraction of G_1 is due to changes in the medium and does not represent some programmed event intrinsic to the myoblast.

The pre-fusion (day 2) pattern can be elicited from the older (day 5) cultures by substituting the same medium (fresh medium) used to chase pre-fusion cultures. The converse is also true, that is, the production of the day 5 pattern in pre-fusion cultures chased with medium withdrawn from the older, day 5 cultures (see Fig. 9). As one might predict, fusion is initiated earlier when day 2 cultures are fed with

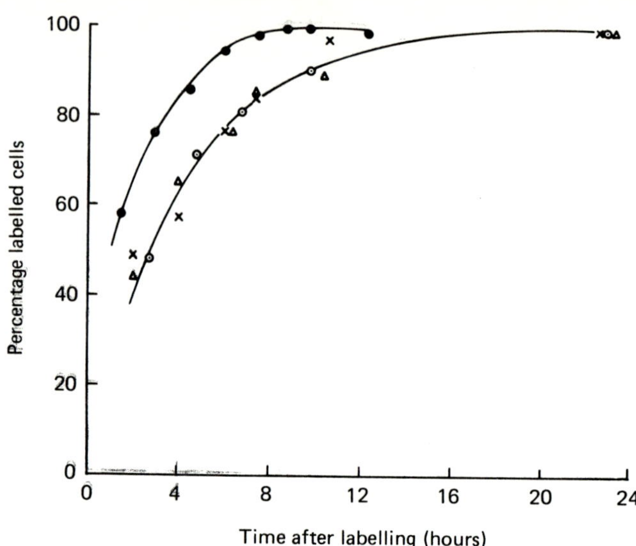

Fig. 9. Graph of the increase of percentage labelled cells under conditions of continuous labelling. On successive days from day 2 through 5, the medium was withdrawn from a set of cultures and replaced with an equal volume of medium (collected from cultures the same age) supplemented with tritiated thymidine (0.5 µCi/ml). At intervals throughout the day, sample cultures were fixed. Following autoradiography and staining, the percentage of mononucleated cells with label over the nucleus was scored. Each point represents at least 200 mononucleated cells scored. The points on the graph represent the following values: solid circles = day 2, crosses = day 3, triangles = day 4, and open circles = day 5. The rate of increase of the percentage of labelled cells on days 3, 4 and 5 is much slower than that on day 2 (see text). (After Buckley, & Konigsberg, 1974a.)

medium collected from the older cultures and cell proliferation is stimulated when day 5 cultures are fed with fresh medium.

Thus the culture medium which is altered by the cell population it supports affects two parameters of culture development: (1) the initiation of fusion, and (2) the protraction of the G_1 phase of the myoblast cell cycle as well as the break in the rate of DNA accumulation.

If these two phenomena were causally related we would expect to find the protraction of G_1 and the initiation of fusion to be closely linked in time.

THE TEMPORAL COINCIDENCE OF THE PROTRACTION OF G_1 AND FUSION

To determine how early protraction of the cell cycle can be detected we measured cell cycle times at daily intervals between days 2 and 5

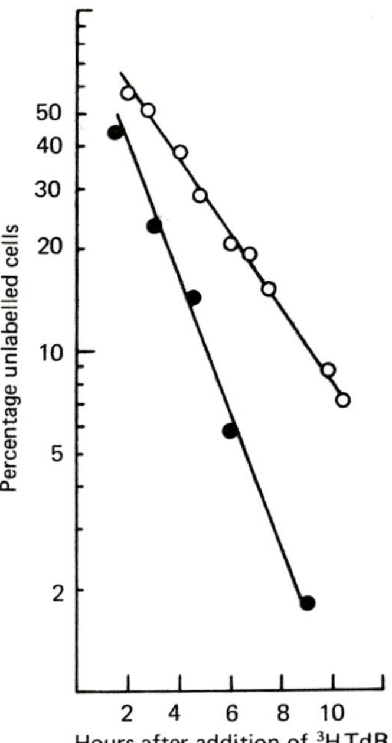

Fig. 10. Semilogarithmic plot of transitional probability of entry into S in the continuous presence of tritiated thymidine. The data presented in Fig. 9 have been converted to $\log_{10}$ of the percentage unlabelled cells and plotted against time.

The steeper curve (solid circles) represents continuous labelling on day 2 of culture; the shallower curve (open circles), the combined data of days 3, 4 and 5. The slope of percentage unlabelled myoblasts on day 2 is -0.220 and on days 3, 4 and 5 is -0.110, this difference being significant below the 0.1% level.

(Buckley & Konigsberg, 1974a). Since we already knew that the increase in generation time can be entirely accounted for by the extension of G_1, we used the simpler, continuous labelling technique (Cleaver, 1967).

At the beginning of each day, from day 2 through day 5, groups of replicate cultures were fed medium withdrawn from sister cultures of the same age supplemented with tritiated thymidine. Cultures were fixed at regular intervals during the ensuing 24 hour period and the percentage of unfused, single cells whose nuclei have incorporated label was determined. In Fig. 9 one sees two curves. Both reach the same maximum (virtually 100%) but at different times and at different rates.

In order to determine the rate of entry into S of the asynchronous populations on day 2 as compared to days 3, 4 and 5, combined, a semilogarithmic plot was constructed of the percentage unlabelled mononucleated cells remaining against time (see Fig. 10). Each of these plots of the probability of entering S follow first order kinetics with slopes that are significantly different ($p < 0.01$). Extrapolating the transitional probability plots to the time axis, maximum labelling is reached at 9.1 h on day 2 but not until 18.2 h on days 3, 4 and 5. Protraction of G_1 occurs as early as day 3, the same day that initiation of the rapid fusion stage is observed in our cultures.

This continuous labelling study not only determines the earliest time in culture that protraction of G_1 can be detected but also clearly indicates that in fusing cultures (days 3, 4 and 5) no sizeable pool of myoblasts withdraw from the cell cycle for any significant period of time before fusing. Virtually all of the unfused cells in both pre-fusion and fusing cultures label within a period of time in good agreement with the average G_T for the respective stage.

PROGRESSION THROUGH G_1 AND FUSION
– A HYPOTHESIS

Our experiments suggest that during the purely proliferative phase in culture, the cells deplete the medium of one or more macromolecular, growth-promoting components. Below some threshold level, although all of the cells continue to cycle, they do so more slowly; G_1 times in the population becoming longer and more variable. Since myoblasts when they fuse do so in G_1, it is reasonable to assume that some activity (or activities) required for fusion is restricted to G_1 specifically. If G_1 were protracted, the probability would be increased that these G_1-restricted activities attain some critical threshold either cumulatively throughout G_1 or during a window late in the protracted G_1 phase.

If our concept of the relationship between the cell cycle and fusion is correct, we would predict that the average terminal G_1 of myoblasts which fuse is significantly longer than the average G_1 of those myoblasts in the culture which are still cycling.

TIME-LAPSE ANALYTICAL TEST OF THE HYPOTHESIS

To test this prediction, we combined time-lapse cinematographic and radioautographic techniques to measure these values in individual cells in pre-fusion and fusing clonal cultures (Konigsberg, Sollman & Mixter, 1978). Knowing the framing rate, the length of the G_1 phase terminating in fusion (the terminal G_1) was derived by counting the

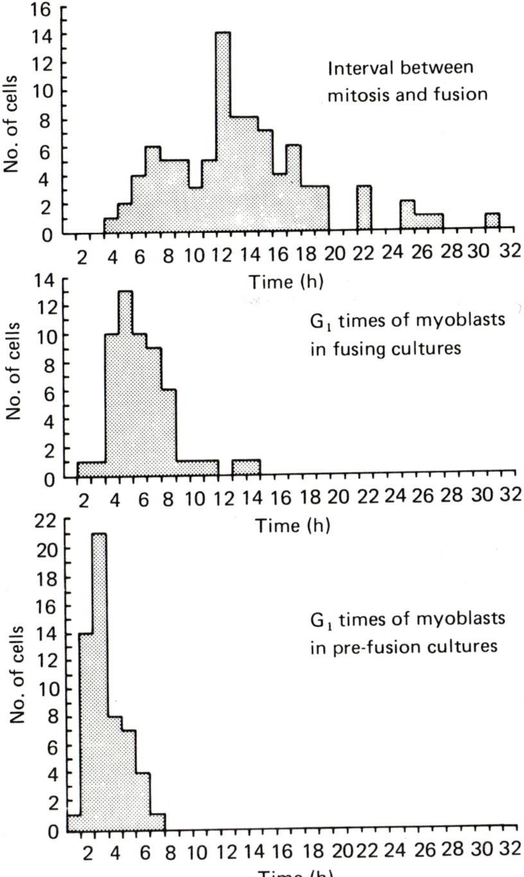

Fig. 11. The distribution of the terminal G_1 times of myoblasts which fused during filming (upper curve) compared to the distribution of G_1 times of myoblasts in the same cultures which were still cycling (middle curve). The lower curve is the distribution of G_1 times in cultures filmed before the initiation of fusion.

Two other measurements in addition to $G_1 + 1/x$ S were made. These were generation time (M to M) in each clone and the av. LI of 10 clones (including the cine-recorded clone) per clonal culture.

From these values (and M, determined from the cine records) the cycle parameters were determined using a derivation of the Stanners & Till (1960) treatment of the distribution of cycling populations.

number of frames which separate a mitotic event from the subsequent fusion of a daughter cell born in that division. The length of G_1 in myoblasts still cycling was measured in cultures which were filmed continuously for 48 to 72 hours and then pulsed with ^{3}H TdR during

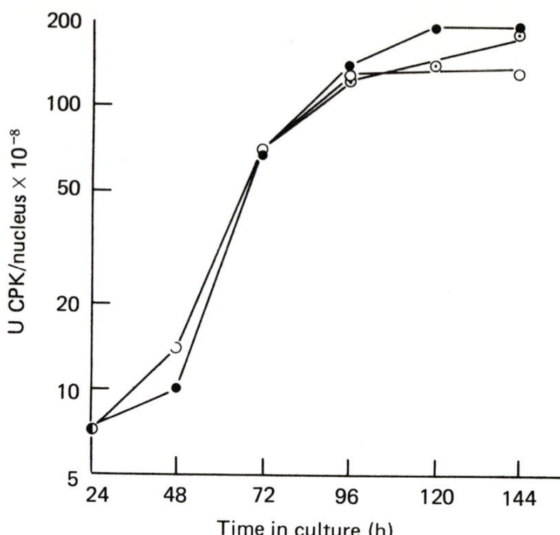

Fig. 12. Accumulation of CPK activity (per nucleus) as a function of time under fusion-permissive and fusion-blocked culture conditions. In all three curves conditioned low-growth media (CM 1–10) was employed, only [Ca^{++}] was varied. Open circles indicate fusion-permissive levels of Ca^{++}; closed circles – fusion-blocking levels of Ca^{++}. The three points marked by dotted circles indicate that these cultures were maintained under fusion-blocked conditions until 72 h and then switched to fusion-permissive medium.

the last 15 min of filming. Using radioautographs of the fixed, processed clones the labelled cells which were in S during the pulse were located on the last frame of the film. Running the film in reverse, each of these cells was traced to the mitosis from which it arose and once again frame counts were converted into actual lapsed time between M and some point in S (viz. $G_1 + 1/x$ S). These values were then corrected (see legend, Fig. 11) to yield actual G_1 times.

Since fusion is cell density dependent, a microscopic field is normally so crowded, by the time fusion is first observed, that it is extremely difficult to track individual cells. Two measures were taken to circumvent this difficulty. First, filming of individual clones of myoblasts was started on the second or third day after inoculating the plates. At this time clones consist of only from 8 to 30 cells. In addition, in order to promote the initiation of fusion at low cell density, cultures were fed, before filming, with medium collected from mass cultures in which fusion had already started. To collect parallel observations on G_1 times in pre-fusion stage clones, cultures were refed with fresh high-growth medium before filming.

The distribution of the terminal G_1 and the G_1 periods of cycling cells which re-entered S in pre-fusion and in fusion stage clones are plotted in Fig. 11. Not only is the average G_1 of myoblasts in fusing clones (6.9 $\pm$ 0.34 hours) twice that of myoblasts in pre-fusion clones (3.1 $\pm$ 0.15) but, comparing the two distributions, there is clearly greater variability of G_1 in fusing cultures as was suggested earlier by our pulse-chase data (Buckley & Konigsberg, 1974a).

Comparing the distribution of the lengths of the terminal G_1 of the 93 cells which fused during filming with the G_1 times of cycling myoblasts in the same clones confirms the prediction that those cells which fuse have longer G_1 times than the cycling myoblasts. The average of the terminal G_1 times (13.5 $\pm$ 0.05 hours) is twice the average G_1 of cycling myoblasts (6.9 $\pm$ 0.34 hours). However, though the difference is significant ($p < 0.001$), there is considerable overlap of these two distributions. Approximately two-thirds of the fusing cells have G_1 times which fall within the range of values of the cycling myoblasts. Furthermore, although the average G_1 of proliferating myoblasts in fusing clones is 6.9 hours, these data indicate that a myoblast may spend as long as 15 hours in G_1 and still re-enter S.

It is also apparent from the distribution of terminal G_1s that a myoblast must spend a minimum of 4 hours in G_1 before fusion can occur. Of the 93 fusion events observed the shortest terminal G_1 was 4.1 hours. This is consistent with the idea that the potential to fuse and initiate muscle-specific protein synthesis is latent, at least, during every G_1 phase of the cycle but is realized only when the length of G_1 is protracted beyond the 4 hour minimum. On inspection of the distributions of G_1 times in Fig. 11, it is obvious that the majority of myoblasts in fusion stage clones (43/52 or 82.7%) have G_1 times exceeding 4 hours and the G_1 time of relatively few myoblasts in pre-fusion clones (8/57 or 14.0%) exceed the minimum. If one calculates the initial probability of encounters between two fusion-competent (viz. $G_1 > 4$ hours) myoblasts, one finds that the probability of this event occuring at fusion stage is 95 $\times$ greater than in pre-fusion populations (see legend, Table 2). These calculations assume that cell movement is random and that initial distances between cells is not limiting at the range of cell densities attained in clones. Both of these assumptions are compatible with the observations made during the analysis of these films.

Such a stochastic mechanism could account for the differentiative transition in myogenic populations without invoking a rigidly timed

Table 2. *Distribution-corrected values* of cell cycle lengths*

Pre-fusion clones	Fusion stage clones	p (difference between stages)
G_1 3.1 ± 0.14	G_1 6.5 ± 0.35	<0.001
S 4.0 ± 0.10	S 3.2 ± 0.11	>0.9
G_2 1.4 ± 0.08	G_2 1.1 ± 0.11	>0.4
M 0.3	M 0.3	—
G_T 8.8	G_T 11.1	

* See legend, Fig. 11.

Calculation of Probability of Contact of Fusion Competent G_1 Cells:
The values above were used to determine the probability of two fusion-competent myoblasts (viz. $G_1 > 4$ h [see text]) making contact at pre-fusion and fusion stages, as follows:

(1) Fraction of the culture population in G_1 at any one time $= G_1/G_T$

Pre-fusion $\qquad$ Fusion

$31/88 = 0.352 \qquad 6.5/11.1 = 0.590$

(2) The proportion of all cells in G_1 whose G_1 times exceed 4 h can be read from the distributions in Fig. 11, these are:

Pre-fusion $\qquad$ Fusion

$8/57 = 0.140 \qquad 43/52 = 0.827$

(3) The fraction of all cells in the population in G_1 for times exceeding 4 h is the sum of the fraction in (1) and the fraction in (2) (above):

Pre-fusion $\qquad$ Fusion

$0.352 \times 0.140 = 0.050 \qquad 0.590 \times 0.827 = 0.488$

(4) Probability is a scalar quantity varying from zero to 1, by definition. The probability of finding one cell which has been in G_1 for an excess of 4 h is equal to the fraction of this type of cell in the population. The initial probability of two such cells making contact is the square of said fraction:

Pre-fusion $\qquad$ Fusion

$0.050^2 = 0.0025 \qquad 0.488^2 = 0.2381$

$p(\text{Fusion} - \text{Stage})/p(\text{Pre-fusion}) = 0.2381/0.0025 - 95.2$

intrinsic programme which, in the past, has served more as a *deus ex machina* than a testable working hypothesis.

All of our data suggest that the regulatory event which channels a myoblast back into another round of cell division or into the differentiative pathway occurs late in G_1 and that the distribution of rates of progression through G_1 can be manipulated in a predictable fashion by controlling the external milieu.

The mechanisms by which the changes in the environment affect cell cycle progression, that is the nature of the mitogen (or mitogens), whether they act on the cell surface or are processed intracellularly, the metabolic pathways affected, etc., are by no means trivial questions. Our major interest at the moment, however, is the nature of the switching mechanism in G_1 itself. Our immediate goal is to dissect the G_1 of the myoblast cell cycle under conditions which permit us to control the rate of progression through G_1. To facilitate such studies we needed a system which allowed us to impose synchrony with respect to both cell cycle phase and cellular differentiation in populations of single cells, uncomplicated by fusion. These requirements are uniquely met by the use of low Ca^{2+}, fusion-blocked myoblasts.

Fusion is one of a spectrum of cellular events which characterize myogenic differentiation. Concomitant with fusion the synthesis of a variety of skeletal muscle specific proteins is coordinately initiated and occurs at rates which accelerate with similar kinetics (Devlin & Emerson, 1978). It has also been amply demonstrated that cell fusion can be dissociated from the initiation of cell-type specific synthesis by lowering the calcium ion concentration ($[Ca^{2+}]$) in the medium to prevent fusion (Shainberg, Yagil & Yaffe, 1970; Ozawa, 1972; Paterson & Strohman, 1972).

Under such fusion-blocking conditions, skeletal muscle myosin synthesis is initiated and proceeds at the same rate (per nucleus) as in fused myofibres (Emerson & Beckner, 1975). We have examined the dissociability of myoblast fusion and the detection of cell-type specific synthesis using creatine phosphokinase (CPK) as a marker (Sutherland, 1977). Under fusion-permissive conditions CPK activity per nucleus increases by greater than 60-fold showing a precipitous rise corresponding closely to the time and course of cell fusion (percentage syncytial cells) (see Fig. 12). This rise in CPK activity in culture is in agreement with studies conducted in other laboratories (Morris, Cooke & Cole, 1972; Dym, Turner, Eppenberger & Yaffe, 1978) as is the demonstration that the increase in enzyme activity is due primarily to the appearance and accumulation of the muscle specific isoenzyme (MM-CPK). Since pre-fusion myoblasts contain low levels of enzyme activity due to the presence in such cells of the non-specific isozyme (similar in electrophoretic mobility to brain CPK) it was necessary first to determine base-line levels of CPK activity per nucleus in pure populations of both proliferating myoblasts and fully-fused muscle fibres. Such information could only be obtained using individual clones derived from single myoblasts. Fig. 13 and 14 contrast the appearance of single clones of proliferating myoblasts and fused muscle fibres respectively. In such preparations it is possible to ascertain whether all

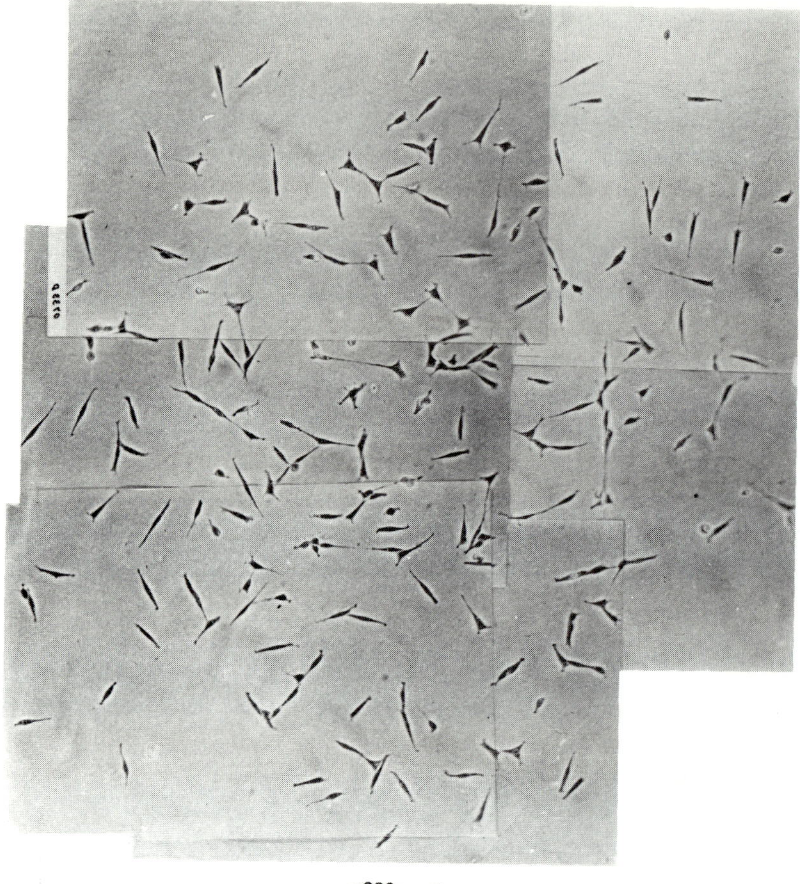

Fig. 13. Clone consisting exclusively of proliferating mononucleated myoblasts. Such cultures were inoculated with 100 cells per petri plate in high-growth medium (FM 10–15) and re-fed with the same medium after 24 h. This montage was photographed at the end of the third day in culture before freeze-drying.

of the nuclei are in single cells or, conversely, are all included in multinucleated fibres and actually count the number of nuclei per clone. In Table 3, CPK activity was determined using an ultramicro-fluorimetric assay and is expressed per nucleus to normalize activity in proliferating myoblasts and multinucleated muscle fibres. Enzyme activity per nucleus is approximately 60 × greater in muscle fibres than in proliferating single myoblasts, and the values of CPK activity per nucleus at each of the two developmental stages are virtually identical whether measured in pure clones or in mass cultures. This

Fig. 14. Clone consisting almost entirely of multinucleated muscle fibres. The few mononucleated myoblasts present have the typical long bipolar morphology of differentiating skeletal myocytes.

close agreement between clonal and mass cultures indicates that the degree of contamination, whether of mass cultures of proliferating myoblasts with small numbers of multinucleated cells or of cultures of fully fused cells with proliferating myoblasts, is negligible. During the transition from purely proliferative to fully fused myogenic cultures intermediate values of CPK per nucleus are observed as expected (see Fig. 12, 48 to 120 h) reflecting the progressive change in the ratio of single myoblasts to syncytial cells.

Table 3 also compares CPK per nucleus in myoblast clones switched

Table 3. *Creatine phosphokinase levels in multinucleated muscle fibres and in proliferating myoblasts*

	Units CPK/nucleus	
	Single cells	Multinucleated cells
Mass cultures	1.9×10^{-6} (low growth; low Ca^{++})	1.4×10^{-6}
Mass cultures	3.8×10^{-8} (proliferating)	
Single clones	5.6×10^{-6} (low growth; low Ca^{++})	
Single clones	3.5×10^{-8} (proliferating)	2.2×10^{-6}

to a maintenance medium ('low growth') under fusion-blocking conditions ("low Ca^{2+}") to both clones of proliferating myoblasts and clones of muscle fibres. With time following the 'step-down' switch, myoblasts become extremely elongated; levels of CPK per nucleus in such clones are indistinguishable from pure clones of muscle fibres. Isozyme patterns from companion experiments using mass cultures clearly demonstrate that the increase in CPK per nucleus in these clones of fusion-blocked 'myocytes' is due to the initiation of synthesis of the muscle-specific (M) CPK subunit (see Fig. 15). This observation is in good agreement with the results obtained for muscle-specific myosin synthesis in fusion-blocked myocytes (Emerson & Beckner, 1975) and demonstrates that myogenic fusion *per se* is not a precondition for cell-type specific protein synthesis.

The one property, however, common to myogenic cells which differentiate under both fusion-permissive and fusion-blocked conditions is that the nuclei of both are blocked in the G_1 phase of the cell cycle. When mass cultures of low-Ca^{2+}, fusion-blocked myocytes are exposed to the continued presence of ^3H TdR for 24 hours periods from the 96th to the 144th hour of culture, at which time CPK per nucleus has reached the same level as in fused muscle cultures (see Fig. 12), fewer than $2-3\%$ of the cells incorporate the label into their nuclei (Sutherland, 1977). It seems reasonable to assume that in fusion-blocked cultures, as under fusion-permissive conditions, the cells also deplete the medium of a heterologous mitogen, lag in G_1, and eventually effectively withdraw in this phase of the cell cycle.

If this assumption were correct one would predict that, given two sets of myoblast cultures, with media containing the same levels of

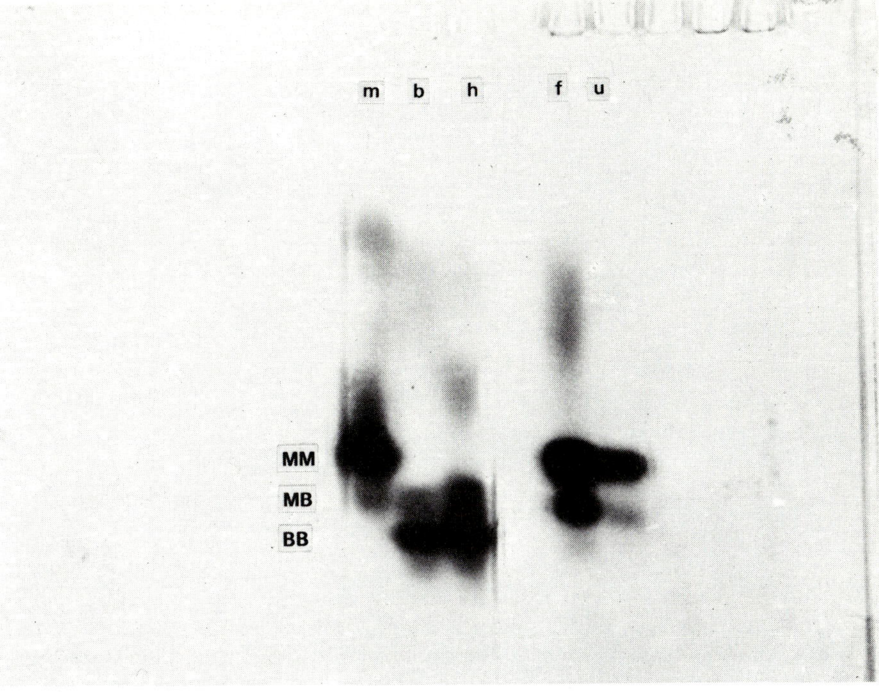

Fig. 15. Isozyme patterns of creatine phosphokinase from highly fused and fusion-blocked muscle cell cultures. MM designates the skeletal muscle specific homodimer, BB the so-called 'brain-type' homodimer and MB the heterodimer of creatine phosphokinase.

The letters above each lane indicate the tissue or culture source of the extract used. The letter m indicates skeletal muscle; b, brain; and h, heart. All three tissues were obtained from adult quail.

The two lanes to the right contained extracts of highly fused (f) cultures and unfused myocyte cultures (u). Total enzyme activity in sample f was 5 × greater than in u.

heterologous mitogens (viz. serum and EE) but which vary in [Ca^{2+}], media-depletion would follow the same course and the kinetics of accumulation of CPK per nucleus would be similar in both sets. This prediction has been tested, and the data plotted in Fig. 12 show that the increase in CPK activity per nucleus does follow the same time course in fusion-blocked and fusing cultures. Conversely one would anticipate that using media of different growth-potentiating properties, both of which contain fusion-impermissive levels of Ca^{2+}, the lower growth promoting medium should result in the earlier accumulation of CPK activity (per nucleus). This is seen in Fig. 16; CPK activity in cultures

                     I. R. Konigsberg

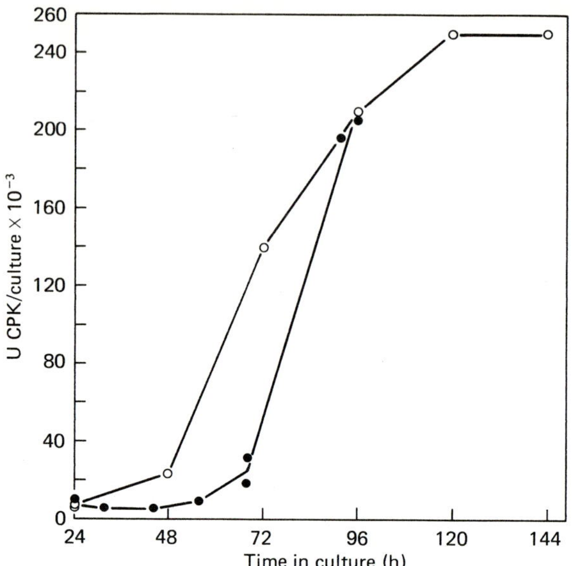

Fig. 16. Accumulation of CPK activity per culture as a function of time in media varying in growth-promoting activity. Culture conditions are as described in Fig. 5. Open circles represent CPK activity of cultures in low-growth, fusion-permissive medium (CM 1–10); closed circles, high-growth, fusion-permissive medium (FM 10–15). Note the lag in initiation of CPK activity between the two sets of cultures.

maintained in low-growth medium increases 24 hours prior to a similar rise exhibited by the cultures in high-growth medium.

These comparisons of biochemical differentiation under fusion permissive and restrictive conditions are compatible with our earlier studies of the control of fusion. Under conditions which preclude the occurrence of the fusion event mononucleated myoblasts initiate nevertheless the synthesis of skeletal-muscle specific proteins. Initiation occurs during a protracted G_1 in mononucleated myocytes which, if not actually withdrawn from the cell cycle, are traversing this phase extremely slowly.

Based on all of these data our working hypothesis is that there is a programmed sequence of biosynthetic events in G_1 leading to the re-initiation of DNA synthesis or, alternatively, into differentiative expression of the determined state of these myogenic cells. We assume that the two alternative fates are controlled at some switch point, late in G_1, so late that it is rarely reached under culture conditions which favour rapid proliferation. Although we have postulated a unitary programme we cannot exclude the possibility that, in fact, two independent programmes exist, one pre-replicative and one leading to differentiation, such as have been described in yeast (Hartwell, Culotti, Pringle & Reid, 1974). One might envisage a common start point (viz. mitosis) and the condition that if one programme is completed (such as the initiation of DNA synthesis) the other programme is terminated.

As a first step in testing this hypothesis we have initiated a series of studies to map the time point in G_1 of two differentiative events: the initiation of cell-type specific protein synthesis and the withdrawal of the differentiating cell from the cycle. We are interested in the muscle specific subunit of CPK (M) because the ubiquitous homologue (B subunit) is present in proliferating myoblasts. If our working hypothesis is correct, synthesis of the B subunit should map to some other time point in the cell cycle (early G_1 or a different phase entirely) than does the muscle-specific M subunit. For practical considerations, however, we elected to map first the initiation of muscle specific myosin heavy chain (MHC) synthesis (Devlin, 1979).

This study was conducted using cultures of low Ca^{2+} fusion-blocked myoblasts in which the time of initiation of MHC synthesis was scored on a per cell basis. The inoculum employed was a myoblast population synchronized in mitosis (M) by mechanical shake-off from secondary mass cultures. These M-cells were collected in low-growth medium and synchronous cultures established in low-growth medium containing fusion-impermissive levels of Ca^{2+}. Under these conditions, the mitotic cells complete division and traverse G_1 in from $2\frac{1}{2}$ to 3 hours (as determined by the rapid increase in percentage labelled cells following 15-minute pulses with ^{3}H TdR at regular intervals following plating). Within 3 hours after detection of the first labelled nuclei, more than 90% of the cells incorporate label during the pulse. During subsequent pulses, the percentage labelled cells declines rapidly and by 10 hours after plating, 50% of the population has entered G_2. From earlier studies (Buckley & Konigsberg, 1974a) we know that G_2 + mitosis occupy approximately 2 hours. Therefore, 2 hours following the end of S, the vast majority of these synchronized myoblasts should have completed mitosis and have entered G_1 of a second cell cycle. In point of fact, however, after entry into this second G_1 phase no further incorporation of ^{3}H TdR is observed even when 15-minute pulses are administered for as long as 42 hours beyond the end of the first cell cycle (see Fig. 17). In the low-growth, fusion-impermissive medium employed the cells *appear to be* withdrawn from the cell division cycle in what we have operationally defined as the terminal G_1

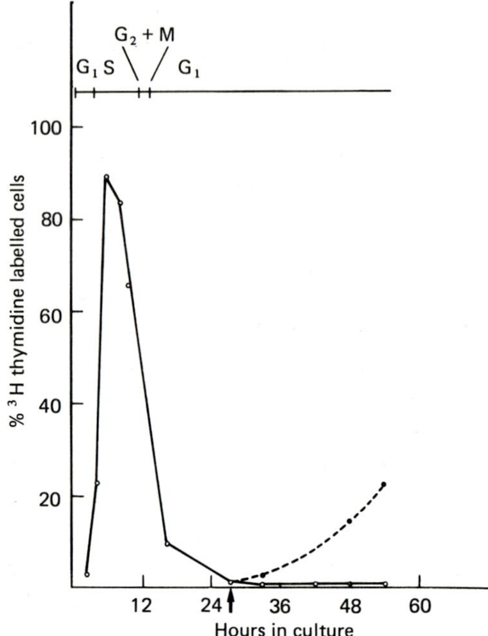

Fig. 17. Acculmulation of ^{3}H TdR labelled M-synchronized cells as a function of time in culture. Following collection by mitotic shake-off, cells were plated into low-growth, fusion-blocking medium. Open circles represent percentage labelled cells pulsed for 15 minutes immediately prior to fixation. Such cells complete one cell cycle following which the labelling index drops to essentially zero.

Re-feeding with a high-growth fusion-blocking medium (marked by arrow) results in a stimulation of proliferation (closed circles and dashed line).

phase. It is worth noting that M-synchronized myoblasts plated in high-growth, fusion-impermissive media, unlike their low-growth controls, do not block in the second G_1. All of these cells continue to cycle, with progressive decay of synchrony, until some time between 36 and 46 hours after plating when the percentage labelled cells in these crowded cultures drops from 98% to 80%. This control is clear evidence that the block in G_1 is a response to the low-growth milieu and not to any prior event or commitment.

To score for the appearance of muscle-specific MHC we have employed an antiserum raised in rabbits and subsequently purified by affinity chromatography against immobilized, purified quail MHC. This reagent as used in an indirect immunofluorescent procedure stains only the A band in differentiated cultured muscle fibres (see Fig. 18). It cross-reacts with neither skin fibroblasts nor with proliferating,

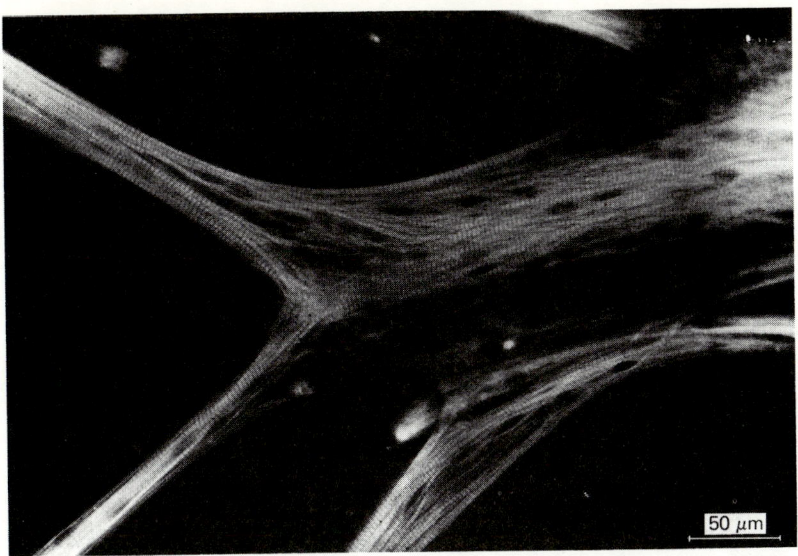

Fig. 18. Immunofluorescent detection of skeletal muscle-specific myosin in muscle fibres of well-fused quail muscle cultures. Cultures fixed in 3% formaldehyde, post-fixed in 1 : 1 ethanol-acetone. Cultures incubated with affinity column purified anti-myosin immunoglobulins (10 μg/ml), rinsed and reacted with fluorescein-labelled anti-rabbit immunoglobulins. Photographed under UV epi-illumination, fluorescence is restricted to A band region.

log phase myoblasts, but stains differentiated, fusion-blocked myocytes brilliantly.

The percentage of myosin-positive (myo+) cells was scored at intervals after entry into the terminal G_1 and the results are plotted in Fig. 19. Myo+ cells are first observed and start to accumulate following 8 hours into the terminal G_1. The percentage of such cells increases rapidly during the next 16 hours, at which time over 90% of the cells are scored as myo+. Thereafter the percentage continues to increase, albeit more slowly, reaching 95% within the following 12 hours.

In summary, synchronized, mitotic myoblasts finish division, traverse a complete cell cycle and enter a second G_1 period. Under the conditions imposed virtually none of the cells re-enters S from this terminal G_1. Instead, during this protracted G_1, the synthesis of MHC is initiated and myo+ cells accumulate with predictable kinetics. These data suggest that skeletal, muscle-specific protein synthesis, like fusion competence, is restricted to a time point late in G_1. They indicate, in fact, that when fusion and MHC synthesis are uncoupled,

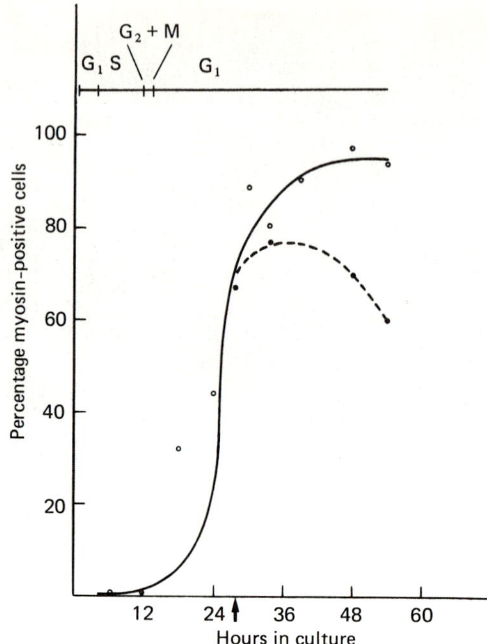

Fig. 19. Percentage of myosin-positive cells scored, in same groups of cells in Fig. 17, prior to radioautography. Myosin-positive cells appear rapidly and synchronously after the completion of the single cell cycle (open circles). A decrease in the percentage myosin-positive cells follows re-feeding (arrow) with high-growth medium (closed circles and broken line).

the initiation of synthesis occurs later in G_1 than fusion-competence under fusion-permissive conditions.

This system, in addition, makes a number of well-defined questions accessible to experimentation for the first time. For example, are myoblasts which are traversing their final G_1 and presumably committed to initiate skeletal muscle specific MHC, withdrawn from the cell cycle? To examine this question, cultures were re-fed with a high-growth potentiating but low-$[Ca^{2+}]$ medium at a time point in terminal G_1 when 50% of the cells are myo+. The curve represented by the dashed line in Fig. 17 describes re-entry into S (percentage cells labelled during a 15-minute pulse) following this change of medium. During the 27 hour period following re-feeding the labelling index rises from 3 to 25%. Clearly a sizeable fraction of the cells traversing what would be their terminal G_1 can be stimulated to re-enter the cell cycle. It is highly unlikely that the cells which re-enter the cycle

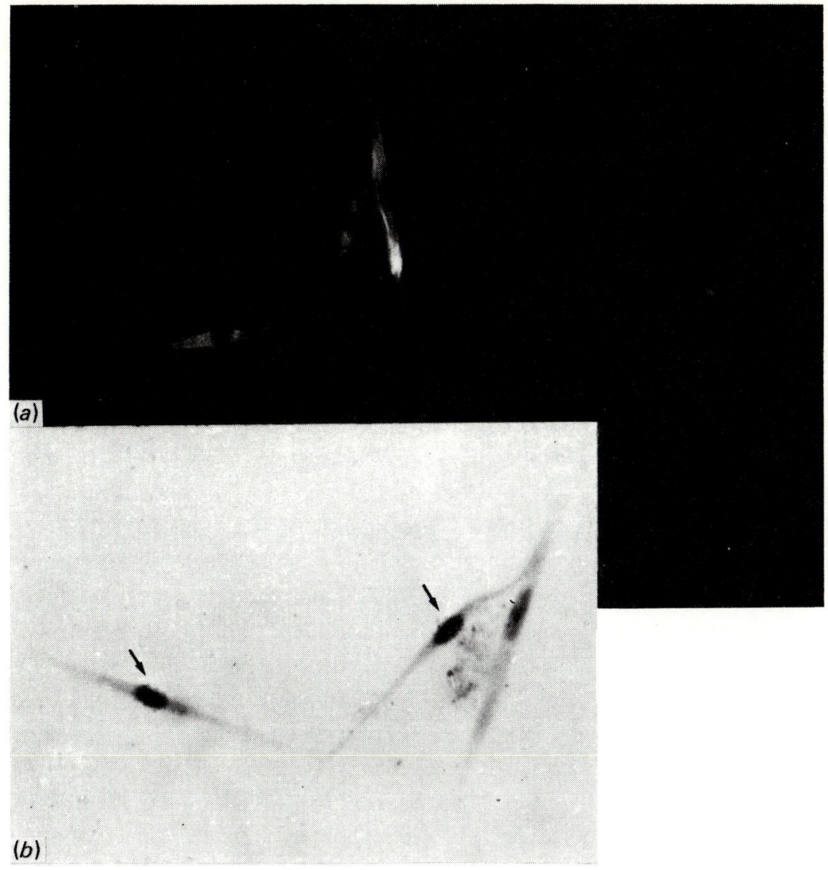

Fig. 20. Incorporation of ^{3}H TdR into myosin-positive cells.

Cells re-fed high-growth, fusion-blocking medium at 15 hours when 50% of the cells
were myo+ were pulsed for 15 minutes with ^{3}H TdR before fixation at 6, 12 and 24
hours after feeding. After processing for immunofluorescent detection of myosin,
randomly selected fields were photographed under UV epi-illumination and the
cultures processed for radioautography and stained with haematoxylin. One such field
is illustrated in (a). In this field the fluorescent emission of the three bipolar cells to the
left clearly exceeded the autofluorescence of that group of six cells, barely visible, to
the right of these myosin-positive cells. Inset (b), below, is a photomicrograph of these
same three immunofluorescent cells following radioautographic processing and
haematoxylin staining. Two of these cells incorporated significant levels of ^{3}H TdR
during the terminal 15-minute pulse (see arrows). The third myosin-positive cell shows
no incorporation either because it was withdrawn from the cycle or was not in
synchrony with the other two. More than 350 cells were examined, 3% of which were
doubly labelled.

Table 4. *Percentage of cells in S at intervals sampled after re-feeding*

		6 hours	12 hours	24 hours
Percentage of dividing cells (37) in S		10%	28%	14%
Percentage of all cells (74) in S		5%	14%	7%
Percentage double labelled	Expected	2.5%	7%	3.5%
	Actual	0.9%	3%	4%

represent contaminating fibroblasts. If such were the case, such non-myogenic cells would overgrow the re-fed cultures. This is not the case, however, since if the cultures are maintained after re-feeding beyond 40 hours, the high-growth medium is depleted, the labelling index declines and the percentage myo+ cells again reaches levels in excess of 90%. A more reasonable hypothesis is that the mitogen-stimulated fraction represents only those myoblasts not engaged in MHC synthesis. This possibility is particularly cogent since the step-up medium change was performed when only 50% of the cells were myo+. To test this alternative we used a double-labelling procedure combining myosin-immunofluorescence and radioautography. Cultures were re-fed as before and fixed at intervals following re-feeding. In these experiments, however, the cells were pulsed with ^{3}H TdR for 15 minutes prior to fixation. Cultures were processed first for immuno-fluorescence, photographed and scored for myo+ cells. Following radioautographic processing the cells in the areas previously scored for immunofluorescently detectable myosin were re-examined for tritium label over the nuclei (see Fig. 20). These comparisons revealed that a small number (approximately 3%) of the myo+ cells had incorporated significant amounts of ^{3}H TdR during the pulse. From the data summarized in Table 4 one observes that the percentage doubly-labelled cells varies somewhat with time after the medium change, being lowest at the shortest time interval. Since the average interval between re-feeding and mitosis (see below) is 1.8 × the total generation time of log phase myoblasts it would not be surprising if few cells had entered S in the first 6 hours after the mitogenic stimulus.

The percentages of doubly labelled cells are small, and one would question the significance of so small a number. These cells might represent a small fraction of cells which have 'escaped' the normal control mechanism. On the other hand, if after myoblasts re-fed in G_1 (or G_0) progress slowly and asynchronously toward S, as do serum starved

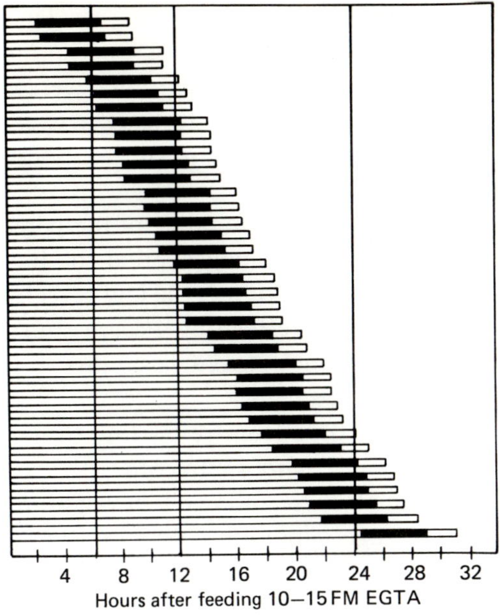

Fig. 21. Determination of the number of cells in S at each time point sampled for myosin immunofluorescent-^{3}H TdR incorporating cells following re-feeding. Each bar in the diagram represents each of the 37 cells which divided (out of a total of 74 cells followed) during filming. Total bar length represents time from re-feeding to mitosis. Black portion represents time in S(4.5 h), open portion of bar following S indicates time in G_2 and mitosis (2 h). Times derived from a previous study (Buckley & Konigsberg, 1974a).

cells of established cell lines after serum replacement (Pardee, Jimenez de Asua & Rozengurt, 1974), one might anticipate that at any one time following re-feeding relatively few cells would be traversing S.

To estimate the percentage of cells in S at each of the time points scored for doubly-labelled cells, the re-feeding experiments were repeated following the same protocol except that time-lapse cinematographic records were made beginning at 2 hours after plating. A total of 74 cells were followed for 32 hours and of these, 37 cells (50%) divided between re-feeding and the termination of filming. A bar graph was constructed representing the 37 cells which divided (see Fig. 21). In comparable studies (Buckley & Konigsberg, 1974a) we found that under differentiation-permissive conditions although G_1 is markedly protracted the lengths of S, G_2 and M are invariant. Using the average length of $G_2 + M$ and S previously determined (Buckley & Konigsberg, 1974a) we located, in time, the S period of each cell which divided on the bar graph (black portion of the bar, Fig. 21). Drawing a perpendicular line at 6, 12 and 24 hours on the abscissa one can determine the percentage of dividing cells which were in S at each time point sampled. Since only half of the cells tracked eventually divided, the percentage of all cells in S at the sampled time points are 2.5, 7 and 3.5% respectively (see Table 4).

These small percentages are in the range of the percentage doubly-labelled myoblasts observed and suggests that the low percentage of double-labelled cells does indeed reflect the low percentage of cells in S during the 12 hour period after re-feeding. Progression towards mitosis is initially slow and asynchronous (note variability in Fig. 21) as in re-fed G_0-blocked serum-starved cells.

In addition the average time from re-feeding to mitosis for the 37 cells which divided is 18.5 ± 6 hours. In contrast, 22 daughter cells of these first divisions which were followed through an additional cell cycle exhibited an average generation time of 10.6 ± 2 hours. The difference between the 'G_T' of the first and second divisions following re-feeding is highly significant (see Table 4) and the average G_T of the second division is in good agreement with previous determinations of G_1 of asynchronous log phase myoblasts in high-growth medium.

Our observations indicate that under the conditions which we have employed the initiation of MHC synthesis is not obligatorily coupled to irreversible withdrawal from the cell cycle. For at least some time, myocytes which have initiated synthesis and accumulated detectable quantities of MHC can re-enter the cell cycle if appropriately stimulated. Since the step-up media change was performed at only one time point, 15 hours into the terminal G_1, we cannot exclude the possibility that competence to re-enter the cell cycle does not decline with the further passage of time in G_1.

The time-lapse records provide additional data which suggest that no such decrement occurs for a period of at least 28 hours in G_1. From the films one can obtain the length of time in G_1 prior to re-feeding (mitosis to medium change). This was done for all 74 cases and the data grouped into three class intervals: less than 9 hours, between 12 and 15 hours and greater than 15 hours in G_1 prior to re-feeding. Approximately 30% of the cases fall into each class interval. Despite the fact that time in G_1 is progressively longer in each class, the same percentage of cells divided (*ca* 50%) in each class. Moreover since the cells in each class have spent increasingly more time in G_1 each successive class should contain a higher percentage of myo+ cells (from $< 15\%$ to $> 70\%$, see Fig. 19). Furthermore, there appears to be no correlation between the length of time a cell spends in the terminal G_1 and the interval between re-feeding and subsequent mitosis since the distribution of these values for each of the classes is similar. These observations are in agreement with the transitional probability hypothesis of progression through the cell cycle in that they demonstrate, for a differentiating cell type in our studies, that the probability of re-entering S is independent of the time spent in G_1. In this sense they also lend support to the hypothesis that in skeletal myocytes engaged in the synthesis of MHC, the probability of re-entering the cell cycle following a mitogenic stimulus is not different from the probability of re-entry of undifferentiated myoblasts.

In summary, all of these results demonstrate that fusion-blocked skeletal myocytes which have synthesized and accumulated MHC can re-initiate DNA synthesis following a simple medium change to a medium which supports rapid proliferation. This suggests that irreversible withdrawal from the cell cycle is not a precondition for contractile protein synthesis nor does cell-type specific synthesis, *per se*, restrict further cell proliferation. We conclude therefore that the irreversible withdrawal which attends myogenic fusion in avian myoblasts is a consequence of some aspect or consequence of the fusion event and is not coupled to contractile protein synthesis.

The original researches described herein were performed under a grant (HD-07083) from the National Institutes of Health, USA and a grant from the Muscular Dystrophy Association of America.

REFERENCES

BOYD, J. D. (1960). Development of striated muscle. In *The Structure and Function of Muscle*, vol. I, ed. G. H. Bourne, pp. 63–85. Academic Press, New York & London.

BUCKLEY, P. A. & KONIGSBERG, I. R. (1974a). Myogenic fusion and the duration of the post-mitotic gap (G_1). *Developmental Biology*, **37**, 193–212.

BUCKLEY, P. A. & KONIGSBERG, I. R. (1974b). The avoidance of stimulatory artifacts in cell cycle determination *in vitro*. *Developmental Biology*, **37**, 186–92.

CLEAVER, J. E. (1967). *Thymidine Metabolism and Cell Kinetics*. John Wiley and Sons, Inc., New York.

COON, H. G. (1966). Clonal stability and phenotypic expression of chick cartilage cells *in vitro*. *Proceedings of the National Academy of Sciences, USA*, **55**, 66–73.

COOPER, W. G. & KONIGSBERG, I. R. (1961a). Succinic dehydrogenase activity of myoblasts in tissue culture. *Experimental Cell Research*, **23**, 576–81.

COOPER, W. G. & KONIGSBERG, I. R. (1961b). Dynamics of myogenesis *"in vitro"*. *Anatomical Record*, **140**, 195–205.

DEVLIN, B. H. (1979). Muscle cell proliferation following activation of myosin synthesis. Dissertation, University of Virginia.

DEVLIN, R. B. & EMERSON, C. P., Jr. (1978). Coordinate regulation of contractile protein synthesis during myoblast differentiation. *Cell*, **13**, 599.

DOERING, J. & FISCHMAN, D. (1974). The *in vitro* cell fusion of embryonic chick muscle without DNA synthesis. *Developmental Biology*, **36**, 225–35.

DYM, H., TURNER, D. C., EPPENBERGER, H. M. & YAFFE, D. (1978). Creatine kinase isoenzyme transition in actinomycin D-treated differentiating muscle cultures. *Experimental Cell Research*, **113**, 15–21.

EMERSON, C. P., JR. (1977). Control of myosin synthesis during myoblast differentiation. In *Pathogenesis of the Human Muscular Dystrophies*, (Fifth International Scientific Conference – MDA), ed. L. P. Rowland, pp. 799–809. Excerpta Medica, Amsterdam-Oxford.

EMERSON, C. P., JR. & BECKNER, S. K. (1975). Activation of myosin synthesis in fusing and mononucleated myoblasts. *Journal of Molecular Biology*, **93**, 431–47.

HARTWELL, L., CULOTTI, J., PRINGLE, J. & REID, B. (1974). Genetic control of the cell division cycle in yeast. *Science*, **183**, 46–52.

HAUSCHKA, S. D. & KONIGSBERG, I. R. (1966). The influence of collagen on the development of muscle clones. *Proceedings of the National Academy of Sciences, USA*, **55**, 119–26.

HOLTZER, H. (1972). The cell cycle, myogenesis and psoriasis. *Journal of Investigative Dermatology*, **59**, 33–34.

HOLTZER, H., ABBOTT, J., LASH, J. & HOLTZER, S. (1960). The loss of phenotypic traits by differentiated cells *in vitro*. I. Dedifferentiation of cartilage cells. *Proceedings of the National Academy of Sciences, USA*, **46**, 1533–42.

HOLTZER, H., RUBINSTEIN, N., FELLINI, S., YEOH, G., CHI, J., BIRNBAUM, J. & OKAYAMA, M. (1975). Lineages, quantal cell cycles, and the generation of cell diversity. *Quarterly Review of Biophysics*, **8**, 523–56.

KONIGSBERG, I. R. (1960). The differentiation of cross-striated myofibrils in short term cell culture. *Experimental Cell Research*, **21**, 414–20.

KONIGSBERG, I. R. (1961a). Cellular differentiation in colonies derived from single cell platings of freshly isolated chick embryo muscle cells. *Proceedings of the National Academy of Sciences, USA*, **47**, 1868–72.

KONIGSBERG, I. R. (1961b). Some aspects of myogenesis *in vitro*. In Biophysics and Biochemistry of the Myocardium (ed. A. P. Fishman). *Circulation*, **24**, (part 2), 447–57.

KONIGSBERG, I. R. (1963). Clonal analysis of myogenesis. *Science*, **140**, 1273–84.

KONIGSBERG, I. R. (1967). The application of clonal techniques to problems of cytodifferentiation. In *Ontogeny of Immunity*, ed. R. T. Smith, P. A. Miescher and R. A. Good. University of Florida Press.

KONIGSBERG, I. R. (1970). The relationship of collagen to the clonal development of embryonic skeletal muscle. In *Chemistry and Molecular Biology of the Intercellular Matrix*, vol. 4, ed. E. A. Balazs, p. 1779. Academic Press, New York.

KONIGSBERG, I. R. (1971). Diffusion-mediated control of myoblast fusion. *Developmental Biology*, **26**, 133–52.

KONIGSBERG, I. R. (1979). Skeletal myoblasts in culture. In *Cell Culture*, ed. W. B. Jakoby and I. Pastan. *Methods in Enzymology 58*, pp. 511–27. Academic Press, New York.

KONIGSBERG, I. R. & HAUSCHKA, S. D. (1966). Cell interactions in the reproduction of cell type. In *Reproduction: Molecular, Subcellular and Cellular*, ed. M. Locke, pp. 243–89. Academic Press, New York.

KONIGSBERG, I. R., MCELVAIN, N., TOOTLE, M. & HERMANN, H. (1960). The dissociability of deoxyribonucleic acid synthesis from the development of multinuclearity of muscle cells in tissue culture. *Journal of Biophysical and Biochemical Cytology*, **8**, 333–43.

KONIGSBERG, I. R., SOLLMAN, P. A. & MIXTER, L. O. (1978). The duration of the terminal G_1 of fusing myoblasts. *Developmental Biology*, **63**, 11–26.

LEBLOND, C. P. (1964). Classification of cell populations on the basis of proliferative behaviour. *Journal of the National Cancer Institute*, **14**, 119–50.

LINKHART, T. A., CLEGG, C. H. & HAUSCHKA, S. D. (1980). Control of mouse myoblast commitment to terminal differentiation by mitogens. In *ICN/UCLA Symposium on Control of Cell Division and Differentiation*, ed. D. Cunningham, *Journal of Supramolecular Structure* (in press).

LIPTON, B. H. (1977). A fine-structural analysis of normal and modulated cells in myogenic cultures. *Developmental Biology*, **60**, 26–47.

MORRIS, G. E., COOKE, A. & COLE, R. J. (1972). Isoenzymes of creatine phosphokinase during myogenesis in vitro. *Experimental Cell Research*, **74**, 582–5.

NADAL-GINARD, B. (1978). Commitment, fusion and biochemical differentiation of a myogenic cell line in the absence of DNA synthesis. *Cell*, **15**, 855–64.

O'NEILL, M. & STOCKDALE, F. E. (1972a). A kinetic analysis of myogenesis *in vitro*. *Journal of Cell Biology*, **52**, 52–65.

O'NEILL, M. & STOCKDALE, F. E. (1972b). Differentiation without cell division in cultured skeletal muscle. *Developmental Biology*, **29**, 410–18.

OZAWA, E. (1972). The role of calcium ion in avian myogenesis *in vitro*. *Biological Bulletin*, **143**, 431–9.

PARDEE, A., JIMENEZ DE ASUA, L. & ROZENGURT, E. (1974). In *Control of Proliferation in Animal Cells*, vol. 1, ed. B. Clarkson and R. Baserga, Cold Spring Harbor Conferences on Cell Proliferation. p. 547.

PATERSON, B. & STROHMAN, R. C. (1972). Myosin synthesis in differentiating cultures of chick embryo skeletal muscle. *Developmental Biology*, **29**, 113–38.

QUASTLER, H. & SHERMAN, F. G. (1959). Cell population kinetics of the intestinal epithelium of the mouse. *Experimental Cell Research*, **17**, 420–38.

REPORTER, M. N., KONIGSBERG, I. R. & STREHLER, B. L. (1963). Kinetics of accumulation of creatine phosphokinase activity in developing embryonic skeletal muscle *in vivo* and in monolayer culture. *Experimental Cell Research*, **30**, 410–17.

ROLSHOVEN, E. (1951). Ueber die Riefungsteilungen bei des Spermatogenese mit einer Kritik des bisheigen Begriffes der Zellteilunge. *Verhandlungen der Anatomischen Gesellschaft*, **49**, 189–97.

SHAINBERG, A., YAGIL, G. & YAFFE, D. (1970). Control of myogenesis *in vitro* by Ca^{++} concentration in nutritional medium. *Experimental Cell Research*, **58**, 163–7.

STANNERS, C. P. & TILL, J. E. (1960). DNA synthesis in individual L-strain mouse

 I. R. Konigsberg

cells. *Biochimica et Biophysica Acta*, **37**, 406–19.

STREHLER, B. L., KONIGSBERG, I. R. & KELLEY, F. E. L. (1963). Ploidy of myotube nuclei developing *in vitro* as determined with a recording double-beam micro-spectrophotometer. *Experimental Cell Research*, **32**, 232–41.

SUTHERLAND, W. M. (1977). Changes in creatine phosphokinase during muscle cell differentiation *in vitro*. Dissertation, University of Virginia.

TURNER, D. C., GMÜR, R., SIEGRIST, M., BURKHARDT & EPPENBERGER, H. (1976). Differentiation in cultures derived from embryonic chicken muscle. 1. Muscle-specific enzyme changes before fusion in EGTA-synchronized cultures. *Developmental Biology*, **48**, 258–83.

The cell cycle and erythroid differentiation

RICHARD A. RIFKIND AND PAUL A. MARKS*

Cancer Center/Institute of Cancer Research, Columbia University, College of Physicians and
Surgeons, New York, New York 10032, USA

INTRODUCTION

Considerable evidence suggests that cell differentiation, detected as the
transition to synthesis of proteins characteristic of the differentiated
state may, under some circumstances, require cell division (Rutter,
Pictet & Morris, 1973). Whether this is true of all developmental
transitions is unknown, as is the role of particular stages of the cell
division cycle in the regulation of cell differentiation (reviewed by
Marks *et al.*, 1977). For terminal erythroid cell differentiation, the
evidence, with respect to the relationship between the cell cycle and
differentiation, falls into two broad, overlapping areas. On the one
hand, there is the relationship between specific stages of the cell cycle
and the transition from latent to expressed erythroid differentiation.
On the other hand, there are the effects of this transition upon cell cycle
kinetics. Evidence with respect to both can be adduced from studies on
normal erythropoietic cell differentiation and on differentiation induced in murine erythroleukaemic cells (MELC) by a variety of chemical
agents (Marks & Rifkind, 1978). This paper reviews both sources in an
attempt to establish the present state of our understanding, and as a
guide to further investigation.

THE CELL CYCLE AND ERYTHROPOIETIN-MEDIATED CELL DIFFERENTIATION

There is considerable evidence that a principal target effect of the
erythropoiesis-regulating hormone, erythropoietin, is the replication
of an erythroid precursor cell (Chui, Djaldetti, Marks & Rifkind, 1971;
Marks & Rifkind, 1972; Cantor, Morris, Marks & Rifkind, 1972) which,
under the influence of the hormone, generates a colony of differentiat-

* Studies from the authors' laboratory were supported, in part, by grants CA-13696
and CA-18314 from the National Cancer Institute, CH-68 from the American Cancer
Society, PCM-75-08696 from the National Science Foundation, and a grant from the
New York Community Trust.

ing erythroid cells (Stephenson, Axelrad, McLeod & Shreeve, 1971; Cooper *et al.*, 1974; Cormack, 1976); these studies and more recent information from a number of laboratories, have recently been reviewed (Axelrad, McLeod, Suzuki, & Shreeve, 1978; Eaves, Humphries & Eaves, 1979). Little is known of the molecular aspects of the regulation of precursor cell proliferation by erythropoietin nor of the relationship between precursor cell proliferation and initiation of the programme of terminal erythroid cell differentiation. Precursor cells respond to erythropoietin with a brisk acceleration of ribosomal RNA synthesis (Djaldetti, Preisler, Marks & Rifkind, 1972) followed by DNA synthesis and subsequently by the onset of accumulation of globin messenger RNA, globin synthesis and the terminal stages of erythropoietic cell differentiation (Ramirez *et al.*, 1975). It has been suggested (Marks *et al.*, 1977) that DNA synthesis, in the presence of erythropoietin, may be required for subsequent globin messenger RNA accumulation and globin synthesis.

It has been suggested that discrete stages of the cell division cycle may be critical for erythropoietin-mediated induction of cell proliferation and terminal erythroid cell differentiation. Both G_1 (Schooley, 1965; Necas & Neuwirt, 1976) and early S phase (Kretchmar, 1966; Monette *et al.*, 1980) have been suggested as possible cell cycle targets for the inductive effect of erythropoietin on erythropoiesis. Interpretation of these experiments is confounded by the problems inherent in *in vivo* studies. It has proved impossible, to date, to obtain sufficient numbers of purified erythroid cell precursors synchronized with respect to the cell cycle in order to study these events under *in vitro* conditions. It is for these reasons that many investigators have turned from the study of erythropoietin-mediated terminal erythroid cell differentiation to the study of chemically induced cell differentiation in the murine erythroleukaemia cell (MELC) system.

THE CELL CYCLE AND INDUCED MELC DIFFERENTIATION

Many laboratories have studied the regulation of expression of the differentiated features characteristic of the terminal stages of erythroid cell differentiation in MELC. A broad spectrum of chemical as well as physical agents induce MELC to express the phenotypic programme of erythroid differentiation (reviewed by Marks & Rifkind, 1978), including both morphogenesis and the synthesis of characteristic gene prod-

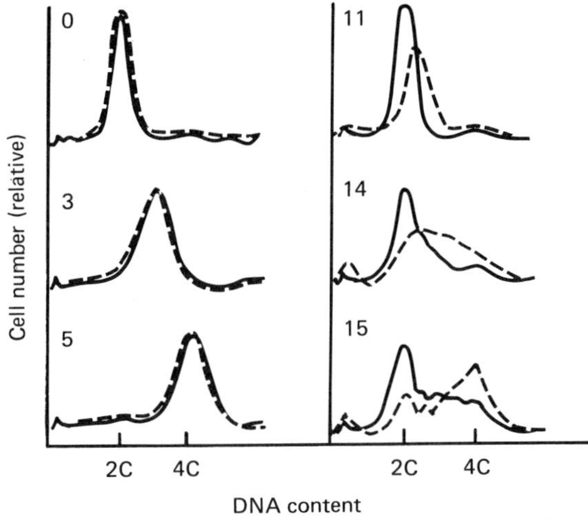

Fig. 1. Cell cycle transit during MELC differentiation induced by HMBA (hexamethylene bisacetamide) (4 mM). The cells were synchronized at the G_1/S boundary by serial exposure to 2 mM thymidine and hydroxyurea (0.5 mM) as previously described (Gambari *et al.*, 1978). After 0, 3, 5, 11, 14 and 15 h culture with (solid line) or without (dashed line) the inducer, aliquots were removed for determination of their position in the cell cycle (measured as relative DNA content per cell) by flow microfluorometry after staining with propidium iodide.

ucts. The ability to grow large numbers of MELC in culture, to synchronize them with respect to the cell division cycle, by chemical and by physical means, and to initiate terminal cell differentiation in a relatively uniform fashion by exposure to inducing chemicals, recommends this cell system for the study of cell cycle-related events during differentiation.

CHANGES IN THE CELLS CYCLE DURING DIFFERENTIATION

The principal cell cycle-related consequence of induced differentiation in MELC is the initiation of terminal division (Singer *et al.*, 1974; Gusella *et al.*, 1976; Fibach, Reuben, Rifkind, & Marks, 1977). Cessation of cell division is characteristic of normal erythropoiesis, as well, and suggests the triggering of a complex programme of events in both normal and transformed erythroid cells. The first sign of a change in cell cycle kinetics is a striking prolongation of the G_1 phase of the cell cycle (Fig. 1), detected within 6–12 h of exposure to inducers (Terada *et al.*, 1977). It is during this prolonged G_1 that an increase in the rate of accumulation of newly synthesized globin messenger RNA is first detected in induced MELC (Gambari *et al.*, 1978). The mechanism for the initiation of this change in cell cycle kinetics has not been elucidated, although the possibility of a cyclic nucleotide-mediated effect is suggested by preliminary evidence (Gazitt *et al.*, 1978) and by analogy with similar effects in other cell systems (Lehnert, 1979).

The initiation of terminal cell division forms one of the critical criteria for the test of commitment of MELC to the programme of induced differentiation (Fibach *et al.*, 1977). An increasing proportion of MELC become committed to differentiate with increasing time of exposure to the inducer in cell culture. Commitment, defined as the ability to express the programme of erythroid differentiation after removal of the inducing agent, can be detected with the first 12–14 h of exposure to inducer; it is virtually complete by 48 h. This assay entails exposure of MELC to the inducer for various periods, washing, transfer to semi-solid medium, then scoring the growing colonies with the benzidine reaction for haemoglobin. The proportion of cells in the culture committed to differentiate is dependent on both the concentration and the duration of exposure to the inducing agent. Committed cells give rise to small colonies (terminated cell division) which contain haemoglobin. Uncommitted cells give rise to large, growing colonies in which no cells contain haemoglobin.

THE CELL CYCLE AND INDUCTION OF DIFFERENTIATION

Three experimental strategies, employing both unsynchronized and cell-cycle synchronized cell populations, provide evidence implicating cell cycle-related events in the induction of erythroid cell differentiation in MELC. These include experiments designed to test the requirement for cell growth for induced differentiation, experiments designed to test, more specifically, the requirement for the cell division cycle for induced differentiation, and finally, studies designed to examine specific phases of the cell cycle with regard to induced differentiation.

Many studies with MELC (Marks *et al.*, 1977; McClintock & Papaconstantinou, 1974; Tabuse, Kawamura & Furusawa, 1976; Parker & Hooper, 1978) reveal a relationship between the extent of proliferation and differentiation in culture with inducer. At higher cell densities, the time required for cell doubling is prolonged and the extent of

Table 1. *Cell proliferation and induced differentiation*

Inoculum ($\times 10^5$ cells/ml)	Cell doublings[a]	B+ Cells[a,b] (%)
1	4.9	89
2	4.2	77
5	3.0	65
10	1.8	58
20	0.9	23
40	0.7	4

[a] Determined after 5 days of culture with 280 mM dimethylsulphoxide as inducer.

[b] Percentage of benzidine-reactive (haemoglobin-containing) cells after 5 days culture with dimethylsulphoxide.

differentiation is decreased (Table 1). Both cell proliferation and differentiation are markedly influenced by conditions of culture (Singer, D., unpublished observations). The higher the rate of cell proliferation, the greater the proportion of cells induced to differentiate.

More direct evidence for the role of the cell cycle in induced MELC differentiation comes from studies with synchronized cell populations. Three different methods of cell synchronization, employing cell cycle inhibitors (2 mM thymidine, Levy, Terada, Rifkind & Marks, 1975; and isoleucine-deficient medium, McClintock & Papaconstantinou, 1974), as well as physical methods of synchronization (Geller, Levenson & Housman, 1978), suggest that events related to the cell cycle may be critical to the commitment of MELC to differentiate along the erythropoietic pathway when exposed to appropriate inducing agents. These conclusions are supported by recent observations with strains of MELC temperature-sensitive for growth (Conkie, Harrison and Paul, personal communication).

Although the bulk of evidence to date is consistent with the notion that some aspect of the cell cycle is critical for induced MELC differentiation, a more precise definition of the phase of the cell cycle implicated is less easy to establish. There is evidence that neither mitosis (Tabuse, Kawamura & Furusawa, 1976), nor cytokinesis (Harrison, 1976; Parker & Hooper, 1978), are essential for chemically induced MELC differentiation. Studies with inhibitors of DNA synthesis (Leder, Orkin & Leder, 1975; Levenson, Kernen, Mitrani & Housman, 1980), have been adduced as evidence that DNA synthesis (the S phase of the cell cycle) is

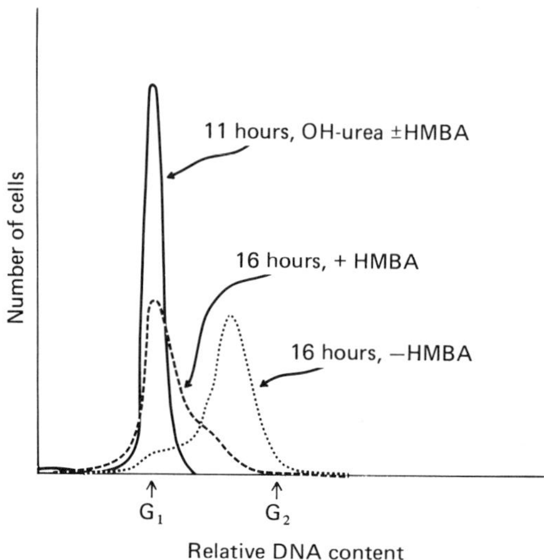

Fig. 2. Effect of inhibition of DNA synthesis in G_1-synchronized cells, on inducer-mediated G_1-arrest. MELC were synchronized in G_1 by exposure to 0.5 mM hydroxyurea for 11 h, without or with inducer (5 mM HMBA). At 11 h the cells were washed and culture resumed, without or with HMBA, as before. Five hours after resumption of cell cycle (16 h total time) the cells were stained for DNA content (propidium iodide) and the cell cycle (DNA content) distribution determined by flow microfluorometry (Gambari, Marks & Rifkind, 1979). At this time the HMBA-treated cells display characteristic G_1 arrest, whereas control cells have moved into S.

not critical to induced MELC differentiation. In neither instance, however, is the inhibition of DNA synthesis absolute, and the participation of critical, discrete, replicative events in the induction process cannot be ruled out. Indeed, recent studies from other laboratories (Brown & Schildkraut, 1979; also Lo, Ross & Mueller, personal communication), examining the sensitivity of the erythroid differentiation programme to inhibition by bromodeoxyuridine, suggest that there may be replicative events early in the S phase which are critical to subsequent expression of the erythroid differentiation programme. The most critical studies, to date, fail to distinguish between events in early S and those which may occur in G_1. Geller, Levenson & Housman (1978), employing unit gravity sedimentation for synchronization, demonstrate that cells exposed to inducing agents starting in the S phase are retarded with respect to commitment, compared to cells exposed from G_2 or G_1. Gambari, Marks & Rifkind (1979), using centrifugal elutriation for synchronization, demonstrate that the initial onset of accumulation of globin messenger RNA occurs during the first prolonged G_1 phase which follows complete traversal of S in the presence of inducer. Comparison of MELC populations exposed to inducers starting in G_1, mid-S or late S/G_2 phases of the cell cycle, provides evidence suggesting that an effect of the inducer during either G_1 or early S is critical to the initiation of accumulation of newly synthesized globin mRNA. More recent, and still preliminary, studies employing hydroxyurea to arrest synchronized (elutriated) G_1 phase MELC at the G_1/S interface, suggest that the events critical to the transient G_1 arrest characteristic of induced MELC differentiation, occur during G_1 or at G_1/S (Fig. 2). The G_1-arrest phenomenon, itself, however, is neither essential nor sufficient for expression of all aspects of erythroid differentiation. Both the phorbol esters (Yamasaki *et al.*, 1977) and dexamethasone (unpublished observations), which inhibit expression of erythroid differentiation, fail to inhibit the G_1-arrest induced by hexamethylene bisacetamide. Hemin, an inducer of globin mRNA synthesis but not of commitment to the complete programme of terminal cell differentiation and cell division, does not initiate a G_1-arrest (unpublished observations).

SUMMARY

Evidence has been cited which suggests that events critical to the initiation of terminal cell differentiation, both of normal erythroid cell precursors under the influence of erythropoietin, and of erythroleukaemia cells, under the influence of a variety of inducing agents, take place during specific phases of the cell division cycle. Exposure to inducers during the sensitive window in the cell cycle appears essential for initiation of the programme of cell differentiation. At present, the best available evidence implicates either G_1 or very early S or perhaps both phases in this process.

REFERENCES

AXELRAD, A. A., McLEOD, D. L., SUZUKI, S. & SHREEVE, M. M. (1978). Regulation of the population size of erythropoietic progenitor cells. In *Differentiation of Normal and Neoplastic Hematopoietic Cells*, ed. B. Clarkson, P. A. Marks & J. E. Till, pp. 155–63. New York: Cold Spring Harbor Laboratory.

BROWN, E. & SCHILDKRAUT, C. L. (1979). Perturbation of growth and differentiation of Friend murine erythroleukemia cells by 5-bromodeoxyuridine incorporation in early S phase. *Journal of Cellular Physiology*, **99**, 261–78.

CANTOR, L. N., MORRIS, A. J., MARKS, P. A. & RIFKIND, R. A. (1972). Purification of the erythropoietin-responsive cell by immune hemolysis. *Proceedings of the National Academy of Sciences, USA,* **69,** 1337–41.

CHUI, D., DJALDETTI, M., MARKS, P. A. & RIFKIND, R. A. (1971). Erythropoietin effects on fetal mouse erythroid cells. I. Cell population and hemoglobin synthesis. *Journal of Cell Biology,* **51,** 585–95.

COOPER, M. C., LEVY, J., CANTOR, L. N., MARKS, P. A. & RIFKIND, R. A. (1974). The effect of erythropoietin on colonial growth of erythroid precursor cells *in vitro. Proceedings of the National Academy of Sciences, USA,* **71,** 1677–80.

CORMACK, D. (1976). Time lapse characterization of erythrocytic colony-forming cells in plasma cultures. *Experimental Hematology,* **4,** 319–27.

DJALDETTI, M., PREISLER, H., MARKS, P. A. & RIFKIND, R. A. (1972). Erythropoietin effects on fetal mouse erythroid cells. II. Nucleic acid synthesis and the erythropoietin-sensitive cell. *Journal of Biological Chemistry,* **247,** 731–5.

EAVES, C. J., HUMPHRIES, R. K. & EAVES, A. C. (1979). In vitro characterization of erythroid precursor cells and the erythropoietic differentiation process. In *Cellular and Molecular Regulation of Hemoglobin Switching,* ed. G. Stamatoyannopoulos & A. W. Nienhuis, pp. 251–78. New York: Grune & Stratton.

FIBACH, E., REUBEN, R. C., RIFKIND, R. A. & MARKS, P. A. (1977). Effect of hexamethylene bisacetamide on the commitment to differentiation of murine erythroleukemia cells. *Cancer Research,* **37,** 440–4.

GAMBARI, R., MARKS, P. A. & RIFKIND, R. A. (1979). Murine erythroleukemia cell differentiation: Relationship of globin gene expression and of prolongation of G_1 to inducer effects during G_1/early S. *Proceedings of the National Academy of Sciences, USA,* **76,** 4511–15.

GAMBARI, R., TERADA, M., BANK, A., RIFKIND, R. A. & MARKS, P. A. (1978). Synthesis of globin mRNA in relation to the cell cycle during induced murine erythroleukemia differentiation. *Proceedings of the National Academy of Sciences, USA,* **75,** 3801–4.

GAZITT, Y., REUBEN, R. C., DEITCH, A. D., MARKS, P. A. & RIFKIND, R. A. (1978). Changes in cyclic adenosine 3′, 5′-monophosphate levels during induction of differentiation in murine erythroleukemia cells. *Cancer Research,* **38,** 3779–83.

GELLER, R., LEVENSON, R. & HOUSMAN, D. (1978). Significance of the cell cycle in commitment of murine erythroleukemia cells to erythroid differentiation. *Journal of Cellular Physiology,* **95,** 213–22.

GUSELLA, J., GELLER, R., CLARKE, B., WEEKS, V. & HOUSMAN, D. (1976). Commitment to erythroid differentiation by Friend erythroid leukemia cells: A stochastic analysis. *Cell,* **9,** 221–9.

HARRISON, P. R. (1976). Analysis of erythropoiesis at the molecular level. *Nature,* **262,** 353–6.

KRETCHMAR, A. L. (1966). Erythropoietin: Hypothesis of action tested by analog computer. *Science,* **152,** 367–70.

LEDER, A., ORKIN, S. & LEDER, P. (1975). Differentiation of erythroleukemic cells in the presence of inhibitors of DNA synthesis. *Science,* **190,** 893–4.

LEHNERT, S. (1979). Changes in morphology and cell cycle traverse induced in CHO cells by methyl isobutyl xanthine. *Experimental Cell Research,* **121,** 383–94.

LEVENSON, R., KERNEN, J., MITRANI, A. & HOUSMAN, D. (1980). DNA synthesis is not required for the commitment of murine erythroleukemia cells. *Developmental Biology*, **74**, 224–30.

LEVY, J., TERADA, M., RIFKIND, R. A. & MARKS, P. A. (1975). Induction of erythroid differentiation by dimethylsulfoxide in cells infected with Friend virus: Relationship to the cell cycle. *Proceedings of the National Academy of Sciences, USA*, **72**, 28–32.

McCLINTOCK, P. R. & PAPACONSTANTINOU, J. (1974). Regulation of hemoglobin synthesis in a murine erythroblastic leukemic cell: The requirement for replication to induce hemoglobin synthesis. *Proceedings of the National Academy of Sciences, USA*, **71**, 4551–5.

MARKS, P. A. & RIFKIND, R. A. (1972). Protein synthesis: Its control in erythropoiesis. *Science*, **175**, 955–61.

MARKS, P. A. & RIFKIND, R. A. (1978). Erythroleukemic differentiation. *Annual Review of Biochemistry*, **47**, 419–48.

MARKS, P. A., RIFKIND, R. A., BANK, A., TERADA, M., MANIATIS, G. M., REUBEN, R. C., & FIBACH, E. (1977). Erythroid differentiation and the cell cycle. In *Growth Kinetics and Biochemical Regulation of Normal and Malignant Cells*, ed. B. Drewinko & R. M. Humphrey, pp. 329–45. Baltimore: Williams & Wilkins.

MONETTE, F. C., KENT, R. B., WEINER, E. J., JARRIS, R. F. JR., OUELLETTE, P. L., THORSON, J. A. & ZELICK, R. D. (1980). Cell-cycle properties and proliferation kinetics of late erythroid progenitors in murine bone marrow. *Experimental Hematology*, **8**, 484–93.

NECAS, E. & NEUWIRT, J. (1976). Study of the effects of hydroxyurea on the erythropoietin-sensitive cells. *Cell Tissue Kinetics*, **9**, 257–66.

PARKER, C. L. & HOOPER, W. C. (1978). Induction of hemoglobin synthesis in dimethylsulfoxide-treated Friend erythroleukemia cells grown in the presence of cytochalasin B. *Leukemia Research*, **2**, 295–303.

RAMIREZ, F., GAMBINO, R., MANIATIS, G. M., RIFKIND, R. A., MARKS, P. A. & BANK, A. (1975). Changes in globin mRNA content during erythroid cell differentiation. *Journal of Biological Chemistry*, **250**, 6054–8.

RUTTER, W. J., PICTET, R. L. & MORRIS, P. W. (1973). Toward molecular mechanisms of developmental processes. *Annual Review of Biochemistry*, **42**, 601–46.

SCHOOLEY, J. C. (1965). Responsiveness of hematopoietic tissue to erythropoietin in relation to the time of administration and duration of action of the hormone. *Blood*, **25**, 795–808.

SINGER, D., COOPER, M., MANIATIS, G. M., MARKS, P. A. & RIFKIND, R. A. (1974). Erythropoietic differentiation in colonies of cells transformed by Friend virus. *Proceedings of the National Academy of Sciences, USA*, **71**, 2668–70.

STEPHENSON, J. R., AXELRAD, A. A., McLEOD, D. L. & SHREEVE, M. M. (1971). Induction of colonies of hemoglobin-synthesizing cells by erythropoietin in vitro. *Proceedings of the National Academy of Sciences, USA*, **68**, 1542–6.

TABUSE, Y., KAWAMURA, M. & FURUSAWA, M. (1976). Induction of haemoglobin synthesis in Friend leukemia cells without the necessity of mitosis. *Differentiation*, **108**, 1–5.

TERADA, M., FRIED, J., NUDEL, U., RIFKIND, R. A. & MARKS, P. A. (1977). Transient inhibition of initiation of S-phase associated with dimethylsulfoxide induction of murine erythroleukemia cells to erythroid differentiation. *Proceedings of the*

National Academy of Sciences, USA, **74**, 248–52.

YAMASAKI, H., FIBACH, E., NUDEL, Y., WEINSTEIN, I. B., RIFKIND, R. A. & MARKS, P. A. (1977). Tumor promoters inhibit spontaneous and induced differentiation of murine erythroleukemia cells in culture. *Proceedings of the National Academy of Sciences, USA*, **74**, 3451–5.

Diversification within embryonic chick somites: *in vitro* analysis of proteoglycan synthesis*

JAMES W. LASH AND CLARISSA M. CHENEY

JWL: Department of Anatomy, School of Medicine/G3, University of Pennsylvania,
Philadelphia, Pennsylvania 19104 USA.
CMC: Department of Biology, The Johns Hopkins University, Baltimore, Maryland 21218 USA.

INTRODUCTION

There has been a steady and growing interest in somite development in recent years. Most investigations have been directed toward the mechanism of somitogenesis whereas diversification within the somite has received much less attention. Initially somites appear as epithelial vesicles. These somitic vesicles are formed from the unsegmented segmental plate by mechanisms poorly understood. It is this transition from an unsegmented mesoderm to an orderly array of bilateral metameric somites that has excited the interest of many investigators in past years. The subsequent differentiation of the somite into a lateral dermatome and myotome, and a medial sclerotome, is no less exciting, but less amenable to experimental analysis. The recent work of Solursh, Fisher, Meier & Singley (1979) suggests that diversification within the somite may be regulated by adjacent tissues, and that the extracellular matrix plays an important role.

It has been shown that the embryonic somites respond to the matrix of the perinotochordal sheath (Kosher & Lash, 1975) by synthesising cartilage matrix. Extracellular matrix materials extracted from cartilage (Kosher, Lash & Minor, 1973) and artificial matrices composed of collagens and/or proteoglycans (Lash & Vasan, 1978) evoke a similar response. The response was measured by the synthesis of cartilage matrix and its attendant macromolecules, particularly proteoglycans. Proteoglycans are ubiquitous glycoconjugates that have been characterised best in cartilage, but have also been reported in cornea (Hassel, Newsome & Hascall, 1979), notochord (Lash & Vasan, 1978), skin (Damle, Kieras, Tzeng & Gregory, 1979), ovarian follicular fluid (Yanagishita, Rodbard & Hascall, 1979), and basement membrane (Hassel *et al.*, 1980). The proportion of carbohydrate to protein varies

* Supported by NIH grant HD 00380 to JWL.

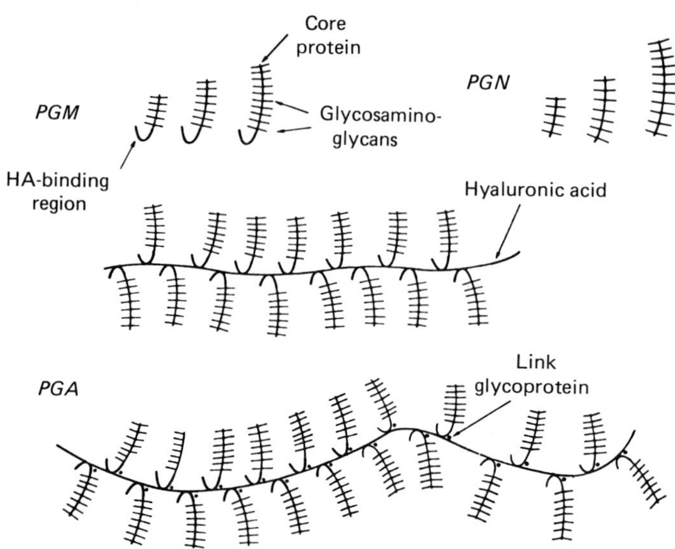

Fig. 1. Diagram of proteoglycan heterogeneity. PGM, proteoglycan monomers, consisting of core protein with hyaluronic acid-binding region and glycosaminoglycans. PGN, proteoglycan not capable of aggregating because HA-binding region is absent. PGA, two sizes of proteoglycan aggregates are shown. The monomers associate with hyaluronic acid to form large macromolecules. Link glycoproteins stabilise the association of the HA-binding region with hyaluronic acid. (Diagram adapted from Cheney & Lash, 1981.)

between tissues, with cartilage proteoglycans having the largest proportion of carbohydrate (90%) to protein (10%). These proteoglycans exist in differentiating cartilage as a very heterogeneous group of molecules with respect to size. The largest molecules are the proteoglycan aggregates (PGA), which are composed of several components. Glycosaminoglycans (primarily chondroitin sulphates) are covalently bound to core proteins to form proteoglycan monomers (PGM). These core proteins have a specific hyaluronic acid-binding region which interacts with hyaluronic acid. These associated molecules (many PGMs for each molecule of hyaluronic acid) constitute the macromolecular PGAs. Although not essential for the interaction, link glycoproteins are known to stabilise the molecular association. Certain non-associated proteoglycans (PGN) lack the HA-binding region of the core protein, and are incapable of interacting with hyaluronic acid. Fig. 1 portrays the heterogeneity of the various proteoglycan molecules.

It is now becoming clear that during chondrogenic differentiation proteoglycans undergo predictable changes in their pattern of complexity, as analysed by molecular sieve chromatography or density gradient centrifugation (Royal & Goetinck, 1977; De Luca *et al.*, 1977; Lash & Vasan, 1978; Ovadia, Parker & Lash, 1980). Precartilaginous tissues synthesise a small proteoglycan which is not capable of binding to hyaluronic acid. During differentiation larger proteoglycans appear until there is eventually a transition to the predominant form of large proteoglycan aggregates and a varying, but smaller proportion of intermediate size proteoglycans. The very small proteoglycans are no longer detectable in differentiated cartilage. Thus, the patterns of proteoglycan synthesised by tissues can be used as an index of their ability to form cartilage. These changing patterns have been shown to be consistent for all cartilages examined in embryonic chicks (Ovadia *et al.*, 1980), including somites (Lash & Vasan, 1978).

In previous studies on somite chondrogenesis there were suggestions from electron micrographs (Minor, 1973) and DNA determinations (Gordon & Lash, 1974) that there is heterogeneity within the somite with respect to cells responding to chondrogenic stimulation. The precartilaginous sclerotome exhibits more cell death *in vitro* than does the adjacent dermamyotome. In addition to having a 'rescue' effect upon sclerotomal cell death, the notochord also stimulates the cells to synthesise cartilage matrix. These results were obtained using stage 17–18 chick somites that were undergoing *in vitro* differentiation into sclerotomal and dermamyotomal tissues (Gordon & Lash, 1974).

To obtain more direct evidence of this seemingly differential response to chondrogenic factors, we took advantage of the fact that there is a narrow time interval (stage $18\frac{1}{2}$–19 see Hamburger & Hamilton, 1951) when the dermamyotome can be surgically isolated from the sclerotome, and the two tissues can be analysed separately *in vitro*. The response of these separate tissues to the inductive influence of the notochord was analysed by determining their ability to form cartilage, and characterising the types of proteoglycans synthesised. Previously, it had been shown that only the sclerotome forms cartilage nodules, and that the sclerotome incorporated more sulphate into sulphated glycosaminoglycans and sulphated proteoglycans (Cheney & Lash, 1981). It was also noted that sclerotomal tissue synthesised large proteoglycan aggregates and intermediate size proteoglycans, whereas the dermamyotome synthesised only intermediate proteoglycans. A characteristic feature of cartilage proteoglycans is that they have a hyaluronic acid-binding region, and are capable of associating with hyaluronic acid to form large macromolecular aggregates. Thus, we set out to determine whether the intermediate-size proteoglycans synthesised by the sclerotome were different from those synthesised by the dermamyotome. With the assays applicable to such small amounts of radioactively labelled molecules, we sought to determine whether morphological diversification within the somite could be correlated with the molecular products synthesised.

MATERIALS AND METHODS

Tissue and method of culture

After removing the extraembryonic membranes, head, tail, and flank tissues of stage $18\frac{1}{2}$–19 (staging series of Hamburger & Hamilton, 1951) chick embryos, the remaining trunk was briefly trypsinised before peeling off the epidermis. Using microdissection knives, the dermamyotomes were peeled back, cut off, and collected in a separate dish of sterile saline. The sclerotomes were then cut away from the spinal cord and collected in a separate dish. After all of the somites that could be separated in this manner were removed, the remaining somites in the posterior portion of the embryo were cut away and also collected. The last dissection procedure was to cut away the notochord, removing as many adhering sclerotomal cells as possible. In all instances there was an unavoidable 'contamination' of both the notochord and dermamyotome with a few sclerotome cells. Although it was impossible to routinely remove all sclerotome cells from the dermamyotome, in one experiment dermamyotomes were meticulously cleaned before culturing them with notochords, also free of sclerotome cells. In this instance there was very little incorporation of radioactive sulphate into sulphated glycosaminoglycans, and no instances of small cartilage nodules in the explants. Routinely, dermamyotomes with a few adhering sclerotome cells form occasional small cartilage nodules. For this reason we concluded that only sclerotome cells form cartilage (see Cheney & Lash, 1981).

The explants were placed upon Nuclepore filters over nutrient medium containing radioactive sulphate, as previously described (Lash & Vasan, 1978; Cheney & Lash, 1981).

Proteoglycan analyses

The explants were cultured in the presence of radioactive sulphate and extracted with a guanidine hydrochloride (GuHCl) solution, as described in Lash & Vasan (1978). The molecular size of the proteoglycans was determined by molecular sieve chromatography using controlled-pore glass beads (CPG-10-2500, mean pore diameter 257.3 nm) as described by Lever & Goetinck (1976), and modified by Lash& Vasan (1978). The extracted proteoglycans were lyophilised, dissolved in 4.0 M GuHCl and dialysed overnight against 0.5 M NaCl. The dialysed material was added to the CPG column and eluted at 4 °C with the eluant solution (0.5 M NaCl). The eluate was collected in 0.7 ml fractions, and transferred to scintillation vials for radioactivity determinations. Determinations were made in a Model 4200 Intertechnique Scintillation counter which was programmed to present the data in bar graph form (IN/US Service Corporation, Fairfield, New Jersey).

Assays for hyaluronic acid binding

The methodology and terminology are derived from the reports of Heinegard (1972), Faltz *et al.* (1979) and Vasan & Lash (1979). Radioactively labelled proteoglycans were extracted from the tissues under conditions which cause dissociation of the proteoglycan monomers from hyaluronic acid (4.0 M GuHCl). These proteoglycans were centrifuged in a caesium chloride gradient under conditions which cause reassociation of the proteoglycans (0.5 M GuHCl). The bottom one-fourth of the tube (the A1, or associated fraction) was collected, dialysed and lyophilised. These purified proteoglycan aggregates (A1) were then centrifuged as above in a caesium chloride gradient under conditions which cause the aggregates to dissociate (4.0 M GuHCl). The bottom two-fifths of the tube was collected as purified proteoglycan monomer fractions (A1-D1). After dialysis and lyophilisation, the A1-D1 fraction was dissolved in 2.5 ml of 4.0 M GuHCl. To this was added 10 ul of hyaluronic acid (stock solution of 1 mg/ml in deionised water) and 0.25 ml of a non-radioactive proteoglycan monomer (250 mg in 4.0 M GuHCl) prepared from 13-day embryonic chick sterna. The mixtures of A1-D1 fractions and hyaluronic acid were mixed on a Vortex and left at room temperature for 1 h. They were then exhaustively dialysed against 0.5 M NaCl at 4 °C, conditions which facilitate the interaction of the proteoglycans with hyaluronic acid (Hardingham & Muir, 1972).

The samples were chromatographed on CPG-10-2500 and the eluted fractions were assayed as described above.

RESULTS AND DISCUSSION

The formation and differentiation of somites is a complex series of events, most of them still poorly understood. The somite's existence begins as it separates from the anterior portion of the segmental plate. Recent evidence has been presented indicating that there is an internal meristic pattern within the segmental plate which precedes segmentation (Meier, 1979). The process of somitogenesis has excited many embryologists in the past, and currently active research offers many promising leads to our understanding of somite formation.

The nascent somite is an epithelial vesicle containing a variable number of core cells. The more posterior forming somites contain more core cells than the anterior somites (Williams, 1910). This epithelial somite subsequently transforms into a structure containing at least three distinct cell types; the chondrogenic sclerotome, the myogenic myotome, and the dermatome (which will migrate to the ectodermal epithelium to contribute to the epidermis). The latter two cell types are tightly joined together as a bilayered dermamyotome (see Fig. 2). The entry of migrating neural crest cells contributes to the cellular heterogeneity of the somite.

It is well established that the somites will respond both *in vivo* and *in vitro* to the inductive influence of the embryonic spinal cord and notochord by undergoing chondrogenesis (see Lash, 1968; Hall, 1978). Previous studies (Lash, 1967) had shown that the posterior, epithelial somites have a greater propensity for *in vitro* chondrification than the more anterior, diversified somites from the same embryo. This suggests that the entire epithelial somite is capable of forming cartilage whereas in the differentiated somite, as we will demonstrate, only a portion has chondrogenic capabilities.

Except for the work of Solursh *et al.* (1979), suggesting that the differentiation of a somite is related to its position with regard to the axial structures, very little is known about the factors influencing somite differentiation. One area of somite differentiation that has been studied extensively is that of somite chondrogenesis. It is now clear that the matrix products of the embryonic notochord (primarily proteoglycans and collagens) stimulate the somites to synthesise cartilage matrix, which also consists primarily of proteoglycans and collagens

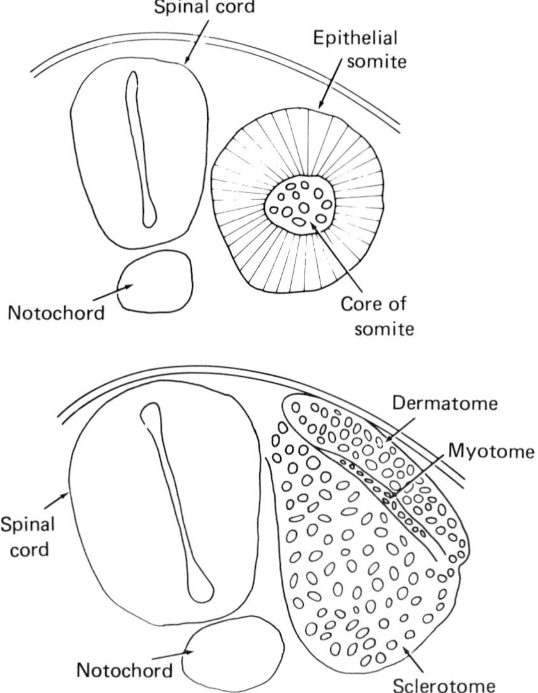

Fig. 2. Drawings of epithelial somite (top) and a somite that has undergone further diversification (bottom).

(Kosher *et al.*, 1973; Lash & Vasan, 1978). This is an interesting instance of a seemingly 'homotypic' induction, where exogenous molecules (proteoglycans and collagens) stimulate the somites to synthesise the same, or similar, type molecules. Although proteoglycans exist in cartilage as a heterogeneous group of molecules, the predominant form in mature embryonic cartilage is the large proteoglycan aggregate, and exogenous aggregates are more effective in stimulating chondrogenesis than smaller proteoglycans (Lash, 1976; Lash & Vasan, 1978; Belsky, Vasan & Lash, 1980).

In experiments testing the whole somite's response to chondrogenic induction, there were suggestions from microscopic analyses (Minor, 1973) and DNA measurements (Gordon & Lash, 1974) that the sclerotome responded differently to both the *in vitro* conditions and to the influence of the notochord. There was more cell death in the cultured sclerotomal portion of the somite, and this cell death was alleviated by the presence of the notochord.

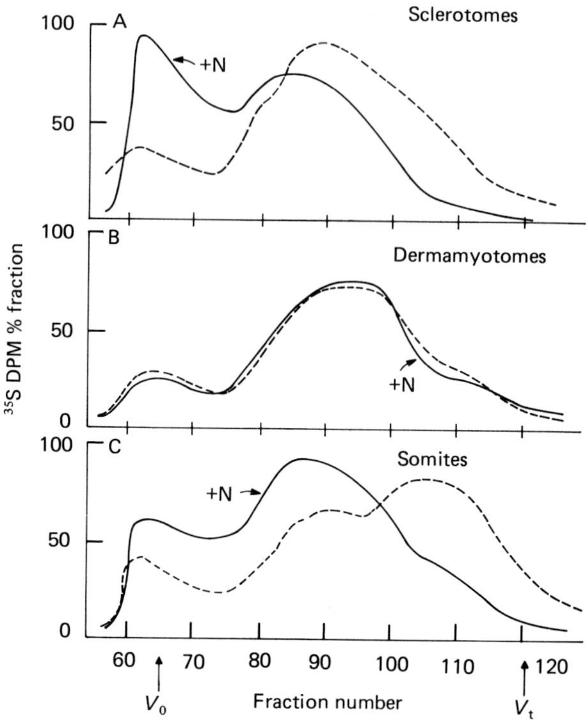

Fig. 3. Molecular size distribution of proteoglycans. Molecular sieve chromatography of proteoglycan extracts was performed on controlled-pore glass beads. Scale is derived from the total disintegrations per minute (DPM) of ^{35}S in the fraction. The fraction with the maximum counts is normalised to 100%, and all other fractions are expressed as percentage maximum counts. Proteoglycans which elute near the void volume mark (V_0) represent proteoglycan aggregates. V_t marks the total volume. (A) Sclerotomes (dotted line), sclerotomes + notochord (solid line). (B) Dermamyotomes (dotted line), dermamyotomes + notochord (solid line). (C) Somites (dotted line), somites + notochord (solid line).

In an effort to test more precisely the behaviour of the sclerotome and the dermamyotome, these tissues were isolated as described previously (Materials and Methods), and cultured with or without the inclusion of notochordal tissue. As seen in Fig. 3, the molecular size of the proteoglycans synthesised by these tissues can be an indication of their chondrogenic potential. There are three main classes of proteoglycans synthesised by these tissues. Large proteoglycan aggregates, eluting from the column in the void volume (V_0). The sclerotome synthesises a greater proportion of aggregates than either the dermamyotome or the complete somite (cf. Fig. 3A). All three types of explants synthesise an intermediate size molecule, which is the predominant form in the dermamyotome (Fig. 3B). The small proteoglycans that elute just before the total volume (V_t) are most prominent in the extract from whole somites (Fig. 3C, dotted line).

In interpreting the chondrogenic potential of these tissues by analysing the types of proteoglycans synthesised (dotted lines in Fig. 3), it is seen that the sclerotomes and dermamyotomes synthesise equivalent populations of proteoglycans with respect to size. Both synthesise a small proportion of aggregates (shown at V_0), and a predominating intermediate size proteoglycan. As mentioned in the Materials and Methods, the small proportion of aggregates synthesised by the dermamyotome are undoubtedly made by the slight amount of sclerotomal tissue adhering to the dermamyotomes. Extracts from the entire somites clearly show three size classes, a small proportion of aggregates, a sizeable amount of intermediates, and a predominating amount of small proteoglycans eluting just before the total volume (V_t). This interpretation emphasises an important fact when trying to correlate chondrogenic potential with only the molecular size of the proteoglycans. There is no obvious difference between the size of the proteoglycans extracted from the dermamyotomes and the sclerotomes, yet it is known that the sclerotomes form cartilage *in vitro* whereas the dermamyotomes do not (Cheney & Lash, 1981).

The tissue's response to chondrogenic induction by the notochord is more significant with respect to the changing patterns of proteoglycans synthesised. It is readily seen that the notochord stimulates the synthesis of larger proteoglycans in both the sclerotomes and whole somites (Fig. 3A, 3C, solid line), but not in the dermamyotome (3B). This suggests that the dermamyotomes either synthesise a different type of proteoglycan, or that the tissues do not respond to the notochord. Clearly, the notochord stimulates the synthesis of aggregates in isolated sclerotomes and (presumably) in the sclerotomal portion of the whole somites.

Since a cardinal feature of cartilage proteoglycans is the ability of the proteoglycan monomers to form large aggregates by binding to hyaluronic acid via the HA-binding region of the core protein (cf. Fig. 1), a further test of the chondrogenic potential of these tissues is to examine the ability of the purified monomers from these tissues to bind with exogenous hyaluronic acid. In Fig. 4 experiments are portrayed where the A1-D1 monomers (see Materials and Methods) from sclerotomes and dermamyotomes were tested for their ability to associate with

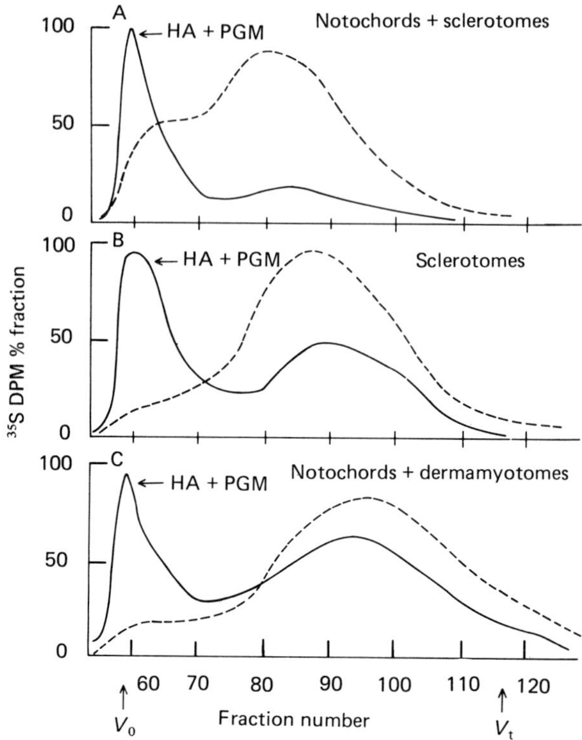

Fig. 4. Molecular size distribution (as in Fig. 3) of A1-D1 proteoglycan monomers before (dotted lines) and after (solid lines) the addition of hyaluronic acid + PGM (see Materials and Methods on hyaluronic acid-binding). (A) Notochords + sclerotomes. (B) Sclerotomes. (C) Notochords + dermamyotomes. The A1-D1 monomers with HA-binding regions associate with hyaluronic acid and elute as large molecules near the void volume (V_0).

hyaluronic acid. A large proportion of the A1-D1 monomers from the sclerotomes are capable of binding to hyaluronic acid (Fig. 4B), indicating the chondrogenic bias of these tissues. In Fig. 4A it is evident that the notochord stimulates the synthesis of even more molecules capable of binding to hyaluronic acid. The proportion of aggregates in this instance is comparable to the molecular size profiles obtained from mature embryonic cartilage (cf. Lash, Ovadia & Vasan, 1978; Ovadia *et al.*, 1980). The contaminating sclerotomal cells in the dermamyotomes complicate the interpretation of Fig. 4C, but assuming the small proportion of aggregates formed are synthesised by the sclerotomal cells, the proteoglycans synthesised by the dermamyotomes are not capable of binding to hyaluronic acid. This agrees with the previous interpretation, viz., that the dermamyotomes synthesise a different proteoglycan than the sclerotomes. One clear difference is the lack of a HA-binding region on the dermamyotome proteoglycans. Given the immense heterogeneity possible in these glycoconjugates, other differences in either the protein core structure or the glycosaminoglycan constitution may also be present. Thus, *in vitro* methods have enabled us to determine that the extracellular matrix products of the notochord can influence the differentiation of somites. Furthermore, the diversification within the somite with respect to morphologically distinct tissue types can be shown to have a molecular basis in the types of molecules synthesised. The chondrogenic potential of the sclerotome can be correlated with its ability to synthesise cartilage-type proteoglycans which are capable of binding to hyaluronic acid to create the large proteoglycan aggregate characteristic of cartilage tissue.

Although not mentioned in this report, there are other extracellular matrix components which undoubtedly effect or regulate somite differentiation. With respect to somite chondrogenesis, collagen type II (see Miller, 1977 for discussion of collagen types) has been implicated in the stimulation of somite chondrogenesis (Lash & Vasan, 1978). Another extracellular component known to affect many cells is fibronectin (see Yamada & Olden, 1978 for review). The addition of fibronectin to chondrocytes has been shown to alter their morphology, the type of collagen made, and decrease the size of the proteoglycans synthesised. The result of these perturbations is a more fibroblastic phenotype (Pennypacker, Hassel, Yamada & Pratt, 1978). Whether fibronectin plays a role in early development in general, or somite formation and differentiation in particular, is unknown. The few studies on fibronectin in early embryos indicate that it is associated with the separation of germ layers and the appearance of mesenchyme (Wartiovaara, Keivo & Vaheri, 1979). It is tempting to hypothesise that some of the complex events in somite formation and differentiation may be regulated by fibronectin as well as the other extracellular matrix components. Immunofluorescence studies have shown that fibronectin is located in the area surrounding the myotome region of 7-day chick embryos and their perinotochordal sheath of both 5-day chick embryos (Linder, Vaheri, Ruoslahti & Wartiovaara, 1975) and 3-day chick embryos (Cheney, Seitz & Lash, in preparation). Fibronectin has been reported in the somites of 2-day quail embryos (Loring, Erickson & Weston, 1977), and we have detected fibronectin in the notochord and somites of stage 9 and 15 chick embryos (Cheney, Seitz

& Lash, unpublished data). Although fibronectin has not been reported in differentiated cartilage (Linder *et al.*, 1975; Stenman & Vaheri, 1978), dissociated chondrocytes can synthesise fibronectin (Dessau *et al.*, 1978). It appears that chondrocytes might utilise an adhesion protein differing from fibronectin, which has been called chondronectin (Kleinman *et al.*, 1979; Hewitt, Kleinman, Pennypacker & Martin, 1980). Collagens are known to bind to fibronectin (Kleinman, McGoodwin & Klebe, 1976; Balian *et al.*, 1979), and an interaction between fibronectin and proteoglycans has been reported (Perkins, Ji & Hynes, 1979).

These findings on adhesion proteins, and our findings on the regulatory role of collagens and proteoglycans in somite differentiation strongly implicate extracellular matrix components either singly or in consort, as playing major roles in both somite formation and somite differentiation. The work of Meier (1979) on meristic patterns preceding somite formation, and that of Solursh *et al.* (1979) on somite differentiation support our contention that these events have a molecular mechanism, and the basis of the mechanism may be the microheterogeneities of the extracellular matrix.

REFERENCES

BALIAN, G., CLICK, E. M., CROUCH, E., DAVIDSON, J. M. & BORNSTEIN, P. (1979). Isolation of a collagen-binding fragment from fibronectin and cold-insoluble globin. *Journal of Biological Chemistry*, **254**, 1429–32.

BELSKY, E., VASAN, N. S. & LASH, J. W. (1980). Extracellular matrix components and somite chondrogenesis. A microscopic analysis. *Developmental Biology*, **79**, 159–80.

CHENEY, C. M. & LASH, J. W. (1981). Diversification within embryonic chick somites: Differential response to notochord. *Developmental Biology*, **81**, 288–98.

DAMLE, S. P., KIERAS, F. J., TZENG, W-K. & GREGORY, J. D. (1979). Isolation and characterization of proteochondroitin sulfate from pig skin. *Journal of Biological Chemistry*, **254**, 1614–20.

DE LUCA, S., HEINEGÅRD, D., HASCALL, V., KIMURA, J. H. & CAPLAN, A. L. (1977). Chemical and physical changes in proteoglycans during development of chick limb bud chondrocytes grown *in vitro*. *Journal of Biological Chemistry*, **252**, 6600–8.

DESSAU, W., SASSE, J., TIMPL, R., JILEK, F. & VON DER MARK, K. (1978). Synthesis and extracellular deposition of fibronectin in chondrocyte cultures. Response to the removal of extracellular cartilage matrix. *Journal of Cell Biology*, **79**, 342–55.

FALTZ, L. L., REDDI, A. H., HASCALL, G. K., MARTIN, D., PITA, J. C. & HASCALL, V. C. (1979). Characteristics of proteoglycans extracted from the swarm chondrosarcoma with associative solvents. *Journal of Biological Chemistry*, **254**, 1375–80.

GORDON, J. S. & LASH, J. W. (1974). *In vitro* chondrogenesis and differential cell viability. *Developmental Biology*, **36**, 88–104.

HALL, B. K. (1978). *Developmental and Cellular Skeletal Biology*, 304 pp. New York: Academic Press.

HAMBURGER, V. & HAMILTON, H. L. (1951). A series of normal stages in the development of the chick embryo. *Journal of Morphology*, **88**, 49–92.

HARDINGHAM, T. E. & MUIR, H. (1972). The specific interaction of hyaluronic acid with cartilage proteoglycans. *Biochimica et Biophysica Acta*, **279**, 401–5.

HASSEL, J. R., NEWSOME, D. A. & HASCALL, V. (1979). Characterization and biosynthesis of proteoglycans of corneal stroma from rhesus monkey. *Journal of Biological Chemistry*, **254**, 12346–54.

HASSEL, J. R., ROBEY, P. G., BARRACH, H-J., WILCZEK, J., RENNARD, S. I. & MARTIN, G. R. (1980). Isolation of a heparin sulfate-containing proteoglycan from basement membrane. *Proceedings of the National Academy of Sciences, USA*, **77**, 4494–8.

HEINEGÅRD, D. (1972). Extraction, fractionation and characterization of proteoglycans from bovine tracheal cartilage. *Biochimica et Biophysica Acta*, **285**, 181–92.

HEWITT, A. J., KLEINMAN, H. K., PENNYPACKER, J. P. & MARTIN, G. R. (1980). Identification of an adhesion factor for chondrocytes. *Proceedings of the National Academy of Sciences, USA*, **77**, 385–8.

KLEINMAN, H. K., HEWITT, A. T., MURRAY, J. C., LIOTTA, L. A., RENNARD, S. I., PENNYPACKER, J. P., McGOODWIN, E. B., MARTIN, G. R. & FISHMAN, P. H. (1979). Cellular and metabolic specificity in the interaction of adhesion proteins with collagens and with cells. *Journal of Supramolecular Structure*, **11**, 69–78.

KLEINMAN, H. K., McGOODWIN, E. B. & KLEBE, R. J. (1976). Localization of the cell attachment region in Types I and II collagens. *Biochemical and Biophysical Research Communications*, **72**, 426–32.

KOSHER, R. A. & LASH, J. W. (1975). Notochordal stimulation of *in vitro* somite chondrogenesis before and after enzymatic removal of perinotochordal materials. *Developmental Biology*, **42**, 362–78.

KOSHER, R. A., LASH, J. W. & MINOR, R. R. (1973). Environmental enhancement of *in vitro* chondrogenesis by exogenous chondromucoprotein. *Developmental Biology*, **35**, 210–20.

LASH, J. W. (1967). Differential behavior of anterior and posterior embryonic chick somites *in vitro*. *Journal of Experimental Zoology*, **165**, 47–56.

LASH, J. W. (1968). Somitic mesenchyme and its response to cartilage induction. In *Epithelial-Mesenchymal Interactions*, ed. R. Fleischmajer, pp. 165–72. Baltimore: Williams & Wilkins.

LASH, J. (1976). Extracellular matrix products and differentiation: Somite chondrogenesis. In *In Vitro Tests of Teratogenicity*, ed. J. D. Ebert & M. Marois, pp. 367–73. Amsterdam: North-Holland.

LASH, J. W., OVADIA, H. & VASAN, N. S. (1978). Microheterogeneities, non-equivalence and embryonic induction. *Medical Biology*, **56**, 333–8.

LASH, J. W. & VASAN, N. S. (1978). Somite chondrogenesis *in vitro*. Stimulation by exogenous extracellular matrix components. *Developmental Biology*, **66**, 151–71.

LEVER, P. L. & GOETINCK, P. F. (1976). Molecular sieve chromatography of proteoglycans: A comparative analysis. *Analytical Biochemistry*, **75**, 67–76.

LINDER, E., VAHERI, A., RUOSLAHTI, E. & WARTIOVAARA, J. (1975). Distribution of fibroblast surface antigen in the developing chick embryo. *Journal of Experimental Medicine*, **142**, 41–9.

LORING, J., ERICKSON, C. & WESTON, J. (1977). Surface proteins of neural crest, crest-derived and somite cells *in vitro*. *Journal of Cell Biology*, **75**, 71a.

MEIER, S. (1979). Development of the chick embryo mesoblast. Formation of the embryonic axis and establishment of the metameric pattern. *Developmental Biology*, **73**, 25–45.

MILLER, E. J. (1977). Biochemistry of the extracellular matrix. In *Cell and Tissue Interactions*, ed. J. W. Lash & M. M. Burger, pp. 71–86. New York: Raven Press.

MINOR, R. R. (1973). Somite chondrogenesis. A structural analysis. *Journal of Cell Biology*, **56**, 27–50.

OVADIA, M., PARKER, C. H. & LASH, J. W. (1980). Changing patterns of proteoglycan synthesis during chondrogenic differentiation. *Journal of Embryology and Experimental Morphology*, **56**, 59–70.

PENNYPACKER, J. P., HASSEL, J. R., YAMADA, K. & PRATT, R. M. (1978). The influence of fibronectin on chondrogenic expression. *In vitro*. *Journal of Cell Biology*, **79**, 149a.

PERKINS, M. E., JI, T. H. & HYNES, R. O. (1979). Cross-linking of fibronectin to sulfated proteoglycans at the cell surface. *Cell*, **16**, 941–52.

ROYAL, P. D. & GOETINCK, P. F. (1977). *In vitro* chondrogenesis in mouse limb mesenchymal cells: Changes in ultrastructure and proteoglycan synthesis. *Journal of Embryology and Experimental Morphology*, **39**, 79–95.

SOLURSH, M., FISHER, M., MEIER, S. & SINGLEY, C. T. (1979). The role of the extracellular matrix in the formation of the sclerotome. *Journal of Embryology and Experimental Morphology*, **54**, 75–98.

STENMAN, S. & VAHERI, A. (1978). Distribution of a major connective tissue protein, fibronectin, in normal human tissues. *Journal of Experimental Medicine*, **147**, 1054–64.

VASAN, N. S. & LASH, J. W. (1979). Monomeric and aggregate proteoglycans in the chondrogenic differentiation of embryonic chick limb buds. *Journal of Embryology and Experimental Morphology*, **49**, 47–59.

WARTIOVAARA, J., KEIVO, I. & VAHERI, A. (1979). Expression of the cell surface-associated glycoprotein, fibronectin, in the early mouse embryo. *Developmental Biology*, **69**, 247–57.

WILLIAMS, L. W. (1910). The somites of the chick. *American Journal of Anatomy*, **11**, 55–99.

YAMADA, K. M. & OLDEN, K. (1978). Fibronectins-adhesive glycoproteins of cell surface and blood. *Nature*, **275**, 179–84.

YANAGISHITA, M., RODBARD, D. & HASCALL, V. C. (1979). Isolation and characterization of proteoglycans from porcine ovarian follicular fluid. *Journal of Biological Chemistry*, **254**, 911–20.

Levels of complexity in limb-mesoderm cell culture systems

D. A. EDE

Department of Zoology, University of Glasgow, Developmental Biology Building, 124 Observatory Road, Glasgow G12 9LU

For the morphogeneticist the interest of the cell as a unit is the problem of how its activities are so regulated as to integrate its activity with that of several thousands of other cells to produce a complex developing embryonic structure. And the interest of *in vitro* culture is that some of the constraints which act upon cells in the embryo are relaxed or absent, allowing them to reveal more of their capacity to behave as semi-autonomous individuals than in their normal situation. This autonomy is never complete and even in the simplest cell cultures, so long as cell to cell interaction is not prevented, some sort of social organization is set up which leads to particular supracellular arrangements and ultimately to rudimentary structures of a distinct type, e.g. in the swirling patterns described by Elsdale & Foley (1969) in cultures of human lung fibroblasts. Moreover, the conditions of culture may be varied from such extremely simple ones to others in which more of the embryonic constraints are introduced, grading finally into *in vivo* culture situations where only some of these constraints are absent.

One thing which such a series may suggest is that while the cell certainly is a fundamental unit of development, it is not the only one, and that aggregates of cells may form more complex units which have their own organizational potentials, and these will then become involved in more complex units still. It is as well to remember, since embryonic development is not static, that these will be units of activity as well as of structure. The termination of one phase of activity will lead into the beginning of another.

THE DEVELOPMENT OF PATTERN AND SPATIAL ORGANIZATION IN THE LIMB BUD

The developing limb bud is an object of great interest and currently, especially in the chick embryo, of analytical observation and experiment. It is also full of surprises: for example, at one time, through a misinterpretation of various pathological phocomelias, it was thought to produce its distal structures first, but Saunders (1948) showed that its development proceeded from proximal base to distal tip; then it was thought to consist of a homogenous mesodermal population of identical mesenchyme cells, but Dienstman, Biehl, Holtzer & Holtzer (1974) showed that separate chondrogenic and myogenic cell lineages could be established *in vitro*; or again, it was thought that the mesenchyme cells were immovably packed alongside their original neighbours, but it is now known through the work of Christ, Jacob & Jacob (1977) and Chevallier, Kieny & Mauger (1977) that the myogenic progenitor cells migrate into the limb bud from the somites and are there somehow attracted, guided and spatially organized into their ultimate positions by the somatopleural cells.

The most striking developmental feature of the limb bud is the origin of the skeletal structures within it, first as cartilage rudiments, which later become transformed into bones. Each cartilage rudiment carries its own programme of further development from an early stage, such that if it is isolated in organ culture it will grow into a bone of characteristic size and shape – fibula, 4th digit, or whatever. These skeletal rudiments are arranged in the distinctive tetrapod form, though the original pentadactyl pattern is modified considerably in the avian embryo, particularly in the case of the wing. Nevertheless, a basic pattern is evident: a series of sets of more or less cylindrical cartilage rudiments arranged proximo-distally along the limb bud – one (humerus or femur) in the first set, two (radius/ulna or tibia/fibula) in the next set, more (wrist/ankles and digits) in the next sets. Though this system is the product of a long evolutionary history, Hinchliffe (1977) has shown by a careful analysis of the developing patterns of chondroitin sulphate synthesizing, i.e. chondrifying, areas, that the avian limb shows a fundamental resemblance to the skeletal patterns of the crossopterygian fish limb. This involves a series of successive bifurcations: in the bird embryo the first is at the humerus (femur) head, with two further bifurcations beyond the ulna (fibula), the first of which gives digits II and III, and the second digits IV and V. The initial tetrapod skeletal pattern appears to have been rather simple, and evolutionary modifications have occurred through alterations in the size and growth rates of the various rudiments.

POSITIONAL INFORMATION AND PRECARTILAGE CONDENSATIONS AS DEVELOPMENTAL UNITS

The limb bud provides one of the most convincing demonstrations of pattern determination through response of cells in an embryonic field to positional information, following Wolpert's (1969) hypothesis that there exists in embryos a special system of chemical gradients whose function is solely to supply each cell with a grid reference, i.e. to create a specific change in the cell state according to its position. This will lead through interpretation at the level of the cell's genetic machinery to subsequent cytodifferentiation in a manner appropriate to that position and thence to a spatial organization of the cells to form the structures of the embryo. Tickle, Summerbell & Wolpert (1975) have shown how determination of the antero-posterior sequence of digits in the chick wing is consistent with the hypothesis that a region at the posterior border of the wing bud, the zone of polarizing activity (ZPA), produces a diffusible morphogen which provides the gradient of positional information on which this particular pattern is based. The experiments on which those conclusions are based involve altering the form of the supposed gradient by grafting the ZPA into different sites and noting that the digits produced (II, III or IV) conform to the new calculated gradient level.

One of the questions not answered by such experiments is 'How fine-grained is the pattern so produced?' At the time of its determination the limit would be that each cell was assigned a different state, but the information given by the experiments is only that cells in certain regions, defined as the intervals between fixed gradient thresholds, contribute to the formation of rudiments which will give rise to particular digits. Yet these rudiments – the limb cartilages – have a characteristic internal structure from their first appearance in the mesenchyme as precartilage condensations. The question arises: does this internal structure reflect more fine-grained positional information within each rudiment, or does it indicate some other mode of pattern determination? In either case, this internal structure suggests that these condensations form developmental units in the sense described above. Furthermore, they are not only produced within the limb bud, but also arise in limb mesenchyme cells cultured *in vitro*, where certain aspects of their development can be better studied.

CHONDROGENESIS WITHIN THE LIMB BUD

The earliest description of chondrogenesis in the limb bud was by Fell (1925), who described the appearance of precartilage condensations in which the mesenchyme cells were more closely packed than in the surrounding regions. The limb bud mesenchyme (in the leg) up to

Hamilton and Hamburger Stage 22 consists of widely separated polymorphic cells, in contact with each other by fine cytoplasmic filopodia, but from this stage (at the base of the bud) electron microscope studies by Searls, Hilfer & Mirow (1972) and Thorogood & Hinchliffe (1975) show that cells within condensations are more rounded and have broader intercellular contacts. Some early EM investigations (Gould, Day & Wolpert, 1972; Gould, Selwood, Day & Wolpert, 1974) suggested that the increase in cell density observed in the light microscope was illusory, but subsequent analyses (Thorogood & Hinchliffe, 1975) have established that with careful attention to the fixatives employed its occurrence can be confirmed. Between stages 25 and 27 Fell & Canti (1935) noted an increase in metachromasis in the intercellular matrix, indicating cartilage differentiation.

A sequence of biochemical events has been established, accompanying these visible changes. At stage 22 chondroitin sulphate synthesis begins to be localized in the condensations, indicated by ^{35}S uptake (Searls, 1965). Fibronectin and type I collagen are distributed evenly through the intercellular spaces of the limb mesenchyme up to stage 22–23, but from then until stage 25 reach a maximum in the condensations (Dessau, Mark, Mark & Fischer, 1980). At stage 24 there is a decrease in the amount of hyaluronic acid and a parallel increase in hyaluronic acid in the prechondrogenic area (Toole, 1972). Cartilage differentiation is marked from stage 25 by the appearance of type II collagen (Linsenmayer, Toole & Trelstad, 1973) and synthesis of cartilage proteoglycan (Vasan & Lash, 1979). Fibronectin and type I collagen are both transient, appearing at their maximum at the stage of precartilage condensation and disappearing at the time of cartilage differentiation (Dessau *et al.*, 1980). Another transient event at the precartilage condensation stage is an increase in cyclic AMP levels (Elmer, Smith & Ede, in preparation).

CONDENSATIONS *IN VIVO* AND *IN VITRO*

In transverse sections the condensations show a characteristic concentric arrangement of the mesenchyme cells, first described by Anikin (1929) in amphibians. The pattern suggests in the first place that chondrogenesis is initiated centrally and spreads centrifugally. This is confirmed in observations of metachromasia in histological sections. In the second place it suggests the possibility that the condensation is produced by movement of cells towards the centre, perhaps a single or small group of cells. Certainly a similar pattern is produced in *Dictyostelium minutum*, a slime mould in which no streams are set up and in which the plasmodia converge on a central initiator cell. The movement in the limb bud mesenchyme would be of a much reduced order — more of a huddling together towards a central point — and it might not be an active movement, but result from contraction of the filopodia, beginning with those of the central cells, hauling the cells centripetally. An alternative suggestion (Gould *et al.*, 1974) is that the pattern is produced by centrifugal pressure, as the central cells produce cartilage matrix. There is no doubt that this is a factor in later stages, but there is evidence that the pattern appears before any matrix synthesis has occurred.

No direct observations can be made on cell movement within the limb bud, but condensations of an essentially similar type are produced in *in vitro* cultures of limb bud mesenchyme cells (Caplan, 1970), provided they are derived from buds older than stage 17 (Ahrens, Solursh & Reiter, 1977; Solursh, Ahrens & Reiter, 1978). Even at earlier stages the same authors have shown that addition of dibutyryl cAMP to the medium allows chondrogenesis to occur. The cultures must also have grown beyond confluence, so that two or three layers of cells are present on the plastic. At early stages cells from the peripheral areas of the limb bud which would not normally produce cartilage are capable of chondrogenesis, but after stage 25–26 this capacity is lost.

The visible events of chondrogenesis, the sequence of biochemical changes, and the production of an Anikin pattern as the condensations form in the culture, are essentially similar *in vitro* to the events in the limb bud. Many experiments have been made by altering the medium in which the cells are cultured. For example, the significance of the changing concentrations and distribution of hyaluronate in the limb bud has been investigated by Toole, Jackson & Gross (1972) who found that the presence of hyaluronate in the medium inhibits chondrogenesis in culture. These authors believe that cell to cell interaction is necessary for chondrocyte differentiation; the presence of hyaluronate molecules in the extracellular matrix prevents this occurring and only when this is eliminated through the local increase of hyaluronidase can chondrogenesis proceed.

CELL ADHESIVITY CHANGES IN CHONDROGENESIS

There is a great deal of evidence which supports the necessity for close contact between cells for the initiation of chondrogenesis, and the active movement of mesenchyme cells towards a central initiator cell, as proposed by Ede & Agerbak (1968), would bring this about. Drews & Drews (1973) studied condensation in *in vitro* culture with this possibility in mind, and in particular looked at the locomotory activity of cholinesterase-positive and cholinesterase-negative cells in cultures from stage 23–24 limb buds. They found that in a first phase of culture there was no cholinesterase activity, and all the mesenchyme cells behaved as fibroblasts, exhibiting contact inhibition behaviour and therefore moving away from any cell aggregates. In the second phase, ChE-positive cells appeared and detached themselves from the substrate, crawling onto neighbouring cells. In the centre of cell aggregates, the ChE-positive cells moved about very little, leading to over-layering of cells. In a third phase, the ChE-positive cells stopped moving and differentiated as chondrocytes, after which ChE production ceased. The transient appearance of ChE activity in some cells, which is also found *in vivo* (Drews & Drews, 1972), is therefore accompanied by changes in the adhesivity of the cells, which may contribute to the formation of condensations.

Other lines of evidence suggesting a connection between the precartilage condensation process and local adhesivity changes in the mesenchyme cells are numerous.

Thus in the mouse limb (Hewitt & Elmer, 1976; Hewitt & Elmer, 1978) and in the chick limb (Paulsen & Finch, 1977) chondrogenic progenitor cells exhibit a transient clustering of lectin-binding sites at the cell surface.

Again, the transient appearance of fibronectin reported by Dessau *et al.* (1980) is likely to be associated with changes in its interactions with collagen (Klebe, 1974) and therefore to its interactions with neighbouring cells. Differentiated chondrocytes do not normally produce fibronectin, but Dessau, Sasse, Timpl, Jilek & Mark (1978) have shown that dissociation of the cartilage matrix and release of chondrocytes in culture reinstates the synthesis of fibronectin, which comes to connect cells within condensations by short intercellular strands.

IS ACTIVE CELL MOVEMENT INVOLVED IN THE PRECARTILAGE CONDENSATION PROCESS?

Many strands of evidence suggest therefore that regulation of cell to cell adhesion is important in chondrogenesis and it may be that the increasing adhesion of cells to each other, plus the changing shape of differentiating chondrocytes, beginning at the condensation field centre, is sufficient to produce the characteristic Anikin pattern.

The alternative is that there is some active movement towards the centre, and here a combination of observations on the condensation process in *in vitro* culture and *in vivo* provides some clues.

In addition to the observations of Drews & Drews (1973), based on serial still photographs, time-lapse ciné observations by Ede, Flint, Wilby & Colquhoun (1977) confirm that in condensations in culture some cells are moving, often in a zig-zag path, centripetally.

Transmission electron microscope studies by Eguchi & Okada (1971) have been made on late condensations in culture which reveal the shape changes which occur as the chondrocytes become differentiated. A clearer picture of the changing form of the condensation and the shape of the cells which form it can be obtained from scanning electron microscope studies and I have carried these out in a series of condensations arising in cultures of dissociated mesenchyme cells from stage 25 chick embryos. Cells were cultured at a density of $10^5/20$ μl on 13 mm diameter glass coverslips in 35 mm Falcon plastic dishes. Each coverslip was covered with 20 μl of the suspension and left in the incubator for 2–3 h in a humidified atmosphere. 3 ml of medium (Ham's F12 + 10% FCS) was then introduced into the Falcon dish and the culture incubated for 1–6 days, with a medium change every 2 days.

The fixation procedure was as follows:
1. Wash medium off with Tyrode.
2. Place in 2% glutaraldehyde in 0.2 M cacodylate buffer for 1 h.
3. Wash well with 0.2 M cacodylate buffer.
4. Place in 2% osmium in cacodylate buffer for 1 h.
5. Wash well with 0.2 M cacodylate buffer.
6. Dehydrate through ascending grades of acetone – 15 minutes in each.
7. 2–3 washes in analar acetone.
8. Critical point dry using Polaron critical point drying unit.
9. Mount on stubs and coat with gold on Polaron sputter coater for 1 min.

Some cracking in the preparation is inevitable, since there is bound to be some shrinkage in the tissue, while the glass to which it is attached remains rigid, but sufficiently large areas remain undamaged to include complete condensations together with surrounding mesenchyme.

The SEM pictures show that in the earliest stage of prechondral condensation (Fig. 1) a clearly concentric arrangement of cells is to be seen, centred upon a few or often a single cell which is rounded up and shows no elongate filopodia but a number of bleb-like cytoplasmic

D. A. Ede

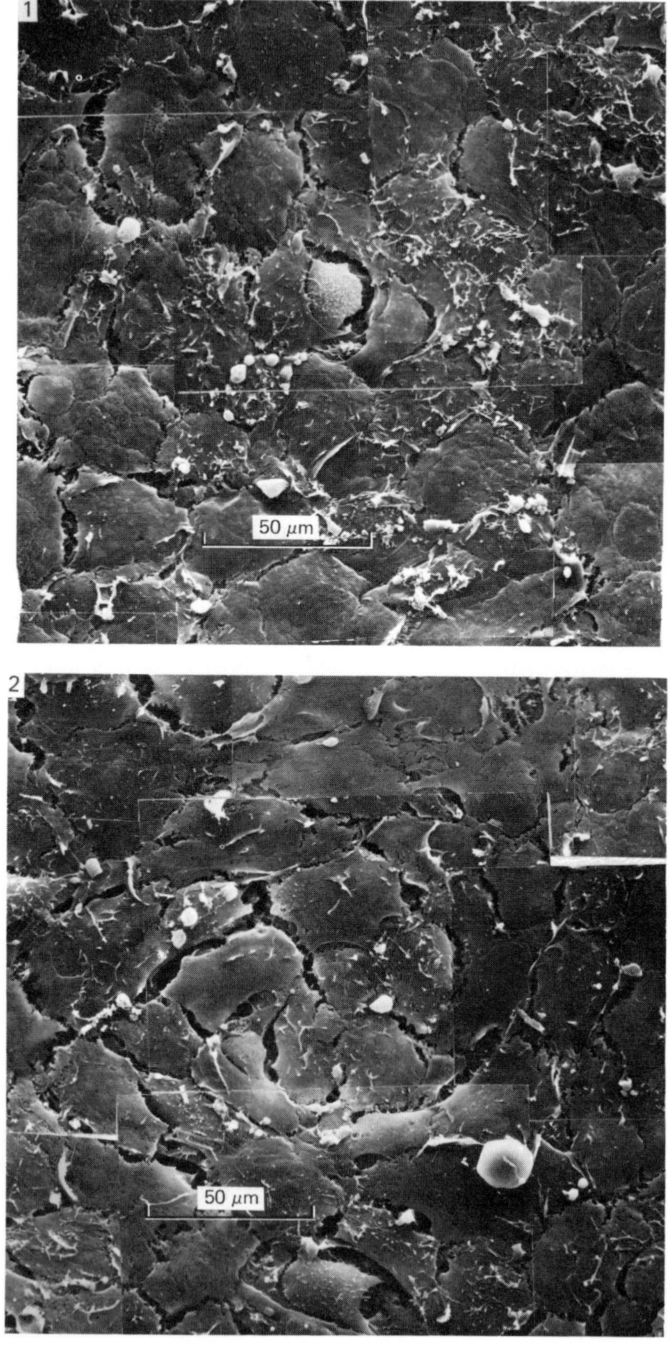

Figs. 1–4. SEM pictures of successive stages in the development of cartilage condensations in normal limb-mesenchyme cell cultures.

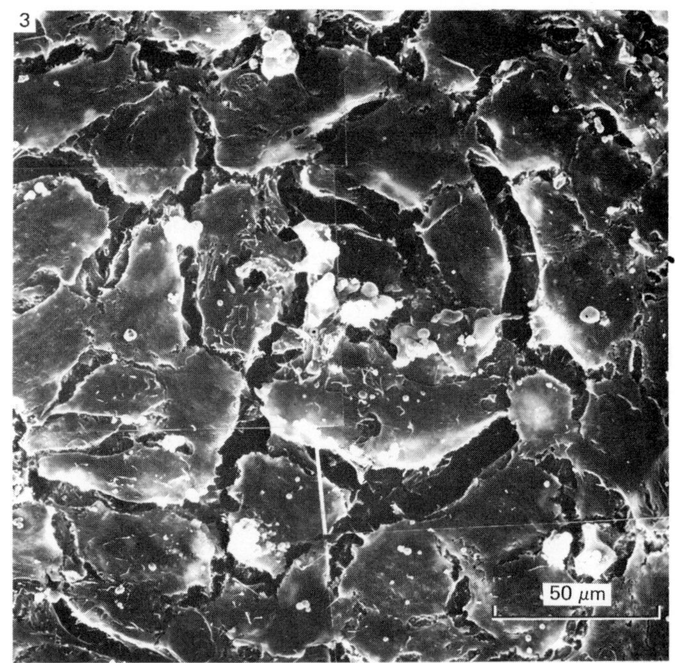

protrusions. There is no disproportionate intercellular space around this cell which would indicate that it had begun cartilage matrix secretion. The cells around it give the appearance of being in an active state, if not of motion, at least with many elongate filopodia reaching out to neighbouring cells. Their appearance is fibroblast-like and they show some orientation towards the central cell (Fig. 2). There follows a phase in which the central cells become separated by clear intercellular spaces (Fig. 3), indicating that cartilage matrix secretion has begun, and this cell separation spreads centrifugally. Finally, the more peripheral cells become stretched tangentially, retaining most filopodial contacts at the narrower ends of the cells (Fig. 4).

ORIENTATION OF CONDENSATION CELLS IN THE LIMB BUD

Some evidence that movement of mesenchyme cells in condensation formation is active rather than passive is found in analysis of Golgi apparatus orientation. The discovery by Trelstad (1977) that in cells of the developing vertebral cartilage the Golgi apparatus in the moving somatic cells was generally located in the trailing edge of the cell, provided an indicator of polarity in cells which otherwise showed no apparent orientation. Ede *et al.* (1977) and Holmes & Trelstad (1980) applied this technique, in which the Golgi apparatus is made visible by a silver-fixation technique, to the developing condensations in the limb bud. In a recent analysis (Ede & Wilby, 1980) of cells in the developing phalangeal condensation in stage 28 wing buds (Figs. 5, 6), three phases can be distinguished (Fig. 7). The first is a precondensation phase, in which there is no visible sign of a condensation but in which the angular distribution indicates that the Golgi is predominately orientated away from the centre, which may indicate movement towards the centre (Fig. 8a). In the second phase, at the condensation tip, a second population of cells is apparent in addition to the first, with the Golgi found primarily towards the centre, which suggests there may be some cell sorting and movement away from the condensation of non-chondrogenic cells in this phase (Fig. 8b). In the third phase, in which cartilage matrix secretion has clearly begun at the centre, Golgi orientation is predominantly tangential, randomly to either tangential pole. This analysis correlates well with the SEM observations on condensation-formation *in vitro*, with the first and second Golgi phases corresponding to the prematrix phase in the cultures. The third phase

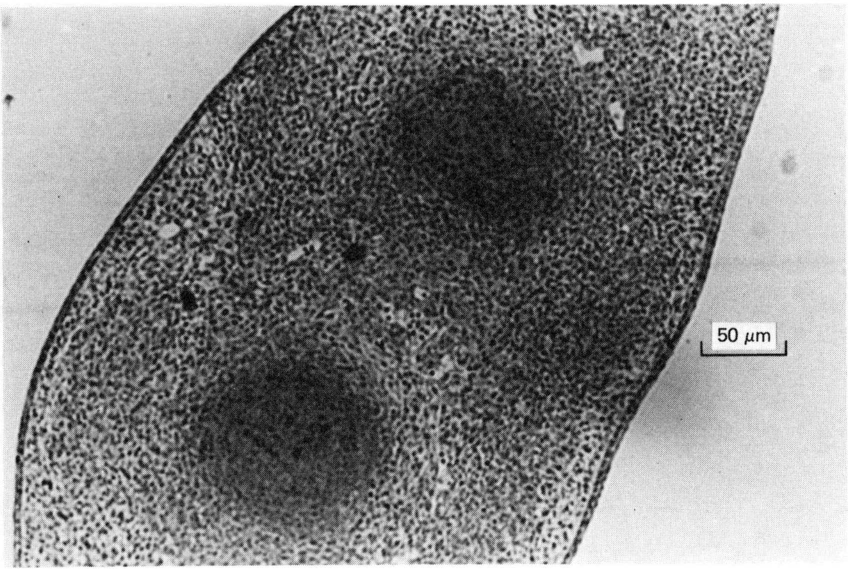

Fig. 5. TS stage 28 chick wing bud at level of metacarpal rudiments. 5 μm, stained alcian blue/chlorantine fast red.

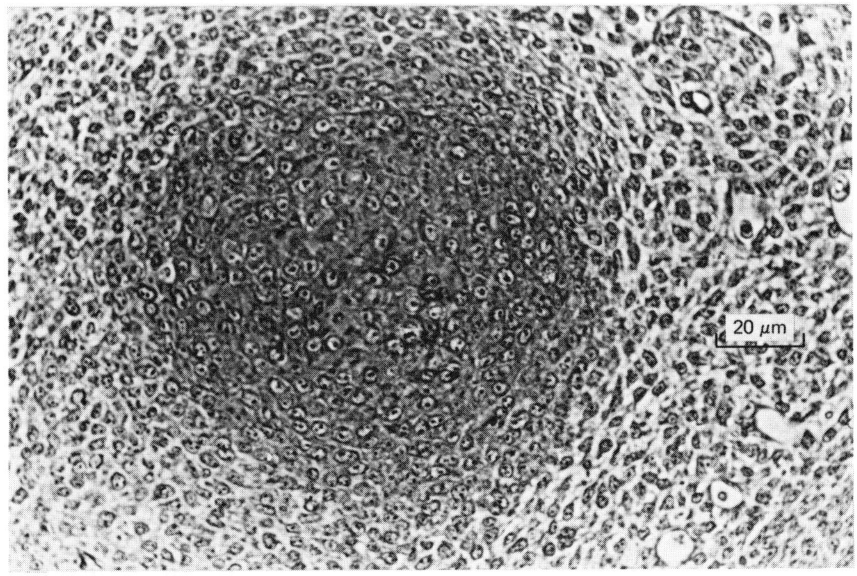

Fig. 6. Detail from Fig. 5, showing metacarpal rudiment at midcondensation stage.

D. A. Ede

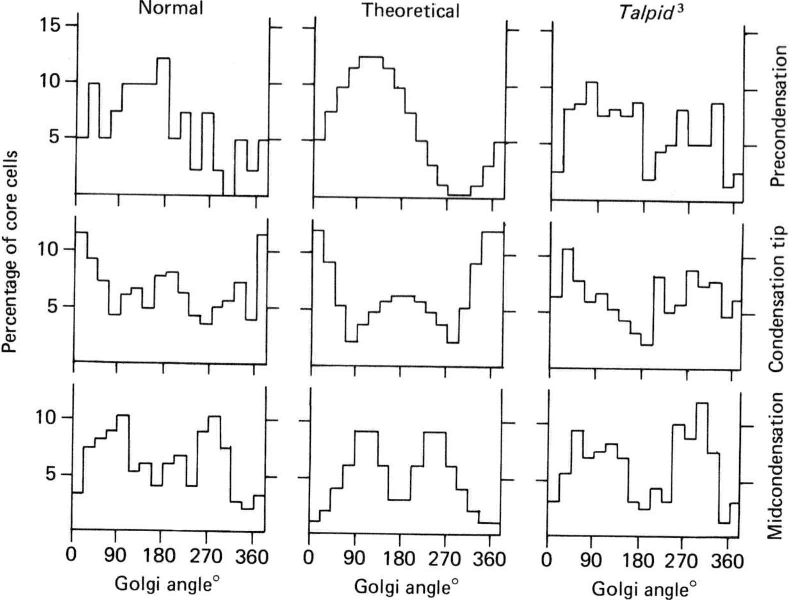

Fig. 7. Histograms showing distribution of Golgi angles in precondensation, condensation tip and midcondensation stage cartilage condensations in normal and *talpid*[3] mutant chick metacarpal rudiments, and corresponding theoretical distributions.

of Golgi orientation clearly parallels the phase in the cultured cells in which the cells towards the periphery of the condensation are being stretched tangentially and forced out centrifugally (Fig. 8*c*).

CONDENSATION IN THE *TALPID*[3] MUTANT *IN VITRO* AND IN THE LIMB BUD

In the *talpid*[3] lethal chick mutant embryo the limb bud is fan-shaped and the skeletal rudiments are polydactylous. In addition the condensations are not clearly separated, so that adjacent rudiments tend to be fused except at the tip of the bud where the digital rudiments are more widely separated.

The effect upon the condensation process can be seen clearly in cultivated *talpid*[3] mesenchyme cells. In normal cultures the substrate becomes covered with condensations, randomly but fairly evenly distributed. Secondary condensations often appear satellite-fashion around the first-formed ones (Figs. 9, 10). In *talpid*[3] cultures the condensation centres appear in the same fashion, but the condensa-

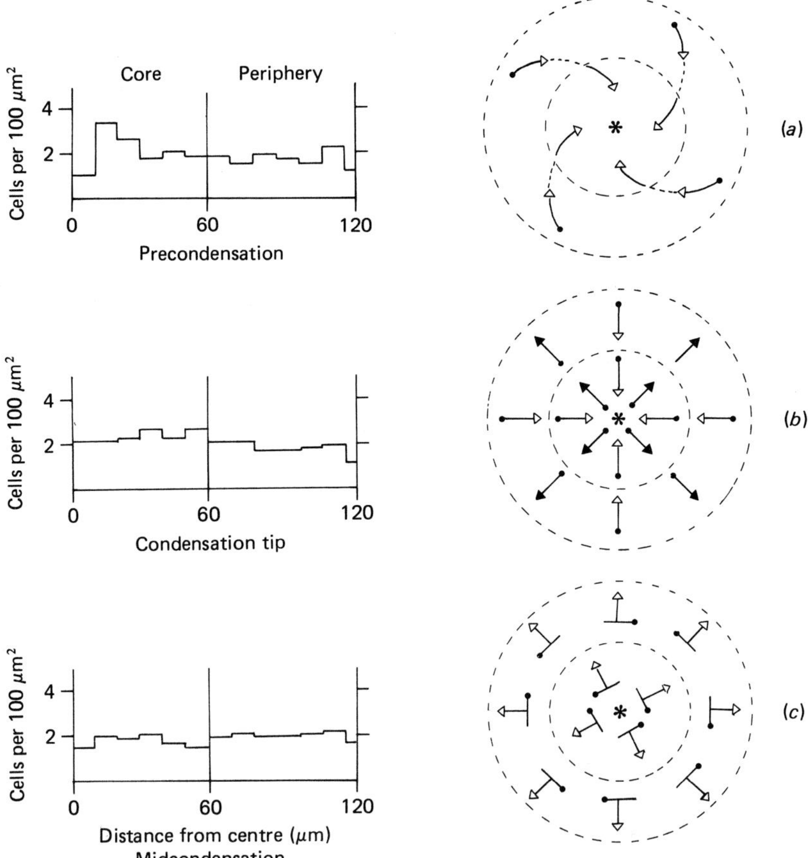

Fig. 8. Histograms showing cell density in precondensation, condensation tip and midcondensation cartilage condensations in the metacarpal region of normal chick wing buds, and corresponding models (*a, b, c*) of cell movements within the condensations at these stages.

tions themselves do not become clearly separated from each other, so that a meshwork of chondrified tissue is produced and no clear Anikin patterns are formed (Figs. 11, 12). It has been shown by Ede & Agerbak (1968) and Ede & Flint (1975) that the *talpid*[3] mesenchyme cells are more adhesive and less motile than normal cells, and, if cell movement is a significant activity in condensation formation, this may account for the failure of condensations to become separated in the *in vitro* cultures and in the limb bud.

Both Golgi orientation analysis of the limb bud condensations and

 D. A. Ede

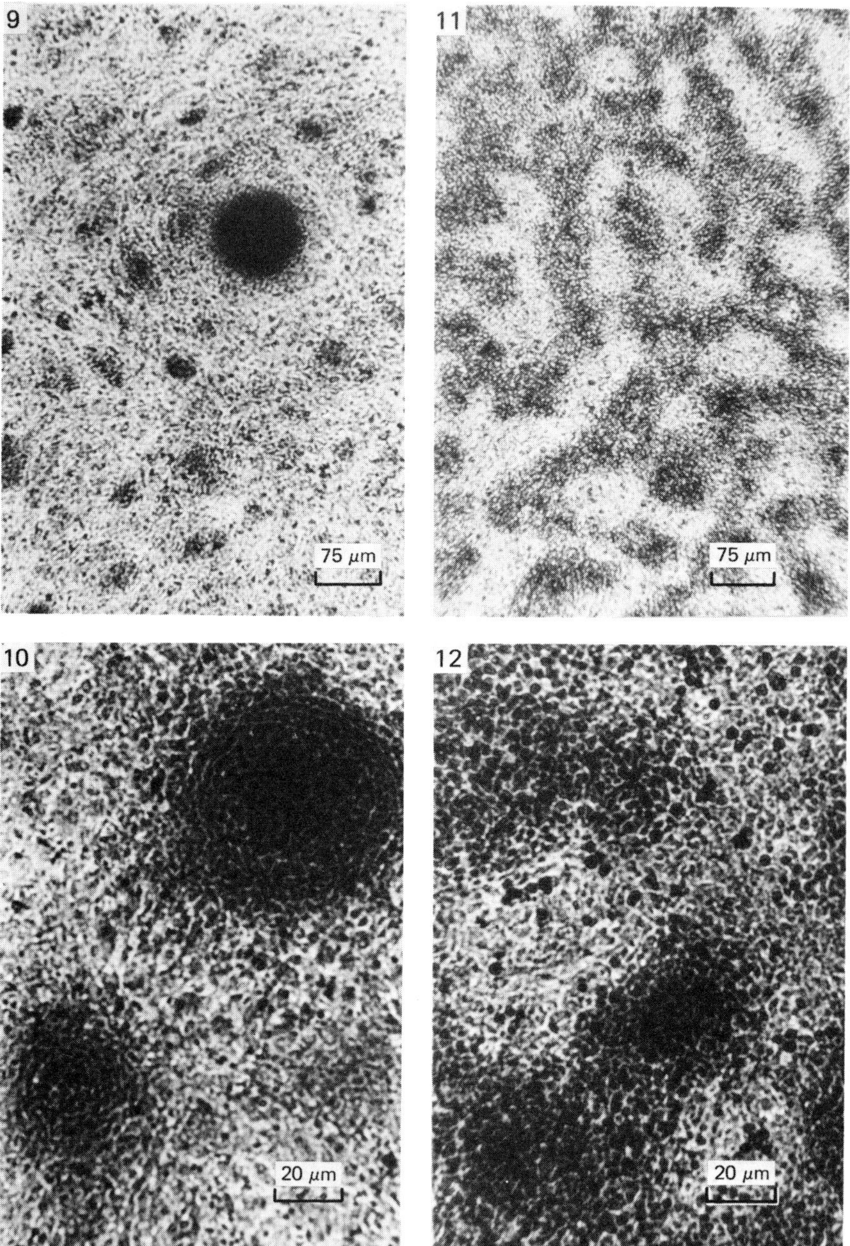

Figs. 9–12. Normal (Figs. 9, 10) and *talpid*[3] mutant (Figs. 11, 12) chick wing-bud mesenchyme cells from stage 23–24 embryos, plated out on plastic at post-confluent densities (1.5×10^6 cells per dish) after 3 days of culture.

SEM studies of *in vitro* condensations lend support to this suggestion. No clear orientation of cells towards or away from the centre of the condensation is revealed in the Golgi analysis, though the subsequent tangential orientation, which is passively produced, does appear (Fig. 7). The SEM pictures show that *in vitro talpid*[3] condensations show much less clearly any centre cell or cell group (Fig. 13), and there is no clear pattern of cell arrangement until the tangential stretching produced by cartilage matrix build-up appears (Fig. 14).

It appears likely, from these observations, that there is some movement of cells towards the centre in the formation of condensations, and that this is disturbed in the *talpid*[3] mutant. If there is such a movement, it may conceivably depend upon cartilage progenitor cells being attracted towards an initiator cell by chemotaxis. It is known (Postlethwaite, Seyer & Kang, 1978) that human dermal fibroblasts are chemotactically attracted to peptide breakdown products of collagen *in vitro*, and Holmes & Trelstad (1979) have shown that small molecular weight hydroxyproline peptides are present in the limb mesenchyme, though there is at present no information about their distribution in any sort of pattern.

INTEGRATION OF THE CONDENSATION UNIT INTO THE SKELETAL PATTERN

The condensation units in the flat field of *in vitro* culture show no pattern beyond a generalized spacing, reminiscent of Claxton's (1964) observations on the spacing of hair follicles in the sheep dermis. A slightly higher level of organizational complexity is introduced when limb mesenchyme cells are reaggregated in rotation culture (Ede & Flint, 1972). In this case spherical condensations appear, distributed through the interior of the aggregate, but never at the surface. Again, an Anikin pattern of cells is produced in each condensation, but this time as concentric spheres rather than concentric circles as in cultures on plastic or glass, or cylinders as in the limb bud.

The next level of complexity studied by us (Ede & Shamslahidjani, in preparation) is in dissociated mesenchyme cells from stage 23 leg buds, cultured not *in vitro* but in ectodermal jackets isolated from other limb buds, and grafted to the dorsal limb bud base of other embryos. Control limbs, developed from undissociated limb-bud grafts, are shown in Figs. 15–18. In the case of grafts with dissociated mesoderm, mesenchyme cells from a normal embryo produce a limb which is in all

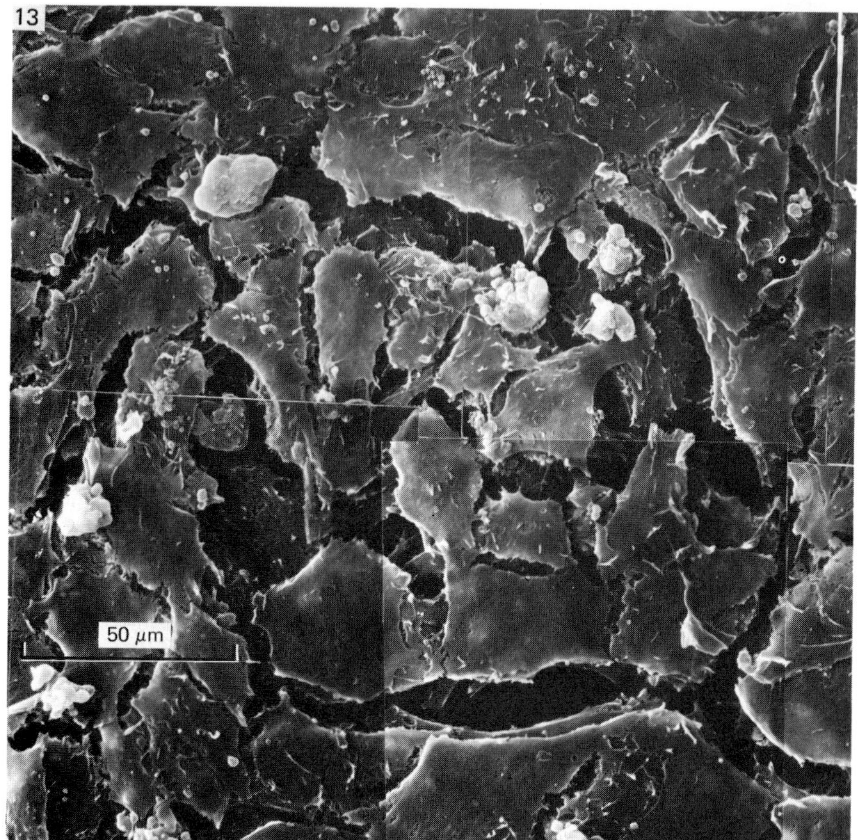

Figs. 13, 14. SEM picture of early and late stages in the development of cartilage condensations in cultures of *talpid*[3] mutant limb mesenchyme cells.

major respects normal in its proximo-distal development, and which produces differentiated skeletal elements in the antero-posterior axis, including digits (Fig. 19). However, these digits are not distinguishable as particular digits, i.e. no polarized pattern of digits is produced as in the control grafts or in unoperated embryos.

What is particularly interesting is the basic resemblance which this pattern bears to the pattern of digits in a *talpid*[3] mutant limb, both *in situ* and when the *talpid*[3] cells have been disaggregated and cultured in ectodermal jackets (Fig. 20).

If it is the case (see above) that A-P polarity in the limb bud depends upon the presence of a morphogen-gradient produced and generated by a cell mass (ZPA) in the posterior mesenchyme, it is not surprising

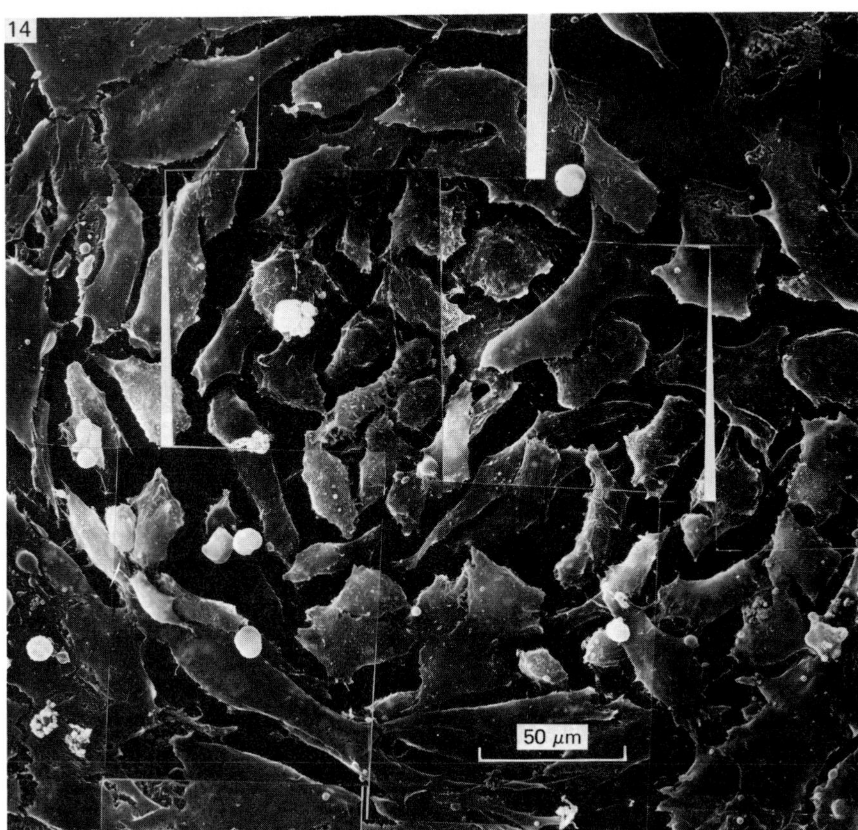

that when the cells of the ZPA are scrambled throughout the newly constituted mesenchyme cultured in the jacket the polarity fails to develop. We have found that introducing normal ZPA tissue into the appropriate site in the reaggregated normal mesoderm leads to significantly better development of the digits and to some indication of polarity (Fig. 21). No such effect is produced with *talpid*[3] tissue of the ZPA region, either in *talpid*[3] or normal mesenchyme (Fig. 22). Nor does *talpid*[3] mesenchyme respond in any way to grafted normal ZPA (Fig. 23). Similar results have been reported for *talpid*[2] by MacCabe & Abbott (1974).

What is more interesting is that in both *talpid*[3] *in situ*, and in normal mesenchyme developing without localized ZPA, though there is no polarity, separate digits are produced. Their differentiation pattern appears to have a periodic basis, the digits being roughly equidistant, with the number of elements at any level related to the width of the

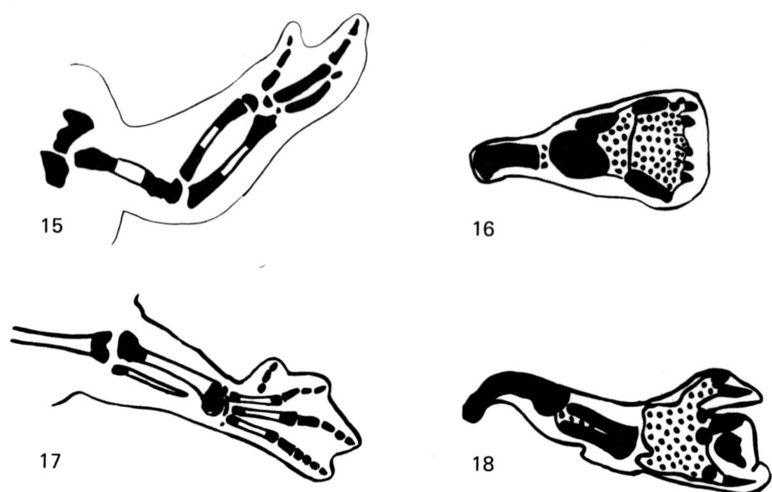

Figs. 15–18. Drawings of alcian-blue cleared 10-d limbs produced from control grafts of stage 18 chick limb buds to normal host limbs. Fig. 15 – normal wing. Fig. 16 – *talpid*[3] wing. Fig. 17 – normal leg. Fig. 18 – *talpid*[3] leg.

limb bud at that level. The most developed digits are those at the anterior and posterior extremities. There is often bifurcation (see Hinchliffe's observations above), and intercalated digits appear as the space between previously established digits widens. A model producing a wave-form gradient which gives rise to appropriate prepatterns in simulations has been suggested by Wilby & Ede (1975) and a model of basically similar type, based upon morphogen diffusion dynamics, by Newman & Frisch (1979), who have more recently proposed that fibronectin (see above) may actually be the morphogen involved (Newman, Frisch, Perle & Tomasek, 1980). Both models find support from the polydactyly found in *talpid* mutants, where the increased number of digits is related to the expanded distal tip of the limb bud. Positional information theory would predict regulation for the normal number, though of broader, digits in this situation. It is also noteworthy that the initiation of each condensation unit, and the beginning of cartilage matrix secretions within it, is central, which the prepattern hypotheses would predict whereas positional information theory, unless the patterns specified were extremely fine-grained, would predict uniformity across the rudiment. Thus there appears to exist in the limb a primitive, periodic system of pattern determination, to which has been added a more elaborate system of positional information as a 'fine-tuning' mechanism.

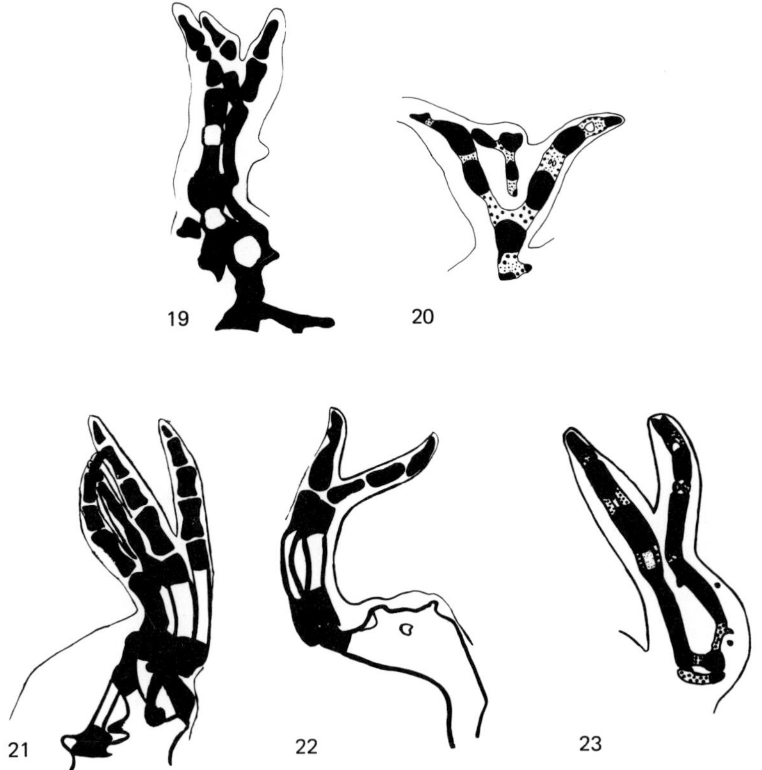

Figs. 19–23. Drawings of alcian-blue cleared 10-d limbs produced from grafts made up of leg-bud ectodermal jackets packed with dissociated and reaggregated leg-bud mesoderm cells. Fig. 19, normal dissociated mesoderm in a normal ectodermal jacket. Fig. 20, *talpid*[3] dissociated mesoderm in a *talpid*[3] ectodermal jacket. Fig. 21, normal dissociated mesoderm in a normal ectodermal jacket, with a piece of normal ZPA at its posterior border. Fig. 22, normal dissociated mesoderm in a normal ectodermal jacket, with a piece of *talpid*[3] ZPA at its posterior border. Fig. 23, *talpid*[3] dissociated mesoderm in a normal ectodermal jacket, with a piece of normal ZPA at its posterior border.

EVOLUTION OF PATTERN-CONTROL MECHANISMS

Just as patterns of differentiation in culture become more complex with increasing levels of complexity in the culture system, so increasingly sophisticated and specialized mechanisms of pattern control have been developed in the course of evolution, replacing the old mechanisms which depend entirely on direct cell to cell interactions, which may have served quite other purposes in early metazoan organization, or supplementing them as fine-tuning mechanisms, with much more elaborate and specialized grid-reference systems depending on the cells' response to cues originating at some considerable distance. The first type will have predominated in the crossopterygian fish ancestors of the tetrapods; the second must have arrived with the evolution of the pentadactyl limb and persisted in a remarkably stable way, producing the same basic developmental pattern in animals whose fully-developed limbs are amazingly diverse.

In the example described here we have seen that in the simplest cell culture systems limb mesenchyme produces a developmental unit, the cartilage condensation, which forms the basis in culture systems of higher level, and in the limb itself, of the skeletal system. These condensation units, formed randomly in the simplest systems, are themselves organized into patterns in systems of higher complexity. In limbs without localized ZPA, as in the *talpid* mutants and in the crossopterygian fish, a comparatively simple periodic diffusion system leads to a skeletal pattern without polarity. Finally, in the limb developing with all its normal embryonic constraints, some system of positional information imposes on these periodic elements specific determination pathways leading to the production of particular rudiments with unique properties of further growth and development.

Acknowledgement

The author wishes to thank the Science Research Council for financial support for the original work reported in this review.

REFERENCES

AHRENS, P. B., SOLURSH, M. & REITER, R. S. (1977). Stage-related capacity for limb chondrogenesis in cell cultures. *Developmental Biology*, **60**, 69.

ANIKIN, A. W. (1929). Das morphogene feld der knorpelbildung. *Wilhelm Roux's Archiv für Entwicklungsmechanik Organum*, **114**, 549–78.

CAPLAN, A. I. (1970). Effects of the nicotinamide (NAD)-sensitive teratogen 3-acetylpyridine on chick limb cells in culture. *Experimental Cell Research*, **62**, 341–55.

CHEVALLIER, A., KIENY, M. & MAUGER, A. (1977). Limb-somite relationship: origin of the limb musculature. *Journal of Embryology and Experimental Morphology*, **41**, 245–58.

CHRIST, B., JACOB, H. J. J. & JACOB, M. (1977). Experimental analysis of the origin of the wing musculature in avian embryos. *Anatomia Embryologia*, **150**, 171–86.

CLAXTON, J. H. (1964). The determination of patterns with special reference to that of the central primary follicles in sheep. *Journal of Theoretical Biology*, **7**, 302–17.

DESSAU, W., V.D. MARK, H., V.D. MARK, K. & FISCHER, S. (1980). Changes in the patterns of collagens and fibronectin during limb bud chondrogenesis. *Journal of Embryology and Experimental Morphology*, **57**, 51–60.

DESSAU, W., SASSE, J., TIMPL, R., JILEK, F. & V.D. MARK, K. (1978). Synthesis and extra-cellular deposition of fibronectin in chondrocyte cultures. Response to the removal of extra-cellular cartilage matrix. *Journal of Cell Biology*, **79**, 342–55.

DIENSTMAN, S. R., BIEHL, J., HOLTZER, S. & HOLTZER, H. (1974). Myogenic and chondrogenic lineages in developing limb buds grown *in vitro*. *Developmental Biology*, **39**, 83–95.

DREWS, U. & DREWS, U. (1972). Cholinesterase in the development of the chick limb. I. Phases of cholinesterase activity in the early bud and in the demarkating cartilage and muscle anlagen. *Wilhelm Roux's Archivs für Entwicklungsmechanik Organum*, **169**, 70–86.

DREWS, U. & DREWS, U. (1973). Cholinesterase in the development of the chick limb. II Enzyme activity and locomotory behaviour of the presumptive cartilage cells in vitro. *Wilhelm Roux's Archivs für Entwicklungsmechanik Organum*, **173**, 208–27.

EDE, D. A. & AGERBAK, G. S. (1968). Cell adhesion and movement in relation to developing limb pattern in normal and *talpid*[3] mutant chick embryos. *Journal of Embryology and Experimental Morphology*, **20**, 81–100.

EDE, D. A. & FLINT, O. P. (1972). Patterns of cell division, cell death and chondrogenesis in cultured aggregates of normal and *talpid*[3] mutant chick limb mesenchyme cells. *Journal of Embryology and Experimental Morphology*, **27**, 245–60.

EDE, D. A. & FLINT, O. P. (1975). Cell movement and adhesion in the developing chick wing bud: studies on cultured mesenchyme cells from normal and *talpid*[3] mutant embryos. *Journal of Cell Science*, **18**, 301–14.

EDE, D. A., FLINT, O. P., WILBY, O. K. & COLQUHOUN, P. (1977). The development of precartilage condensations in limb-bud mesenchyme of normal and mutant embryos *in vivo* and *in vitro*. In *Vertebrate Limbs and Somite Morphogenesis*, ed. D. A. Ede, J. R. Hinchliffe & M. Balls, pp. 161–79. Cambridge University Press.

EDE, D. A. & WILBY, O. K. (1980). Golgi orientation and cell behaviour in the developing pattern of chondrogenic condensations in chick limb bud mesenchyme. *Histochemical Journal* (in press).

EGUCHI, G. & OKADA, T. S. (1971). Ultrastructure of the differentiated cell colony derived from a singly isolated chondrocyte in *in vitro* culture. *Development, Growth and Differentiation*, **12**, 297–312.

ELSDALE, T. & FOLEY, R. (1969). Morphogenetic aspects of multilayering in petri dish cultures of human fetal lung fibroblasts. *Journal of Cell Biology*, **41**, 298–311.

FELL, H. B. (1925). The histogenesis of cartilage and bone in the long bones of the embryonic fowl. *Journal of Morphology and Physiology*, **40**, 417.

FELL, H. B. & CANTI, R. G. (1935). Experiments on the development *in vitro* of the avian knee joint. *Proceedings of the Royal Society of London*, **B116**, 316–49.

GOULD, R. P., DAY, A. & WOLPERT, L. (1972). Mesenchymal condensation and cell contact in early morphogenesis of the chick limb. *Experimental Cell Research*, **72**, 325–36.

GOULD, R. P., SELWOOD, L., DAY, A. & WOLPERT, L. (1974). The mechanism of cellular orientation during early cartilage formation in the chick limb bud and regeneration of amphibian limb. *Experimental Cell Research*, **83**, 287–96.

HEWITT, A. T. & ELMER, W. A. (1976). Reactivity of normal and brachypod mouse limb mesenchymal cells with con A. *Nature*, **264**, 177–8.

HEWITT, A. T. & ELMER, W. A. (1978). Developmental modulation of lectin-binding sites on the surface membranes of normal and brachypod mouse limb mesenchyme cells. *Differentiation*, **10**, 31.

HINCHLIFFE, J. R. (1977). The chondrogenic pattern of chick limb morphogenesis: a problem of development and evolution. In *Vertebrate Limb and Somite Morphogenesis*, ed. D. A. Ede, J. R. Hinchliffe & M. Balls, pp. 293–309. Cambridge University Press.

HOLMES, L. B. & TRELSTAD, R. L. (1979). Identification of rapidly labelled, small molecular weight hydroxyproline-containing peptides in developing mouse limbs *in vitro* and *in vivo*. *Developmental Biology*, **72**, 41–9.

HOLMES, L. B. & TRELSTAD, R. L. (1980). Cell polarity in precartilage mouse limb mesenchyme cells. *Developmental Biology* (in press).

KLEBE, R. J. (1974). Isolation of a collagen-dependent cell attachment factor. *Nature*, **250**, 248–51.

LINSENMAYER, T. F., TOOLE, B. P. & TRELSTAD, R. L. (1973). Temporal and spatial transitions in collagen types during embryonic chick limb development. *Developmental Biology*, **35**, 232–9.

MacCABE, J. A. & ABBOTT, U. K. (1974). Polarizing and maintenance activities in two polydactylous mutants of the fowl: diplopodia and talpid. *Journal of Embryology and Experimental Morphology*, **31**, 735–46.

NEWMAN, S. A. & FRISCH, H. L. (1979). Dynamics of skeletal pattern formation in developing chick limb. *Science*, **205**, 662–8.

NEWMAN, S. A., FRISCH, H. L., PERLE, M. A. & TOMASEK, J. J. (1980). Limb development: aspects of differentiation, pattern formation, and morphogenesis. *Ann Arbor Symposium on Pattern in Development* (in press).

PAULSEN, D. F. & FINCH, R. A. (1977). Age- and region-dependent concanavalin A reactivity of chick wing-bud mesoderm cells. *Nature*, **268**, 639–41.

POSTLETHWAITE, A. E., SEYER, J. M. & KANG, A. H. (1978). Chemotactic attraction of human fibroblasts to type I, II, and III collagens and collagen-derived peptides. *Proceedings of the National Academy of Sciences, USA*, **75**, 871–5.

SAUNDERS, J. W. (1948). The proximo-distal sequence of origin of the parts of the chick wing and the rôle of the ectoderm. *Journal of Experimental Zoology*, **108**, 363–403.

SEARLS, R. L. (1965). An autoradiographic study of the uptake of ^{35}S-sulphate during the differentiation of limb bud cartilage. *Developmental Biology*, **11**, 155–68.

SEARLS, R. L., HILFER, S. R. & MIROW, S. M. (1972). An ultrastructural study of early chondrogenesis in the chick wing bud. *Developmental Biology*, **28**, 123–37.

SOLURSH, M., AHRENS, P. B. & REITER, R. S. (1978). A tissue culture analysis of the steps in limb chondrogenesis. *In Vitro*, **14**, 51.

THOROGOOD, P. V. & HINCHLIFFE, J. R. (1975). An analysis of the condensation process during chondrogenesis in the embryonic chick hind limb. *Journal of Embryology and Experimental Morphology*, **33**, 581–606.

TICKLE, C., SUMMERBELL, D. & WOLPERT, L. (1975). Positional information and specification of digits in chick limb morphogenesis. *Nature*, **254**, 199–202.

TOOLE, B. P. (1972). Hyaluronate turnover during chondrogenesis in the developing chick limb and axial skeleton. *Developmental Biology*, **29**, 321–9.

TOOLE, B. P., JACKSON, G. & GROSS, J. (1972). Hyaluronate in morphogenesis: Inhibition of chondrogenesis *in vitro*. *Proceedings of the National Academy of Sciences, USA*, **69**, 1384–6.

TRELSTAD, R. L. (1977). Mesenchyme cell polarity and morphogenesis of chick cartilage. *Developmental Biology*, **59**, 153–63.

VASAN, N. S. & LASH, J. W. (1979). Monomeric and aggregate proteoglycans in the chondrogenic differentiation of embryonic chick limb buds. *Journal of Embryology and Experimental Morphology*, **49**, 47–59.

WILBY, O. K. & EDE, D. A. (1975). A model generating the pattern of cartilage skeletal elements in the embryonic chick limb. *Journal of Theoretical Biology*, **52**, 199–217.

WOLPERT, L. (1969). Positional information and the spatial pattern of cellular differentiation. *Journal of Theoretical Biology*, **25**, 1–47.

Morphogenetic studies in plant tissue cultures

G. G. HENSHAW, J. F. O'HARA AND K. J. WEBB
Department of Plant Biology, The University of Birmingham, P.O. Box 363,
Birmingham B15 2TT

The concept of cellular totipotency was an integral part of the 'cell theory' as first propounded by Schleiden (1838) and Schwann (1838). Schleiden (1849), for example, basing his ideas on observations of the considerable regenerative powers of plants, commented on 'the possibility of each cell, in any situation, on occasion, going through all the phases of cell life, and becoming developed in any way that the circumstances under which it is placed render necessary'. Although Schleiden took this to indicate the virtual independence of the cells within the organism – a point of view which is no longer acceptable – his statement would not look out of place in a modern text book. It was, however, more than one hundred years before the direct proof of the totipotency of higher plant cells was obtained by the regeneration of plants from isolated cells which had been grown in culture (Muir, 1953). Nevertheless, by the end of the nineteenth century the considerable further regeneration studies with plant organs by workers such as Vöchting (1878) had led Pfeffer (1900) among others to believe that the failure to form a new plant from a single isolated cell was 'obviously due to the difficulty of presenting such a cell with a supply of nutriment of appropriate quality and quantity' and Haberlandt (1902) made his famous prediction that isolated cells should be capable of developing as 'artificial embryos'. The present situation is rather similar in that extensive regeneration studies and, more recently, cell culture studies have led to the wide acceptance of the view that diploid somatic cells in plants are totipotent, despite the fact that this has been difficult to demonstrate with cells from certain species, even though they have been the subject of very intensive investigations. In fact, the more recent comments of workers such as Steward, Ammirato & Mapes (1970), to the effect that 'present failures to rear them [all normally diploid cells] into plants merely present the challenge to find the right conditions for their development', strongly echo the earlier views of Pfeffer and Haberlandt. The real issue now is not whether some higher plant cells may not be totipotent – because such a negative proposition is difficult to prove – but whether the present acceptance of totipotency has led to an oversimplified and distorted view of plant development, which could have important implications both from the fundamental and the applied points of view.

There can be no doubt that plant cell culture successes in recent years have had a considerable influence on the debate about the respective roles of cell and organism in the control of development. This situation was forecast by White (1954) in a comment on contemporary attitudes, in which he pointed out the impracticality of studying integrated wholes only and the practicality of the concept of the cell as an elementary functional unit. The strength of this reductionist argument was quickly confirmed when it was shown that isolated plant cells could be cultured successfully (Muir, 1953) and that complete plants could be regenerated from cultures derived from somatic cells, either directly via an embryogenic pathway (Reinert, 1958; Steward, Mapes & Mears, 1958) or indirectly via an organogenic pathway (Skoog & Miller, 1957). The fact that the morphogenetic behaviour of the cells was strongly influenced by exogenous factors, especially hormones, and that cytodifferentiation was also shown to be influenced by similar factors (Wetmore & Rier, 1963), quickly led to the widely held view that plant cells, in contrast to animal cells, remain relatively uncommitted and that their states are largely maintained by factors which are extrinsic to the cells; isolation from those factors was supposed to lead to the cell's reversion to some sort of base state from which it could readily redifferentiate in accordance with external factors. The obvious practical implications of such a situation led to a large expansion of plant cell culture studies with a considerable emphasis on the 'application of suspected morphogenetically active chemicals to hopefully totipotent cells' (Halperin, 1969). More recently, Wareing (1978) has also commented upon the botanist's fascination by the evidence for totipotency and the way that this has led to an explanation of development being almost automatically sought in terms of factors which are extrinsic to the cell. This recalls the earlier warning of Sinnott (1946) – made at a time when hormone studies were beginning to have a considerable impact on attitudes towards plant development – that 'for developmental problems a study of substances is important but ... the ultimate question is rather the organised system in which this substance operates'.

Clearly the fact that the removal of organisational constraints by isolation leads to a change in the behaviour of plant cells, and possibly reversion towards some sort of base state, affects the way in which *in vitro* methods might be employed to investigate developmental processes. Although it might no longer be so widely assumed that the act of isolation completely removes developmental constraints, there have been remarkably few attempts to investigate systematically the influence of *in vivo* cell states on the subsequent morphogenetic behaviour of the cells *in vitro*. The reason, presumably, is that the empirical approach, based largely on the manipulation of extrinsic factors, especially growth regulators, has had considerable success and the list of species from which cells have been induced to express their totipotency *in vitro* is now extensive (Murashige, 1978). There have however been failures, notably among some of the important crop species (Thomas & Wernicke, 1978) and, further, it is well known that where such an ability does exist it might only be transient. A greater understanding of the effects of cell states on the response to supposedly morphogenetic substances can only help with the more pragmatic aspirations of *in vitro* studies, at the same time as contributing towards a more realistic view of the ways in which cells might operate *in vivo*. Unfortunately, a conceptual basis for the description of cell states in plants has not been established and, in particular, it is not certain whether the terminology employed in animal developmental studies is applicable.

COMPETENCE, DETERMINATION AND TOTIPOTENCY

Heslop-Harrison (1967) has pointed out that the phenomenon of cell competence, which has long been investigated by animal embryologists, has not been greatly emphasised in plant literature, despite the fact that it seems to be widely encountered. He also suggested that the distinction between the acquisition of the competence to react to a stimulus directing differentiation and the response to that stimulus has profound importance for the understanding of control mechanisms. Halperin (1969), in an analysis of morphogenetic studies with plant cell cultures, drew attention to the limitations of present techniques which are frequently based on an assumption that the raw material consists of a disorganised mass of cells which should be completely responsive to the appropriate morphogenetic signals. In particular, he emphasised that this tendency to concentrate on supposedly specific morphogenetic signals overlooks the difference between totipotency and competence and that the cells in a culture which are competent to respond are usually few in number and generally not identifiable in advance.

It would be inaccurate to say that cell culture studies still take no account of cell states carried over from the plant. Murashige (1974), for example, strongly emphasised the importance of choosing a suitable explant for particular morphogenetic responses in culture and in 1977 he stated that the 'fundamental misconception has been that tissue cultures ... are composed of undifferentiated cells'. The real problem now, therefore, is not one of recognising that all cells in culture systems are not equal but one of understanding the reasons for the inequalities. This would appear to be a question of cell competence but it is important that botanists should examine the situation carefully to see whether it warrants the use of this term in the same sense that it has been employed by animal embryologists such as Waddington (1932, 1962, 1968, 1970).

As defined, the concept of competence is intimately related to the concepts of determination and potency, in the sense that a competent system which receives a developmental signal becomes determined and, as a consequence, loses potency. The terms were used in the context of embryo development to describe the ability of a tissue to undergo a specific morphogenetic change in response to 'embryonic induction', that is to say, in response to a signal produced by another tissue in the embryo. When a competent tissue responds to such a morphogenetic signal it becomes determined, a state in which, according to the stage of development, it either becomes competent to react to further specific signals, or in which it remains until it becomes 'activated' by permissive conditions, so that the morphogenetic change is produced. Embryo development is seen, therefore, to involve the sequential appearance and loss of states of competence in the tissues. This is believed to operate in a hierarchical manner, so that tissues are initially competent to develop in a number of different ways according to the signals that are received and with each determinative event their degrees of freedom are reduced, until eventually they become determined with respect to a single fate.

The states of competence and determination are both cryptic states but, whereas competence is a transient condition which is probably not 'cell-heritable' and which might disappear spontaneously even though a signal has not been received, determination is a more stable condition which is known to be inheritable through many cell generations (Waddington, 1968). Little is known about the mechanisms by which cells become competent and how they might be controlled but the fact that specificity seems to reside in the competent cells rather than in the signal has led to suggestions that the 'priming' of a number of sets of genes is involved and that at determination one of those sets is selected to become dominant in the future development of the cell. Following this argument, 'activation' would then involve the production of the specific proteins relating to the structural genes in the selected set. Alternatively, the competent state may simply depend on the presence of a particular set of compounds in the cells, but this is, perhaps, less likely in view of the apparently specific nature of the state.

The close relationship between the concepts of competence and determination might appear to present some problems if the terminology is to be used in this strict sense to describe developmental events in plants, since the question of whether determined states exist at the cell level is somewhat controversial. Wareing (1978, 1979) has recently discussed at length the question of determination in plant development, suggesting that there are two main reasons for the botanist's reluctance to utilise the concept. Firstly, the 'canalisation' of development, leading to progressive specialisation of cells and their loss of potential which has been described for animal embryos is not so evident in plants because of their indeterminate pattern of growth involving apical meristems. Secondly, the fact that the state of determination, by definition, involves a loss of developmental potential and is 'cell-heritable' conflicts strongly with the widespread assumption that somatic plant cells remain totipotent. Wareing argues that these are not necessarily good reasons for avoiding the use of the concept for plants since the evidence that all living mature cells retain their full potential is not conclusive and, further, there are examples of stability at the higher organisational levels in plants which are more readily explained in terms of intrinsic differences in the component cells than in terms of extrinsic factors controlling a population of uncommitted cells. The fact that such committed cells might still be capable of expressing their full potential if, for example, they are induced to grow in culture is explained on the grounds that the changed conditions lead to de-differentiation so that former states, which involved the selective masking of the genome are obliterated.

These points might be valid and, certainly, there is increasing evidence of organ- and tissue-specific states persisting in plant cell cultures (Raff, Hutchinson, Knox & Clarke, 1979; Khavkin, Markov & Misharin, 1980). Nevertheless there appears to be a real difference between cell states in plants and animals although this could be a reflection of technical problems. In higher animals there are good examples of intrinsic states which are stable in culture and, although dedifferentiation might occur, it is not apparently accompanied by 'dedetermination' and the expression of full potential. In some examples, the stability of the states of determination has been demonstrated by the fact that the specific cell phenotypes are restored by permissive conditions (Holtzer & Abbott, 1968). In higher plants, however, there is no doubt that dedifferentiation can lead to the expression of full potential, at least in some cells. Further, most of the examples of cell states which are apparently stable *in vitro* have not yet been subjected to the rigorous test of single-cell cloning and, as Heslop-Harrison (1967) pointed out, there is always the possibility that the states might be maintained by factors extrinsic to the cells.

There is a real biological issue here and not simply one of semantics. There can be no doubt that determination, associated with loss of potential, occurs in all multicellular organisms. As soon as the initial cell divides, to produce a multicellular structure which behaves as an organism, determination has taken place, but not necessarily at the cellular level. The important general point is that determination can occur at any organisational level within an organism and determination at one level does not necessarily imply determination at lower levels of the organisational hierarchy. On the contrary, it is presumably essential that more degrees of freedom should be maintained at the lower organisational levels, otherwise the regulatory mechanisms which tend to maintain the developmental norm at any organisational level, and which are characteristic of all organisms, could not operate. There will be conflict between the need for stability at the higher organisational levels and flexibility at lower levels which will be resolved in a particular organism according to the demands of its life style; a more complex organisational structure will most probably impose greater constraints on the degrees of freedom at the lower organisational levels, or conversely, a need for greater flexibility at the lower levels might impose limits on the overall organisational complexity.

The real issue, therefore, is not *whether* determination occurs in plants in the same sense as it occurs in animals but, rather, the level at which it *does* occur. In higher animals, it is quite clear that determination occurs at all levels in the organisational hierarchy down to the individual cell, although the nucleus of a determined cell might be able to express its full potential when released from its cytoplasmic constraints (Gurdon, 1976). In plants, determination occurs at the organ and tissue levels but it is obvious that much greater flexibility can be retained at the cell level. This must be an advantage for an organism which because of its life-style cannot move to avoid environmental stress and damage caused by other organisms, but it might impose limits on its overall organisational complexity. One way that greater organisational complexity might be achieved in an organism is by developmental flexibility at the cell level being restricted to certain regions, such as the indeterminate meristem, but that occurs in all plants. Many plant species, however, also retain developmental flexibility in a high proportion of the somatic cells, which is released when the higher organisational states are disturbed. The practical question is whether in some plant species that sort of flexibility has been sacrificed at the expense of greater organisational complexity. Are the totipotent cells much more localised in their distribution in the more highly organised plant species?

This situation in plants can be illustrated by reference to the development of the various foliar organs or phyllomes, which include the floral appendages as well as a range of vegetative structures. In the higher plants these organs are regarded as being morphologically homologous and, phylogenetically, they are thought to have been derived from a shoot system (Esau, 1953). *In vitro* studies have shown that these structures are clearly determined from an early stage of development. In the ferns there seems to be a short phase when the young leaf primordium is completely undetermined so that it either develops into a bilaterally-symmetrical leaf structure, if it is in contact with the shoot apex, or into a radially-symmetrical shoot structure if it is isolated. Once the next primordium has been formed, the leaf apex is irreversibly determined as a leaf (Haight & Kuehnert, 1971). The situation in angiosperms is more controversial, with some authors claiming that the primordia are determined as to organ type at the same time as they are initiated and others claiming that the processes of initiation and determination are separated temporarily (see discussion by Halperin, 1978); whichever is correct, the *in vitro* conversion of leaf primordia into shoots has not been demonstrated. Once these organs are determined, they are extremely stable when isolated and grown in culture. There is good evidence, therefore, that determination occurs at this level of organisation in plants, and that this is associated with loss of potency, in the sense that these organs cannot be converted into other organs and they cannot reform an entire plant, except indirectly by adventitious regeneration from some of the cells. There is also good evidence that this state of determination is not dependent upon a similar stability at the lower levels of organisation, since the totipotency of at least some of the cells has been demonstrated in many species, by the adventitious production of shoots and roots after excision (see review by Broetjes, Haccius & Weidlich, 1968). Further, leaf mesophyll cells are commonly chosen as a source of protoplasts from which plants are subsequently to be regenerated. It is obvious, therefore, that although there can be considerable stability at the higher organisational levels in plants, this does not seem to demand the same level of commitment in the individual cells as is found in animals.

It would be misleading, however, to suggest that all leaf cells are uncommitted to the extent that they are capable of expressing their full potential as soon as these organisational constraints are removed, since not all species show the same readiness to form buds on isolated leaves and there are some families, such as the Gramineae and the Palmae, in which adventitious bud formation has never been reported (Broetjes, Haccius & Weidlich, 1968). There is also some evidence in other species that organ-specific antigens continue to be produced in cell cultures derived from leaves (Raff *et al.*, 1979). The reasons for this variation in the readiness of leaf cells – and other somatic cells – from different species to express their totipotency are not understood. There might simply be differences in the conditions required to initiate that type of morphogenetic behaviour but, on the other hand, there could be different levels of commitment in the cells, possibly, reflecting different degrees of complexity at the higher organisational levels. It may not be coincidence, for example, that the palm leaf is the most complex of all leaf forms (Periasamy, 1962) and that its cells seem to have a low morphogenetic potential.

With regard, therefore, to the question of whether the related concepts of competence, determination and potency, as originally defined for animal systems, should also be used to describe developmental processes in plants, it might be concluded that the terminology is appropriate. *In vitro* studies in particular have shown, however, that the states do not necessarily exist at the same levels of organisation in plants and animals. Plant cells, for example, might act in a determined manner *in situ* when they are subject to the constraints of the higher levels of organisation, but the state might not be stable when those constraints are removed. Johansen (1950) recognised how some of these problems affect plant embryology, when he described how the cells of a proembryo – as he thought – normally give rise to particular parts of the embryonic body at the same time as remaining totipotent. He used the terms 'actual potentiality' and 'virtual potentiality' to distinguish between a cell's potency under, respectively, normal and modified conditions. In principle, this would be a satisfactory way of dealing with the problem, except that there are practical difficulties in defining 'normal' conditions, especially since it is now recognised that the cells of a plant embryo, even *in vivo*, do not follow such rigidly defined pathways as was once believed (Wardlaw, 1965).

EMBRYOGENIC COMPETENCE

During normal development in most plant species, the egg cell is the only cell which normally demonstrates its embryogenic competence. The state can obviously exist in other cells, as in some species adventive embryogeny occurs in a sufficiently regular manner in adjacent ovular tissues, for it to constitute a mechanism of asexual reproduction. *In vitro* studies have now shown that other somatic cells can give rise to adventive embryos, indicating either that the state of embryogenic competence was simply constrained by the determinative events taking place at higher levels of organisation or that the state was in some way restored by the *in vitro* conditions. The failure to demonstrate embryogenic competence in some somatic cells under *in vitro* conditions could be due to one or more of the following reasons:

(*a*) A permanent loss of totipotency.

(*b*) A failure to remove developmental constraints.

(*c*) A failure to induce embryogenic competence once the developmental constraints have been removed.

(*d*) A failure to provide a specific morphogenetic signal.

(*e*) A failure to provide suitable permissive conditions for embryo development.

The fact that it is virtually impossible to demonstrate the negative proposition of a loss of totipotency, combined with the manifest totipotency of somatic cells from a wide range of species (see review by Narayanaswamy, 1977), has inevitably led to an assumption that this state can exist in most, if not all, mature, living plant cells. Even where there are obvious genetic changes in the cells, detectable at the chromosomal level, it cannot always be assumed that developmental potential has been lost (D'Amato, 1978). Nevertheless, more cryptic genetic changes possibly do occur in certain somatic cells, as well as organisational changes in the genome involving, for example, transposable elements (Esposito & Esposito, 1977). The question of whether any of these mechanisms have a role in development is still debatable, but they could produce states which are essentially irreversible, so that there is a permanent restriction of potency.

Just as it is difficult to be certain whether an apparent lack of embryogenic competence is due to a permanent restriction of potency, it is also difficult to distinguish between the other possible reasons, outlined above. Steward, as a result of his early *in vitro* investigations of carrot embryogenesis, strongly emphasised that if cells are to express their totipotency they should be physically isolated and nourished by a medium which is a near approximation of the chemical environment inside the ovule and which will support their rapid growth and development (Steward, Mapes, Kent & Holsten, 1964). Haberlandt (1902) had predicted that isolated plant cells should be capable of behaving as embryos and it now seemed that this had been confirmed by the technical advances which had eventually permitted isolated cells to be grown *in vitro*. It soon became apparent, however, that this procedure did not always lead to the expression of totipotency and other workers convincingly showed that in the carrot suspension cultures embryogeny occurred largely, if not entirely, in the surface layers of the cell aggregates, rather than in the isolated cells (Halperin, 1966; Sussex, 1972; Jones, 1974; McWilliam, Smith & Street, 1974). It is quite clear, therefore, that although the removal of certain organisational constraints must be one of the important factors which enable somatic cells to express their totipotency under *in vitro* conditions, it is not essential for the cells to be completely isolated. In fact, there is some evidence, also from carrot, suggesting that completely isolated cells will first produce a multicellular aggregate before embryogeny is initiated (Backs-Hüsemann & Reinert, 1970).

Although the list of species, in which embryogeny has been successfully induced *in vitro*, has been steadily growing during the last twenty years, it is still not easy to assign a role to any particular set of conditions which might have been employed. The majority of investigations have, for example, confirmed the observation of Halperin & Wetherell (1964) that exogenous auxin concentration is the major factor controlling embryogeny *in vitro*, in the sense that a reduction in the effective auxin concentration leads to the initiation of the embryogenic process. It is assumed that the auxin has an important function before embryogeny is initiated but the process is far from being understood. A statement to the effect that auxin is required for 'the induction of cells with embryogenic competence' (Kohlenbach, 1978), for example, leaves open the question of whether the auxin is involved in the process which leads to the removal of developmental constraints, or whether it is required to induce embryogenic competence once those constraints have been removed, or whether it provides the signal which evokes the embryogenic response in a suitably competent cell. Steward (1967), however, has suggested that the auxin provides a means of removing, from more mature cells, the developmental constraints which accumulate in the plant; at the earlier stages of development, physical isolation might be sufficient to release the full potential of the cell but, with the more mature cells, more drastic treatments are likely to be required. He made a distinction between those cells which 'merely grow' in culture and others which have the 'maximum ability to develop' and proposed that the auxin converts the former to the latter by inducing a very actively proliferating state. The implications of these ideas are important because they suggest that development in plants produces determined cell states which might be stable with certain types of cell proliferation and not with others.

There are two classes of cell proliferation which might be expected to have differing effects on cell states: those in which the nuclear events are more or less in balance with cytoplasmic events, so that the same cell size is regained between each division, and those in which the two events are out of balance, so that cell size is not regained between each division. Cell states which are dependent upon specific nuclear–cytoplasmic interactions are more likely to be unstable with the latter type of proliferation where cytoplasmic states will probably undergo regressive changes. On this basis, the asymmetrical type of division which is thought to precede many specific types of differentiation in the plant (Bünning, 1952) would presumably be a variant of the regressive type of proliferation in which polarisation produces full regression in one of the daughter cells in one step rather than in several.

The regressive type of cell division occurs particularly after wounding and during the early stages of callus formation where it seems to be associated with the process of dedifferentiation (Gautheret, 1966; Aitchison, Macleod & Yeoman, 1977). Following the phase of rapid cell proliferation or 'regression phase', which tends to produce radial rows of cells at the surface of the tissue, there is a 'differentiation phase' during which cell division and cell expansion are in balance and the orientation of the planes of division alters so that more deep-seated meristematic nodules are produced. What is not certain is whether any of this regression ever leads to the re-establishment of embryogenic competence. Halperin (1967) and Street (1976a) have both suggested that embryogenic competence is only achieved during the special conditions which obtain in the primary explant, and that competence might then be maintained by proliferation of the induced cells. Murashige (1977) seems to go further by suggesting that 'it is not now possible to induce embryogenesis *in vitro*, although it has been possible to achieve an enhancement or repression in normally embryogenic tissues'. The question of whether embryogenic competence is ever restored under *in vitro* conditions should, however, remain open, since there seem to be no *a priori* reasons for assuming that it is not possible and the evidence, either way, is not convincing.

The other questions about embryogenic competence which must also remain open concern the issues of whether it constitutes *the* base state in plant cells which can be achieved by regression alone and what sort of signal is required to cause that type of cell to become embryogenically determined. It does, however, seem certain that once the cell is determined the permissive conditions required to activate and maintain embryogeny are relatively simple. The reported requirements for reduced nitrogen compounds (Wetherell & Dougall, 1976), particular levels of dissolved oxygen (Kessel & Carr, 1972), relatively high levels of potassium ions (Brown, Wetherell & Dougall, 1976) and a particular pH range (Wetherell & Dougall, 1976) seem to fall into the category of permissive conditions. The most characteristic feature of embryogeny *in vitro*, once the process has been initiated and relatively simple permissive conditions have been provided, is the high degree of autonomy in its progress towards the production of a normal plant. This suggests that the term 'embryogenic competence' is being used correctly, since it can be defined accurately in terms of the response – i.e. embryogeny – although the precise morphogenetic signal is unknown, unless it happens to be auxin.

ORGANOGENIC COMPETENCE

Any discussion at the present time about organogenic competence is complicated by the difficulty in establishing the organisational level at which the state might become established. It is widely believed that embryogenesis is usually initiated in a single cell and Street & Withers (1974) actually used that as one of the criteria to define the process, although it has been suggested that, *in vivo*, certain adventive embryos might arise from more than one cell (Maheshwari, 1950). Organogenesis, on the other hand, although sometimes seeming to begin in one cell, is usually seen to involve further cells, eventually, and Stewart (1978), on the basis of studies with chimaeras has suggested that the formation of an apical meristem always involves, at the time of determination, a group of three or more cells, which may or may not have been derived ultimately from a single cell. If this is correct, the states of organogenic competence and embryogenic competence would seem to operate at different organisational levels and, strictly, single cells should not be regarded as being organogenically competent, although they may have the competence to form a tissue which at some point becomes morphogenetically active. The chimerical nature of some organs derived from calluses containing populations of cells from two tobacco species (Carlson & Chaleff, 1975), certainly seems to argue against the suggestion (Davidson, Aitchison & Yeoman, 1976) that that type of co-ordination might only develop in an ordered fashion in a group derived from a single cell. This distinction between the organisational qualities of the states of embryogenic and organogenic competence could have practical implications with regard to the way in which the structure of a callus might influence its morphogenetic behaviour, so that the potential of a friable callus is different from that of a more structured callus.

RELATIONSHIP BETWEEN EMBRYOGENIC AND ORGANOGENIC COMPETENCE

Although a distinction might be made between the development of embryos and organs on the basis of the number of cells from which they are derived, in both cases an important stage is the production of a morphogenetically active group of cells in which pattern-formation mechanisms become established. Flexibility in the way in which these groups of cells can be produced seems to be tolerated, with the embryogenically competent groups being more frequently made up of cells derived, ultimately, from a single cell. The question that arises is whether these groups of cells have multiple competence so that embryos or organs are produced according to extrinsic conditions or whether they are determined from an early stage.

Many plant tissue culture workers seem to favour the view that the first stage in organogenesis, and possibly embryogenesis, is the production of an undetermined primordium. This view derives particularly from Bünning's (1952) concept of the 'meristemoid', a cell in which the 'character of embryonic cells' is to some extent regained as the result of an unequal division, although the idea of flexibility in the development of plant organs has a much longer history (see discussions by Peterson, 1975 and Halperin, 1978). Bünning illustrated the meristemoid concept, particularly, by examples of cellular differentiation which could be related to earlier unequal divisions, but he extended his arguments to include higher levels of organisation, suggesting that there is only one type of primordium (anlage) 'which can lead to a great multiplicity of structures depending on special conditions'. Subsequently, the meristematic nodules which are present in many callus cultures have been equated with meristemoids (see Thorpe, 1978a, b), but it is debatable whether such structures should be regarded as uncommitted primordia or anlagen, in Bünning's sense and, particularly, whether they can also become embryos, as suggested by Murashige (1974, 1977). Certainly, organs can arise *from* these structures, but they also arise from meristematic tissues which have a more two-dimensional arrangement, such as at the surface of calluses. Further, these more deep-seated meristematic structures are a regular feature of callus development (see Gautheret, 1966; Aitchison, Macleod & Yeoman, 1977) and frequently they do not appear to be organogenic. It is probably more accurate to regard such structures as providing one of the various means by which an intimate arrangement of competent cells might arise so that organogenesis can be initiated, given appropriate conditions. The question of whether the structures can also be embryogenic is complicated by the difficulty in defining embryogenesis in plants, if the early stages are not readily observed. It is not uncommon for the complementary organs to be formed adventitiously on or in the vicinity of organs which have been initiated in callus and there is no doubt that the term 'embryoid' has been used loosely to describe both genuine adventive embryos and structures which are strictly of organogenic origin. Whether a meristematic nodule which produces both a shoot and root should be regarded as embryogenic depends on how embryogeny is defined but, certainly, such a process bears little resemblance to the process as it occurs *in ovulo*.

It is difficult to decide from direct observation, therefore, whether the meristematic nodules have the competence to be both embryogenic and organogenic. Indirect evidence comes, however, from the observation that although embryogenesis and organogenesis can sometimes be induced in the same tissue, as described for carrot (Halperin, 1966) and *Atropa belladonna* (Konar, Thomas & Street, 1972), more frequently the processes occur independently and rhizogeny, in particular, commonly occurs in tissues which have never shown any competence for embryogeny. Further, where embryogenic and organogenic competence have been monitored in long-term cultures it has generally been found that any decline in the two properties follows independent time courses, as shown with callus lines from two races of *Brachycome lineariloba* by Gould (1978). At the very least, this suggests that the two types of competence are not inseparable. It is not possible to say whether this indicates different competences of the individual cells – a differential 'unmasking' of potency, as suggested by Street (1976b) – or whether structural changes in the tissues alter competence at higher organisational levels.

Further evidence that embryogenic and organogenic competences are distinct states which are not necessarily shared by the same cells, comes from the observation (e.g. Rao, Handro & Harada, 1973; Rao, Bapat & Harada, 1976; Tanimoto & Harada, 1980) that organogenic cultures are sometimes more readily produced in the presence of the weaker auxins, IAA and NAA, and embryogenic cultures in the presence of the more powerful 2,4-D. Just as Steward (1967) argued that the more powerful growth regulator was required to eliminate the developmental constraints which accumulate in more mature carrot cells, it might be argued that the regression brought about by the less powerful growth regulators is incomplete and that it only restores organogenic competence.

The remaining question of whether organogenesis involves the production of an uncommitted primordium which can develop either as a root or shoot according to extrinsic factors will not be answered easily by direct observation of the process in callus tissues where the early stages of the process are not readily distinguishable. Arguments similar to those used in relation to embryogenic and organogenic competence can be used in the sense that rhizogeny and caulogeny frequently do not occur in the same tissues, suggesting that the two processes do not necessarily have a common origin. In this case, however, more direct evidence should become available from studies of organogenesis in leaf epidermal cells from species such as

Nautilocalyx lynchii (Venverloo, 1976), where it has been possible to keep individual organs under observation from the time of their inception. It would seem that, in this case at least, the patterns of division in the two types of organ are different from very early stages, with four or more of the original epidermal cells being involved in the production of a root primordium and as few as one or two of the cells contributing to a shoot primordium. The other major difference between the two types of primordium was the stage at which cell enlargement occurred, the first divisions having led to a reduction in cell size.

Since it would seem that individual epidermal cells could be involved in the production of either roots or shoots, according to their position in the explant and the types and concentrations of growth regulators employed, it might be argued that they have multiple competence. However, since these cells would not normally produce organs *in vivo*, unless the leaf is damaged it can also be argued that as long as they are subject to normal organisational constraints they have no organogenic competence. Such competence only arises after the constraints are removed, but the precise points at which it does arise and at which the determination of organ type occurs are difficult to identify. One possible interpretation is that the first divisions of the cells are 'regressive', in the sense that has already been discussed, and that the observed differences in the balance between cell division and cell enlargement, presumably under the influence of the growth regulators, lead to different degrees of regression and, consequently, to different competences being restored. If this interpretation is correct, it would suggest that in this situation the organ type is determined from the very earliest stages of development, and that the growth regulators do not act as specific organ determinants except in the sense that they influence the removal of developmental constraints as a result of their effects on nuclear-cytoplasmic interactions. There are other possible interpretations, but these particular observations do not seem to provide support for the view that organogenesis involves the initial production of uncommitted primordia. In other situations, possibly, regression is more complete, passing the point at which rhizogenic and caulogenic competence are separable; multiple organogenic competence would then occur and specific determinants might be required.

CONCLUSIONS

Strictly, therefore, for the concepts of competence, determination and potency to be used in a logical manner, it is necessary to define the system for which they are being employed. In practice, the terms are often used loosely and a plant cell might be described as being totipotent even when, under a particular set of conditions, it clearly has no competence to express its full potential. In that case it is the cell's 'virtual potentiality', in Johansen's (1950) sense, which is really being described and the effect is to divert attention from the cell's lack of competence. With regard to the practical aim of regenerating plants from cell culture systems, it is important that the concepts should be used correctly so that the true nature of the problem is understood. If, as a result, it appears that the real problem is one of removing developmental constraints, rather than extending the search for specific morphogenetic substances, the approach might well be altered.

Since plants can be regenerated by both embryogenic and organogenic pathways, it is necessary to examine how the concepts should be applied to each of those processes. When this is done, attention is drawn to the questions of when and how organisational constraints are removed and when determination takes place. This is important if only because it emphasises the need to distinguish between the roles that exogenous growth regulators play in removing constraints and as specific morphogenetic determinants. There is little evidence that the role of exogenous growth regulators in embryogenesis involves more than the removal of constraints and the maintenance of a population of embryogenically competent cells; beyond that their effects on the embryogenic process are generally inhibitory. Similar comments can be made about organogenesis except that there might also be a requirement for an exogenous supply of regulators at critical stages in the organogenic process, since endogenous supplies, normally produced elsewhere in the organism, will not be available. If the role of exogenous growth regulators in morphogenetic studies with cell cultures proves to be largely one of removing developmental constraints, the question that arises is whether there are somatic cell states in plants which are stable enough to resist the influence of such compounds. If that is the case considerably more attention will have to be paid towards the problem of identifying suitably competent cells in the plant.

REFERENCES

AITCHISON, P. A., MACLEOD, A. J. & YEOMAN, M.M. (1977). Growth patterns in tissue (callus) cultures. In *Plant Tissue and Cell Culture*, 2nd edn, ed. H. E. Street, pp. 267–306. Oxford: Blackwell.

BACKS-HÜSEMANN, D. & REINERT, J. (1970). Embryobilding durch isolierte Einzelzellen aus Gewebekulturen von *Daucus carota. Protoplasma*, **70**, 49–60.

BROETJES, C., HACCIUS, B. & WEIDLICH, S. (1968). Adventitious bud formation on isolated leaves and its significance for mutation breeding. *Euphytica*, **17**, 321–44.

BROWN, S., WETHERELL, D. F. & DOUGALL, D. K. (1976). The potassium requirement for growth and embryogenesis in wild carrot suspension cultures. *Physiologia Plantarum*, **37**, 73–9.

BÜNNING, E. (1952). Morphogenesis in plants. *Survey of Biological Progress*, **2**, 105–40.

CARLSON, P. S. & CHALEFF, S. (1975), Heterogeneous associations of cells formed *in vitro*. In *Genetic Manipulations with Plant Materials*, ed. L. Ledoux, pp. 254–61. New York: Plenum Press.

D'AMATO, F. (1978). Chromosome number variation in cultured cells and re-generated plants. In *Frontiers of Plant Tissue Culture 1978*, ed. T. A. Thorpe, pp. 287–95. Calgary: The International Association for Plant Tissue Culture 1978.

DAVIDSON, A. W., AITCHISON, P. A. & YEOMAN, M. M. (1976). Disorganized systems. In *Cell Division in Higher Plants*, ed. M. M. Yeoman, pp. 407–32, London: Academic Press.

ESAU, K. (1953). *Plant Anatomy*. New York: Wiley.

ESPOSITO, M. S. & ESPOSITO, R. E. (1977). Gene conversion, paramutation, and controlling elements: a treasure of exceptions. In *Cell Biology – A Comprehensive Treatise*, vol. 1. ed. L. Goldstein & D. M. Prescott, pp. 59–92. London: Academic Press.

GAUTHERET, R. J. (1966). Factors affecting differentiation of plant tissues grown *in vitro*. In *Cell Differentiation and Morphogenesis*, ed. W. Beermann, pp. 55–95. Amsterdam: North Holland.

GOULD, A. R. (1978). Diverse pathways of morphogenesis in tissue cultures of the composite *Brachycome lineariloba* (2n = 4). *Protoplasma*, **97**, 125–35.

GURDON, J. B. (1976). The pluripotentiality of cell nuclei. In *The Developmental Biology of Plants and Animals*, ed. C. F. Graham & P. F. Wareing, pp. 55–63. Oxford: Blackwell.

HABERLANDT, G. (1902). Kulturversuche mit isolierten Pflanzenzellen. *Sitzungsberichte der Akadamie der Wissenschaften in Wien. Mathematische-naturwissenschaftliche Klasse*, **111**, 69–92.

HAIGHT, T. H. & KUEHNERT, C. C. (1971). Developmental potentialities of leaf primordia of *Osmunda cinnamomea*. VI. The expression of P_1. *Canadian Journal of Botany*, **49**, 1941–5.

HALPERIN, W. (1966). Alternative morphogenetic events in cell suspensions. *American Journal of Botany*, **53**, 443–53.

HALPERIN, W. (1967). Population density effects on embryogenesis in carrot cell cultures. *Experimental Cell Research*, **48**, 170–3.

HALPERIN, W. (1969). Morphogenesis in cell cultures. *Annual Review of Plant Physiology*, **20**, 395–418.

HALPERIN, W. (1978). Organogenesis at the shoot apex. *Annual Review of Plant Physiology*, **29**, 239–62.

HALPERIN, W. & WETHERELL, D. F. (1964). Adventive embryony in tissue cultures of the wild carrot, *Daucus carota. American Journal of Botany*, **51**, 274–83.

HESLOP-HARRISON, J. (1967). Differentiation. *Annual Review of Plant Physiology*, **18**, 325–48.

HOLTZER, H. & ABBOTT, J. (1968). Oscillations of the chondrogenic phenotype *in vitro*. In *The Stability of the Differentiated State*, ed. H. Ursprung, pp. 1–16. Berlin: Springer-Verlag.

JOHANSEN, D. L. (1950). *Plant Embryology: Embryogeny of the Spermaphyte*. Waltham: Chronica Botanica Co.

JONES, L. H. (1974). Factors influencing embryogenesis in carrot cultures (*Daucus carota* L.) *Annals of Botany*, NS, **38**, 1077–88.

KESSEL, R. H. J. & CARR, A. H. (1972). The effect of dissolved oxygen concentration on growth and differentiation of carrot (*Daucus carota*) tissue. *Journal of Experimental Botany*, **23**, 996–1007.

KHAVKIN, E. E., MARKOV, E. YU. & MISHARIN, S. I. (1980). Evidence for proteins specific for vascular elements in intact and cultured tissues. *Planta*, **148**, 116–23.

KOHLENBACH, H. W. (1978). Comparative somatic embryogenesis. In *Frontiers of Plant Tissue Culture 1978*, ed. T. A. Thorpe, pp. 59–66. Calgary: The International Association for Plant Tissue Culture 1978.

KONAR, R. N., THOMAS, E. & STREET, H. E. (1972). The diversity of morphogenesis in suspension cultures of *Atropa belladonna* L. *Annals of Botany*, NS, **36**, 249–58.

MCWILLIAM, A. A., SMITH S. M. & STREET, H. E. (1974). The origin and development of embryoids in suspension cultures of carrot (*Daucus carota*). *Annals of Botany*, NS, **38**, 243–50.

MAHESHWARI, P. (1950). *An Introduction to the Embryology of Angiosperms*. New York: McGraw-Hill.

MUIR, W. H. (1953). Cultural conditions favouring the isolation and growth of single cells from higher plants *in vitro*. Ph. D. Thesis. University of Wisconsin, Madison, USA.

MURASHIGE, T. (1974). Plant propagation through tissue culture. *Annual Review of Plant Physiology*, **25**, 135–66.

MURASHIGE, T. (1977). Manipulation of organ initiation in plant tissue cultures. *Botanical Bulletin Academica Sinica*, **18**, 1–24.

MURASHIGE, T. (1978). The impact of plant tissue culture on agriculture. In *Frontiers of Plant Tissue Culture 1978*. ed. T. A. Thorpe, pp. 15–26. Calgary: The International Association for Plant Tissue Culture 1978.

NARAYANASWAMY, S. (1977). Regeneration of plants from tissue cultures. In *Plant Cell, Tissue and Organ Culture*, ed. J. Reinert & Y. P. S. Bajaj, pp. 179–206. Berlin: Springer-Verlag.

PERIASAMY, K. (1962). Morphological and ontogenetic studies in palms. I. Development of the plicate condition in the palm-leaf. *Phytomorphology*, **12**, 54–64.

PETERSON, R. L. (1975). The initiation and development of root buds. In *The Development and Function of Roots*, ed. J. G. Torrey & D. T. Clarkson, pp. 125–61. London: Academic Press.

PFEFFER, W. (1900). *The Physiology of Plants*, 2nd edn, vol. 1, transl. A. J. Ewart. Oxford: Clarendon.

RAFF, J. W., HUTCHINSON, J. F., KNOX, R. B. & CLARKE, A. E. (1979). Cell recognition: antigen determination of plant organs and their cultured callus cells.

Differentiation, **12**, 179–86.

RAO, P. S., BAPAT, V. A. & HARADA, H. (1976). Gamma radiation and hormonal factors controlling morphogenesis in organ cultures of *Antirrhinum majus* L. c.v. Red Majestic Chief. *Zeitschrift für Pflanzenphysiologie*, **80**, 144–52.

RAO, P. S., HANDRO, W. & HARADA, H. (1973). Hormonal control of differentiation of shoots, roots and embryos in leaf and stem cultures of *Petunia inflata* and *Petunia hybrida*. *Physiologia Plantarum*, **28**, 458–63.

REINERT, J. (1958). Untersuchungen über die Morphogenese an Gewebekulturen. *Berichte der Deutschen Botanischen Geseleschaft*, **71**, 15.

SCHLEIDEN, J. M. (1838). Beiträge zur Phytogenesis. *Archiv für Anatomie, Physiologie und wissen schaftliche Medizin*, **4**, 137–76.

SCHLEIDEN, J. M. (1849). *Principles of Scientific Botany*, transl. E. Lankester. London: Longman, Brown, Green & Longmans.

SCHWANN, TH. (1838). *Mikroskopische Untersuchungen über die Übereinstimmung in der Struktur und dem Wachstume der Thiere und Pflanzen*, Ostwalds Klassiker Nr 176. Leipzig: Engelmann.

SINNOTT, E. W. (1946). Substance or system: the riddle of morphogenesis. *The American Naturalist*, **80**, 497–505.

SKOOG, F. & MILLER, C. O. (1957). Chemical regulation of growth and organ formation in plant tissues cultured *in vitro*. *Symposium of the Society for Experimental Biology*, **11**, 118–30.

STEWARD, F. C. (1967). *Growth and Organization in Plants*. Reading, Mass.: Addison-Wesley.

STEWARD, F. C., AMMIRATO, P. V. & MAPES, M. O. (1970). Growth and development of totipotent cells. *Annals of Botany*, NS, **34**, 761–87.

STEWARD, F. C., MAPES, M. O. & MEARS, K. (1958). Growth and organized development of cultured cells. II. Organization in cultures grown from freely suspended cells. *American Journal of Botany*, **45**, 705–8.

STEWARD, F. C., MAPES, M. O., KENT A. E. & HOLSTEN, R. D. (1964). Growth and development of cultured plant cells. *Science, NY*, **143**, 20–7.

STEWART, R. N. (1978). Ontogeny of the primary body in chimeral forms of higher plants. In *The Clonal Basis of Development*, ed. S. Subtelny & I. M. Sussex, pp. 131–60. London: Academic Press.

STREET, H. E. (1976a). Experimental embryogenesis – The totipotency of cultured plant cells. In *The Developmental Biology of Plants and Animals*, ed. C. F. Graham & P. F. Wareing, pp. 73–90. Oxford: Blackwell.

STREET, H. E. (1976b). Cell cultures: a tool in plant biology. In *Cell Genetics in Higher Plants*, ed. D. Dudits, G. L. Farkas & P. Maliga, pp. 7–38. Budapest: Akadéimiai Kiadó.

STREET, H. E. & WITHERS, L. A. (1974). The anatomy of embryogenesis in culture. In *Tissue Culture and Plant Science 1974*, ed. H. E. Street, pp. 71–100. London: Academic Press.

SUSSEX, I. M. (1972). Somatic embryos in long term carrot tissue cultures:Histology, cytology and development. *Phytomorphology*, **22**, 50–9.

TANIMOTO, S. & HARADA, H. (1980). Hormonal control of morphogenesis in leaf explants of *Perilla frutescens* Britton var. *crispa* Decaisne f. *Viridi-crispa* Makino. *Annals of Botany*, NS, **45**, 321–7.

THOMAS, E. & WERNICKE, W. (1978). Morphogenesis in herbaceous crop plants. In *Frontiers of Plant Tissue Culture 1978*, ed. T. A. Thorpe, pp. 403–10. Calgary: The International Association for Plant Tissue Culture 1978.

THORPE, T. A. (1978a). Physiological and biochemical aspects of organogenesis *in vitro*. In *Frontiers of Plant Tissue Culture 1978*, ed. T. A. Thorpe, pp. 49–58. Calgary: The International Association for Plant Tissue Culture 1978.

THORPE, T. A. (1978b). Regulation of organogenesis *in vitro*. In *Propagation of Higher Plants Through Tissue Culture*, ed. K. W. Hughes, R. Henke & M. Constantin, pp. 87–101. US Department of Energy: Technical Information Center.

VENVERLOO, C. J. (1976). The formation of adventitious organs. III. A comparison of root and shoot formation on *Nautilocalyx* explants. *Zeitschrift für Pflanzenphysiologie*, **80**, 310–22.

VÖCHTING, H. (1878). *Über organbildung in Pflanzenreich*. Bonn: Max Cohen.

WADDINGTON, C. H. (1932). Experiments on the development of chick and duck embryos, cultivated *in vitro*. *Philosophical Transactions of the Royal Society*, **B221**, 179–230.

WADDINGTON, C. H. (1962). *New Patterns in Genetics and Development*. New York: Columbia University Press.

WADDINGTON, C. H. (1968). Theoretical biology and molecular biology. In *Towards a Theoretical Biology*, vol. 1, ed. C. H. Waddington, pp. 103–8. Edinburgh University Press.

WADDINGTON, C. H. (1970). Concepts and theories of growth, development, differentiation and morphogenesis. In *Towards a Theoretical Biology*, vol. 3, ed. C. H. Waddington, pp. 177–97. Edinburgh University Press.

WARDLAW, C. W. (1965). General physiological problems of embryogenesis in plants. In *Encyclopedia of Plant Physiology*, vol. 15 part 1, ed. W. Ruhland, pp. 424–42. Berlin: Springer-Verlag.

WAREING, P. F. (1978). Determination in plant development. *Botanical Magazine*, Tokyo, *Special Issue*, **1**, 3–17.

WAREING, P. F. (1979). What is the basis of the stability of apical meristems? In *Differentiation and the Control of Development in Plants – Potential for Chemical Modification*, ed. E. C. George, pp. 1–9. Wantage: British Plant Growth Regulator Group.

WETHERELL, D. F. & DOUGALL, D. K. (1976). Sources of nitrogen supporting growth and embryogenesis in cultured wild carrot tissue. *Physiologia Plantarum*, **37**, 97–103.

WETMORE, R. H. & RIER, J. P. (1963). Experimental induction of vascular tissues in callus of angiosperms. *American Journal of Botany*, **50**, 418–30.

WHITE, P. R. (1954). *The Cultivation of Animal and Plant Cells*, New York: The Ronald Press Company.

Differentiation within the immune system: the importance of cloning

A. A. CZITROM, N. A. MITCHISON AND YEH MING

Tumour Immunology Unit, Department of Zoology, University College London, Gower Street, London WC1E 6BT

Ideas about the working of the immune system are now at an extraordinarily exciting stage. We are entering a period of crisis in which many old and seemingly well-established views are coming into question, and where there is little general agreement about the conceptual framework which will replace them. This is happening at a time when certain important immunological concepts, which have for decades been the subject of intense controversy, have suddenly become much clearer. At first sight it may seem odd that clarification should generate crisis, but this is really perfectly natural. Only when we think clearly can we perceive the intellectual difficulties.

So first of all let us make clear where in our opinion the clarifications have taken place. After all, not all colleagues agree that this has happened at all, and as usual the biggest obfuscators tend to be those most actively engaged in research on the topics which others think have been clarified. It is their business – and their material interest – to point out that *much remains to be done*. So let us simply point out the two great clarifications: the working of the immunoglobulin genes, and the working of molecules encoded by the major histocompatibility complex. As regards the immunoglobulin genes, it is now generally agreed that each peptide chain which goes to make up an immunoglobulin molecule is comprised of three portions, V (variable), J (junctional) and C (constant) in that order. Each portion is encoded by a separate family of genes which are arranged in sequence on a chromosome. During the differentiation of a B cell (and perhaps also a T cell) one member of each of these three families is brought into juxtaposition, and it is these three juxtaposed genes which are alone expressed. The bringing into juxtaposition is done by DNA deletion. For a review of the evidence for this process, see Tonegawa, Breck, Hirama & Lenhard-Schuller (1979); here, as elsewhere, the most authoritative and most recent reviews have appeared in the Proceedings of the IV International Congress of Immunology.

For the purposes of our discussion here, the main consequence of

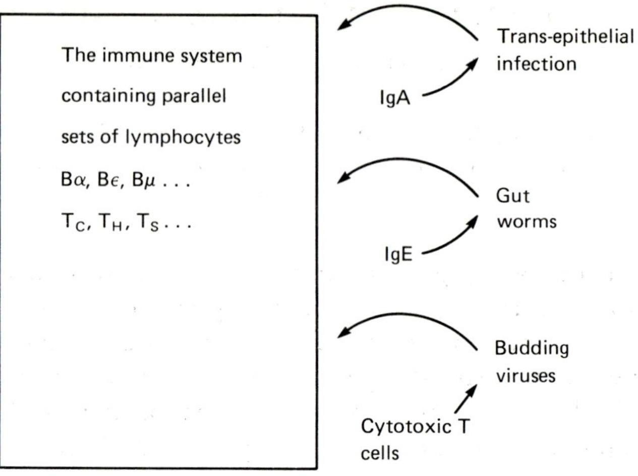

Fig. 1. Appropriate responses.

this genetic process is that it generates parallel sets of lymphocytes. Each set is defined by expression of a single C gene in combination with the full repertoire of V and J genes. The function of the various sets is to operate distinct effector functions. For example cells expressing the Cα gene produce antibody of the IgA class, and thereby protect against trans-epithelial infection in the gut and elsewhere. This in turn poses a major problem in immunoregulation: how does the immune system come to deliver an appropriate response, in terms of activating the appropriate set of lymphocytes, to a particular form of infection? This concept of appropriateness lies at the heart of our discussion here, see Fig. 1, and it must be remembered that it concerns lymphocyte parallel sets and C-regions, not the older immunological concept of specificity which concerns V and J regions. Oddly enough, although this question has been well understood for many years as a problem in B-lymphocyte regulation, it is there still largely unanswered; the best we can do is invoke groups of poorly understood immunoglobulin class-specific regulatory T-lymphocytes. On the other hand for T cells, whose receptors and division into parallel sets are in comparison poorly understood, we now have an excellent theory about the guidance of appropriate responses. One further point is worth making. At the level of DNA this arrangement provides a marvellous example of economy. The combination of tandem spacing and excision permits the same pieces of DNA to be used over and over again in different contexts. At the level of cells, on the other hand, we see profligacy. By representing the same repertoire of antibody-combining sites over and over again, the immune system becomes far larger than might have been thought necessary, and demands appropriately cumbersome regulatory mechanisms. Why not make decisions about immunoglobulin class *after* rather than *before* antigenic stimulation, thus making tremendous savings in number of cells? Like all what-might-have-been speculations, we cannot answer this question with confidence. But it seems likely that the profligacy reflects *either* difficulty in constructing a mechanism of controlled C genes switching, *or* a need for speed in delivering an appropriate response which can be accommodated only by pre-selection of C genes, *or* both, in the sense that controlled switching could be arranged but would take too long.

As regards the second great clarification, over the major histocompatibility complex, we have taken three steps forward. For review see Zinkernagel (1978a, b). The first was to recognize that all of the apparently diverse functions controlled by single 'regions' within the complex are in fact carried out by the same molecule. Thus a single molecule, for example the 44k membrane glycoprotein encoded by H-2K, functions as a restriction element, a serologically defined allo-antigen, a T-cell-defined allo-antigen, and an immune response gene. The second was to recognize that the apparently different functions of 44k molecules and 28/32k molecules (respectively Class I and Class II in J. Klein's nomenclature) are misleading, and that both types of molecule function in the same way but in relation to different T lymphocyte sets. And the third was to recognize that the natural function of these molecules, as distinct from their artificial functions as allo-antigens, is to guide T lymphocyte parallel sets: 44k molecules guide cytotoxic T cells, 28/32k molecules guide helper T cells, and (but the evidence for this is weaker) I-J molecules, of as yet undetermined size, guide suppressor T cells.

Up to a point these parallel sets resemble the parallel sets of B lymphocytes: in both cases each set contains essentially a full repertoire of receptors, subject presumably only to perturbations resulting from antigenic selection. Consequently a given antigen is potentially able to stimulate a subset of cytotoxic, helper, and suppressor T cells, and which it chooses is determined by the MHC molecule which the T cells see it to be associated with. Thus the guidance of appropriate T cells is at least partially understood at the molecular level, something which, as has been mentioned, is still lacking for B cells.

The main gap in our knowledge of T cells is that we still do not know how their receptors work. Evidently they fulfil two functions, recognition of antigens and recognition of MHC guiding molecules. Whether these functions are fulfilled by one receptor or two is unknown. So also is whether an immunoglobulin-like molecule is involved, and so therefore is whether the various T cell parallel sets express distinct C genes. Our guess is that the next important evidence bearing on these questions will come from DNA chemistry. Our contribution, in collaboration with Allan Williamson in Glasgow, is to screen a large number of randomly picked T cell lines with one or two VH probes.

Whatever the molecular nature of the T cell receptor, it functions, as has been often pointed out, in a neat way. Dual recognition of antigen and MHC molecule kills two birds with one stone. The requirement for the MHC molecule ensures that the appropriate T cell parallel set is chosen for activation. At the same time it ensures that T cells are not activated prematurely, by encounter with antigen free in body fluids. This is obviously more important for a cell whose effector function is essentially local, than for a B cell which releases antibody to act at a distance.

Our discussion thus far has been of areas where intellectual strife is settling down. The only place where colleagues might disagree seriously is on the role of I-J, a matter where we have direct experience and hold strong views (Czitrom, Sunshine & Mitchison, 1980). The discussion now turns to the present crisis.

REGULATORY CIRCUITS AND INTEGRATIVE PROPERTIES

The crisis has arisen because of the increase in complexity of the immune system. At its simplest, this complexity is reflected in the proportion of cells of the immune system which are now known to engage in regulatory rather than effector activity. Like a modern army, few individuals are in the front line; more are engaged in administration, supply, and intelligence. The same is true of our nervous system: more neurons act on one another than ever watch the outside world or act upon it. For 60% of lymphocytes which are T cells, we can roughly calculate a proportion of 70% as regulatory (all the 40% of Lyt. 1 cells, plus say half the 10% of Lyt. 2, 3 cytotoxic or suppressor cells, plus say half the remaining 50% of Lyt. 1, 2, 3 cells). For the 40% B cells, say half look at one another's idiotypes. Say 60–70% regulatory cells overall.

How are these regulatory cells divided up? Until recently a reasonable list would have run as follows: T helper cells, divided into three main categories, mediating antigen-specific help, immunoglobulin-specific help, and non-specific help (via factors of the T cell growth factor or allogeneic effect factor type); immunoglobulin specific help would have been further subdivided into idiotype-, allotype-, and isotype-specific help; and for each helper cell there would have been an equivalent suppressor cell, making say ten categories of regulatory T cell in all. Add at least two further categories of regulatory B cell, mediating antigen-concentration and feedback suppression and we have a round dozen regulatory cells. This is already a large number, but one which could still be visualized in terms of a simple balance between two opposing armies, the up-regulators and down-regulators (for a diagrammatic representation of this balance, see Mitchison, 1979a).

Rather suddenly the situation has changed. In the first place, many new regulatory cells have entered the scene. They are identified by new surface markers such as Qal (Cantor *et al.*, 1978a) and the I-J subspecificities (Tada *et al.*, 1979), as well as by their ability on occasion to release various soluble immunoregulatory factors (Feldmann & Kontiainen, 1980). This work has been conducted mainly in culture systems, but others have reached similar conclusions through the use of cell transfers (Herzenberg, Black & Herzenberg, 1980). Functions as far-fetched as suppressor-amplification and contra-suppression have been proposed.

Secondly, the concept of a regulatory circuit has been introduced. The term circuit can be applied loosely to any pattern of interacting cells in which some cells control the activity of others. But it now carries a more restricted connotation, of a circuit which feeds back on itself in a reverberatory manner and therefore has flip-flop properties (Cantor & Gershon 1979; Herzenberg *et al.*, 1980; for further discussion see Mitchison, 1979b). While there is no direct evidence that circuits of this type operate in the immune system, all the necessary components have been isolated separately and the system as a whole has properties which suggest that flip-flops operate.

Understandably this increasingly complex picture of immune regulation has not exactly proved welcome. As ever in biology, there are splitters and lumpers. Concepts in cellular immunology come and go, and the lumpers nurse the hope that some of the more abstruse postulates will blow away. So perhaps it is worth listing the main evidence in support of circuitry.

(1) The trend to complexity. When suppressor cells were first proposed some of us hoped that they would not last. But then came the Lyt markers, which by assigning a special phenotype to suppressor cells as distinct from helpers set them on a firm footing.

(2) Confirmatory evidence from man. The Lyt markers were defined originally as allo-antigens in the mouse. Molecules have now been identified on the membranes of human lymphocytes which resemble them not only in terms of the function of the cells which they mark (Kung, Goldstein, Reinherz & Schlossman, 1979), but also in terms of physical properties. This has recruited many clinicians to the splitters.

(3) There is increasing evidence that immunological diseases are associated with abnormal representation of sub-sets of regulatory T cells. In the mouse, five separate congenital auto-immunities are each associated with distinct T cell perturbations although it is by no means generally accepted that these perturbations actually cause the disease (Cantor *et al.*, 1978b; Theofilopoulos *et al.*, 1979). The prototype examples of auto-immunity in the mouse, the New Zealand strains, have all manner of defects in other parts of their immune system (Knight & Adams, 1978; Knight, Knight & Winchester, 1981). There is also direct evidence of loss of human suppressor cells during auto-immune disease (Strelkauskas *et al.*, 1978).

POSITIONAL INFORMATION: GRADIENTS VERSUS SURFACE STRUCTURES

This volume contains a number of chapters on differentiation in limb buds and somites. These are rather simple systems, at least in terms of the number of molecules affected ($< 10^2$?). The immune system is very complex in this respect ($> 10^6$ molecules?), and the nervous system must be of intermediate complexity. The simple systems would seem characteristically to involve Positionally Informative Gradients, or PIG for short. In contrast the immune system depends characteristically on positionally informative surface structures, as exemplified by MHC molecules. It is quite clear that nothing as complex as the immune system could possibly organize itself by gradients rather than by surface structures, particularly when the cells concerned move about. The interesting question is how the nervous system gets organized: positionally informative surface structures may well turn out to be crucial there, too.

T CELL CLONES

In response to the crisis of complexity many laboratories have turned to cloning. The assumption is that any worthwhile set of regulatory T cells will show up among the clones, and that this will enable it to be adequately examined. Furthermore it is hoped that a library of clones will enable us to do reconstruction experiments on regulatory circuits.

Several experimental approaches have been followed. One of them is simply to collect lymphomas and leukaemias as clonal samples (Greaves, 1975). This has told us much about markers, but little about function. Another is to fuse normal and malignant cells, and then select functionally active hybrids (Hämmerling, 1977; Kontiainen *et al.*, 1978; Nabholz *et al.*, 1980a). Yet another is to transform lymphocytes with virus, and again select for functional activity (Finn, Boniver & Kaplan, 1979). These procedures all have the drawback that only limited function can be recovered, and in particular that the ability of the cell to respond to antigen and other regulatory stimuli is largely lost. Some encouragement can be gleaned from recent work with B cell tumours and hybrids which suggests that responsiveness may not entirely be lost (Abbas, Ratnofsky & Burakoff, 1980; Boyd & Schrader, 1980), but what remains is slight.

More attention has therefore been paid to approaches which leave responsiveness to antigen more-or-less intact. The first long-term cultures which retained this property were simply mixed lymphocyte reactions in which fresh stimulator cells were added after an interval (MacDonald, Engers, Cerottini & Brunner, 1974). They established the principle, which was subsequently extended to long-term cell lines, that functionally active cytotoxic T cells could survive and proliferate provided that they received antigenic stimulation at intervals (Dennert & De Rose, 1976). Since then cloning of T cells has proceeded in two directions: the use of short-term clones to measure the frequency of various T cell sets by limit dilution analysis, and the use of long-term clones to characterize individual properties more fully. Polyclonal mitogens, particularly concanavalin A (Eichmann, Falk, Melchers & Simon, 1980) and T cell growth factor (TCGF) (Gillis & Smith, 1977) have proved useful in initiating and maintaining cultures. The subject has recently been comprehensively reviewed (MacDonald *et al.*, 1980; Nabholz *et al.*, 1980b; Wagner *et al.*, 1980; Schreier *et al.*, 1980).

Thus far clones of regulatory T cells have shown many of the same requirements and activities as the mixed populations from which they originate. These clones and long-term cultures of T helper cells do not respond to antigenic stimulation without the participation of specialized antigen-presenting cells (Julius & Augustin, 1979; Schreier *et al.*, 1980). The only likely exception would be helper cells responding to MHC antigens, and clones do indeed proliferate in response to HLA-D incompatible lymphocytes without any need for additional cells (Inouye *et al.*, 1980); however, these clones have not yet been shown to have helper activity.

There is some confusion about the way in which clones and long-term cultures of T cells exert helper effects. Those which are specific for protein antigens appear to act in a strictly antigen-specific way, in the sense that they will not help the responses of bystander cells (Waldmann, 1977; Julius & Augustin, 1979). With cellular antigens on the other hand helper activity *in vitro* appears to be non-specific, in the sense that once a cell has been stimulated by its own specific antigen it can help bystanders to respond to other antigens, presumably through the release of non-specific factors. This principle applies to the control of both the B cell response to foreign erythrocytes (Schreier *et al.*, 1980) and of the cytotoxic T cell response to allo-antigens (Glasebrook & Fitch, 1980). Yet cloned suppressor T cells responding to foreign erythrocytes release an antigen-specific effector molecule (Kontiainen *et al.*, 1978; McVay-Boudreau, Fresno, Nabel & Cantor, 1980). These differences may be more apparent than real, and may reflect in part the way that *in vitro* assay systems can mislead. It is significant that one T helper clone has displayed a bystander effect when tested *in vitro* but not when tested *in vivo* (Tees & Schreier, 1980; Schreier *et al.*, 1980).

In choosing our own approach to these problems we have been influenced by two considerations. One is the advantage of using MHC molecules as antigens, in that they remove the need for adding specialized antigen-presenting cells. The other is the need for functional assays which (a) can be performed easily *in vitro*, for screening purposes, and which (b) detect physiological regulatory activity, in the sense of intrastructural help (Lake & Mitchison, 1977). We have accordingly chosen to work mainly with allo-H-2I antigens, and have used two assays, the mixed lymphocyte proliferation assay and a transfer system to detect physiological helper activity.

Our system is illustrated in Fig. 2, together with our best data so far. The cell line was established by stimulating *in-vivo*-primed lymphocytes with antigen in combination with TCGF, providing the latter in

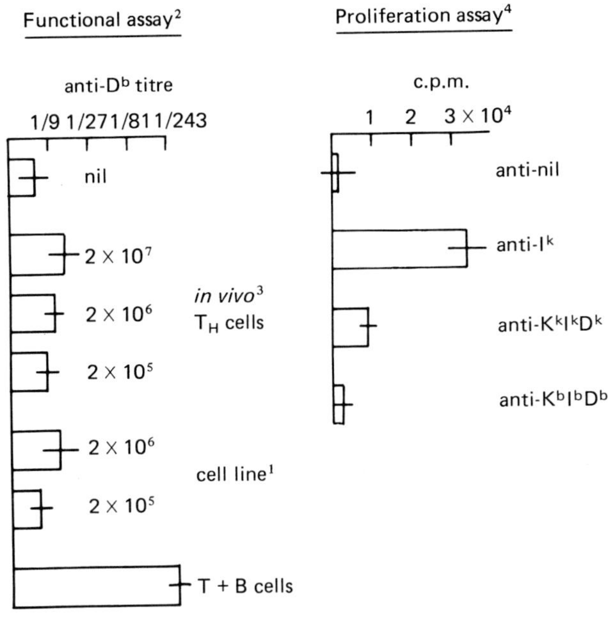

[1] ATH × Balb anti-ATL (anti-I^k), 4 × 7–14 day passages with occasional restimulation.

[2] Adoptive transfer, with B cells (anti-C57, anti-K^{b}I^{b}D^b) plus antigen (2R, K^{k}I^{k}D^b).

[3] 10^7 ATL cell – 7 days.

[4] 3-day mixed lymphocyte reaction: anti-I^k, ATL stimulators; anti-K^{k}I^{k}D^k, CBA stimulators; anti-K^{b}I^{b}D^b, C57 stimulators.

Fig. 2. An anti-H-2I cell line.[1]

the form of supernatant from con-A-stimulated rat lymphocytes. The proliferation data are scattered but clearly show that the line can respond to its specific antigen; we do not regard the difference between the two I^k stimulators as significant. The functional assay follows principles which have been described previously (Lake & Mitchison, 1977). The data are again unimpressive, in this assay because the effects produced by the helper cells are small. Nevertheless they are consistent, and suggest that this cell line can indeed deliver physiological help. Obviously these results are worth little; but they encourage us to believe that this form of physiological help will be amenable to full clonal analysis.

CLONES: HOW FAR WILL THEY TAKE US?

The importance of clones in testing and validating the various hypothetical T cell subsets has already been emphasized. We can add a variety of unexpected or otherwise unobtainable findings which clones have already turned up. The list includes: (1) odd specificities and cross-reactions detected by cytotoxic clones, such as F_1 hybrid antigens (von Boehmer *et al.*, 1979; Nabholz *et al.*, 1980b), (2) subdivision of antigen-specific helper cells, according to their susceptibility to suppression (Eichmann *et al.*, 1980), and (3) very precise testing of the matching requirements of helper cells and B cells for H-2I antigens (Schreier *et al.*, 1980). We can add a list of semi-technical problems which clones are likely to solve: (1) whether cells mediating allogeneic help/suppression (Waterfield, Dennert, Swain & Dutton, 1979; Coutinho & Augustin, 1980) also mediate physiological help, (2) to what extent the MHC molecules which guide particular self-T-cell subsets under normal conditions preferentially activate allo-T-cells of the same subsets (Czitrom *et al.*, 1980), (3) to what extent the immune system sees minor allo-antigens in association with donor or host MHC molecules (Clark, Lake, Mitchison & Nakashima, 1979). These are in fact problems which we hope that our own system as it develops will help answer.

A major question remains, one which any reductionist approach raises. Can we ever understand how a complex system works merely by taking it apart? Flip-flopping of the type which we have discussed is a property of ordered circuits, not of individual clones or random assemblies of clones. Perhaps the future truly lies with the theoreticians, however abstract and unsatisfying their models may feel at present (Bell, Perelson & Pimbley, 1978; Grossman & Cohen, 1980).

REFERENCES

ABBAS, A. K., RATNOFSKY, S. E. & BURAKOFF, S. J. (1980). T lymphocyte-mediated suppression of myeloma function *in vitro*. Evidence for regulation of hapten-binding myeloma by syngeneic hapten-specific cytolytic T cells. *Journal of Experimental Medicine*, **154**, 306–23.

BELL, G. I., PERELSON, A. S. & PIMBLEY, G. H. (eds.) (1978). *Theoretical Immunology.* Dekker, New York.

VON BOEHMER, H., HENGARTNER, H., NABHOLZ, W., LENHARDT, W., SCHREIER, M. H. & HAAS, W. (1979). Fine specificity of a continuous growing killer cell clone specific for H-Y antigen. *European Journal of Immunology*, **9**, 592–7.

BOYD, A. W. & SCHRADER, J. W. (1980). Mechanisms of effector cell blockade. I. Antigen-induced suppression of Ig synthesis in a hybridoma cell line, and correlation with cell-associated antigen. *Journal of Experimental Medicine*, **151**, 1436–51.

CANTOR, H. & GERSHON, R. K. (1979). Immunological circuits: cellular compositions. *Federation Proceedings*, **38**, 2058–64.

CANTOR, H., HUGENBERGER, J., McVAY-BOUDREAU, L., EARDLEY, D. D., KEMP, J., SHEN, F. W. & GERSHON, R. K. (1978a). Immunoregulatory circuits among T cell sets. Identification of a subpopulation of T-helper cells that induce feedback inhibition. *Journal of Experimental Medicine*, **148**, 871–7.

CANTOR, H., McVAY-BOUDREAU, HUGENBERGER, J., NAIDORF, K., SHEN, F. W. & GERSHON, R. K. (1978b). Immunoregulatory circuits among T-cell sets. II. Physiological role of feedback inhibition *in vivo* observed in NZB mice. *Journal of Experimental Medicine*, **147**, 1116–25.

CLARK, E., LAKE, P., MITCHISON, N. A. & NAKASHIMA, I. (1979). Alternative routes of entry for cell surface antigens into the immune system. In *Natural and Induced Cell-Mediated Cytotoxicity*, ed. G. Riethmuller, P. Wernet & G. Cudkowicz, pp. 157–63. Academic Press, New York.

COUTINHO, A. & AUGUSTIN, A. A. (1980). Major histocompatibility complex-restricted and unrestricted T helper cells recognizing minor histocompatibility antigens of B cell surfaces. *European Journal of Immunology*, **10**, 535–41.

CZITROM, A. A., SUNSHINE, G. & MITCHISON, N. A. (1980). Suppression of the proliferative response to H-2D by I-J subregion gene products. *Immunogenetics*, **11**, 97–102.

DENNERT, G. & DE ROSE, M. (1976). Continuously proliferating T killer cells specific for H-2^{b} targets: selection and characterization. *Journal of Immunology*, **116**, 1601–6.

EICHMANN, K., FALK, I., MELCHERS, I. & SIMON, M. M. (1980). Quantitative studies on T cell diversity. I. Determination of the precursor frequencies for two types of Streptococcus A-specific helper cells in non-immune, polyclonally activated spleen cells. *Journal of Experimental Medicine*, **152**, 477–92.

FELDMANN, M. & KONTIAINEN, S. (1980). Antigen specific T cell factors. *Molecular and Cellular Biochemistry*, **30**, 177–93.

FINN, O. J., BONIVER, J. & KAPLAN, H. S. (1979). Induction, establishment in vitro, and characterization of functional, antigen-specific, carrier-primed murine T-cell lymphomas. *Proceedings of the National Academy of Sciences, USA*, **76**, 4033–7.

GILLIS, S. & SMITH, K. A. (1977). Long term culture of tumour specific cytotoxic T cells. *Nature (Lond.)*, **268**, 154.

GLASEBROOK, A. L. & FITCH, F. W. (1980). Alloreactive cloned T cell lines. 1. Interactions between cloned amplifier and cytolytic T cell lines. *Journal of Experimental Medicine*, **151**, 876–95.

GREAVES, M. F. (1975). *Cellular Recognition*. 'Outline Series in Biology'. Chapman and Hall, London.

GROSSMAN, Z. & COHEN, I. (1980). A theoretical analysis of the phenotypic expression of immune response genes. *European Journal of Immunology*, **10**, 633–50.

HÄMMERLING, G. J. (1977). T lymphocyte tissue culture cell lines produced by cell hybridization. *European Journal of Immunology*, **7**, 743–6.

HERZENBERG, L. A., BLACK, S. J. & HERZENBERG, L. A. (1980). Regulatory circuits and antibody responses. *European Journal of Immunology*, **10**, 1–11.

INOUYE, H., HANK, J. A., CHARDONENNES, X., SEGALL, M., ALTER, B. J. & BACH, F. H. (1980). Cloned primed-lymphocyte-test reagents in the dissection of HLA-D. *Journal of Experimental Medicine*, **152**, 143s–55s.

JULIUS, M. H. & AUGUSTIN, A. A. (1979). Helper activity of T cells stimulated in long-term culture. *European Journal of Immunology*, **9**, 671–80.

KNIGHT, J. G. & ADAMS, J. D. (1978). Three genes for lupus nephritis in NZB × N2W mice. *Journal of Experimental Medicine*, **147**, 1653–60.

KNIGHT, J. G., KNIGHT, A. & WINCHESTER, G. (1981). Evidence that autoimmunity in NZB mice is caused by a defect in immune specificity rather than a generalised deficit in tolerance of self antigens. *Cellular Immunology* (in press).

KONTIAINEN, S., SIMPSON, E., BOHRER, E., BEVERLEY, P. C. L., HERZENBERG, L. A., FITZPATRICK, W. C., VOGT, P., TORANO, A., MCKENZIE, I. F. C. & FELDMANN, M. (1978). T cell lines producing antigen-specific factor. *Nature (Lond.)*, **274**, 477–80.

KUNG, P. C., GOLDSTEIN, G., REINHERZ, E. L. & SCHLOSSMAN, S. S. (1979). Monoclonal antibodies defining distinctive human T cell surface antigens. *Science*, **206**, 347–9.

LAKE, P. & MITCHISON, N. A. (1977). Regulatory mechanisms in the immune response to cell surface antigens. *Cold Spring Harbor Symposia on Quantitative Biology*, **41**, 589–95.

MACDONALD, H. R., CEROTTINI, J.-C., RYSER, J.-E., MARYANSKI, J. L., TASWELL, C., WIDMER, M. B. & BRUNNER, K. T. (1980). Quantitation and cloning of cytolytic T lymphocytes and their precursors. *Immunological Reviews*, **51**, 93–123.

MACDONALD, H. R. H., ENGERS, H. D., CEROTTINI, J.-C. & BRUNNER, K. T. (1974). Generation of cytotoxic T lymphocytes. II. Effect of repeated exposure to alloantigens on the cytotoxic activity of long-term mixed lymphocyte cultures. *Journal of Experimental Medicine*, **140**, 718–30.

MCVAY-BOUDREAU, L., FRESNO, M., NABEL, G. & CANTOR, H. (1980). Functional analysis of Ig-specific suppressor molecules synthesised by T-suppressor clones. *Proceedings of the 4th International Congress of Immunology*. Academic Press, London.

MITCHISON, N. A. (1979a). Regulation of the response to cell surface antigens. In *Current Trends in Tumour Immunology*, ed. S. Ferrone, S. Gorini, R. B. Herberman & R. A. Reisfeld, pp. 111–18. Garland STPM Press, New York.

MITCHISON, N. A. (1979b). Regulatory circuits initiated by highly specialised antigen-presenting cells: the lesson for transplantation. *Behring Institut Mitteilungen*, **63**, 52–5.

NABHOLZ, M., CIANFRIGLIA, M., ACUTO, O., CONZELMANN, A., HAAS, W., VON BOEHMER, H., MACDONALD, H. R., POHLIT, H. & JOHNSON, J. P. (1980a). Cytolytically active murine T-cell hybrids. *Nature (Lond.)*, **287**, 437–40.

NABHOLZ, M., CONZELMANN, A., ACUTO, O., NORTH, M., HAAS, W., POHLIT, H., VON BOEHMER, H., HENGARTNER, H., MACH, J.-P., ENGERS, H. & JOHNSON, J. P. (1980b). Established murine cytolytic T-cell lines as tools for a somatic cell genetic analysis of T-cell functions. *Immunological Reviews*, **51**, 125–56.

SCHREIER, M. H., ISCOVE, N. N., TEES, R., AARDEN, L. & VON BOEHMER, H. (1980). Clones of killer and helper T cells: growth requirements, specificity and retention of function in long-term culture. *Immunological Reviews*, **51**, 315–36.

STRELKAUSKAS, A. J., CALLERY, R. T., MCDOWELL, J., BOREL, Y. & SCHLOSSMAN, S. F. (1978). Direct evidence for loss of human suppressor cells during active autoimmune disease. *Proceedings of the National Academy of Sciences, USA*, **75**, 5150–4.

TADA, T., NONAKA, M., OKUMURA, K., TANIGUCHI, M. & TOKUHISA, T. (1979). In *Cell Biology and Immunological Function of Leucocytes*, ed. M. Quastel, p. 353. Academic Press, New York.

TEES, R. & SCHREIER, M. H. (1980). Selective reconstitution of nude mice with long-term cultured and cloned specific helper T cells. *Nature (Lond.)*, **283**, 780.

THEOFILOPOULOS, A. N., EISENBERG, R. A., BOURDON, M., CROWELL, J. S. & DIXON, F. J. (1979). Distribution of lymphocytes identified by surface markers in murine strains with systemic lupus erythematosus-like syndromes. *Journal of Experimental Medicine*, **149**, 516–34.

TONEGAWA, S., BRECK, C., HIRAMA, M. & LENHARD-SCHULLER, R. (1979). Somatic recombination and structure of an immunoglobulin gene. In *Cells of Immunoglobulin Synthesis*, ed. B. Pernis & H. J. Vogel, pp. 15–32. Academic Press, New York.

WAGNER, H., HARDT, C., HEEG, K., PFIZENMAIER, K., SOLBACH, W., BARTLET, R., STOCKINGER, H. and ROLLINGHOFF, M. (1980). T-T cell interactions during cytotoxic T lymphocyte (CTL) responses: T cell derived helper factor (interleukin 2) as a probe to analyze CTL responsiveness and thymic maturation of CTL progenitors. *Immunological Reviews*, **51**, 215–55.

WALDMAN, H. (1977). Conditions determining the generation and expression of T helper cells. *Immunological Reviews*, **35**, 121–45.

WATERFIELD, J. D., DENNERT, G., SWAIN, S. L. & DUTTON, R. W. (1979). Continuously proliferating allospecific T cells. I. Specificity of cooperation with allogeneic B cells in the humoral antibody response to sheep erythrocytes. *Journal of Experimental Medicine*, **149**, 808–14.

ZINKERNAGEL, R. M. (1978a). Speculations on the role of major transplantation antigens in cell-mediated immunity against intramolecular parasites. *Current Topics in Microbiology and Immunology*, **82**, 113–38.

ZINKERNAGEL, R. M. (1978b). Thymic and lymphohaemopoietic cells: their role in T cell maturation in selection of T cells' H-2-restriction-specificity and in H-2 linked IR gene control. *Immunological Reviews*, **42**, 224–70.

Differences between cell differentiation *in vitro* and *in vivo*

L. WOLPERT

Department of Biology as Applied to Medicine, The Middlesex Hospital Medical School,
London W1P 6DB, UK

Cell differentiation is a major process in development, but it is not the only important process. Pattern formation, which is the spatial organisation of cellular differentiation, and changes in form – morphogenesis in its strict sense – are just as important (Wolpert, 1971). Cellular differentiation, particularly as studied *in vitro*, is concerned with the change with time of the synthesis of particular molecules, together with changes in cell structure, which characterise cell function. Thus, the differentiation of striated muscle is characterised by cell fusion, the appearance of characteristic striations and the synthesis of specific actins and myosins. For red blood cell differentiation, the key marker is the synthesis of haemoglobin. Attention is thus largely directed towards a well-determined sequence of events. This can be regarded as cell maturation: the cell being committed to a particular pathway of differentiation which it then carries out. A great deal of development is probably of this kind, carrying out an invariant programme and it probably characterises most of development, in vertebrates at least, from the late neurula onwards. This type of development must be contrasted with early development in which there is considerable lability, and the pathways of differentiation are not yet specified. Here the problem is to unravel the processes involved in this specification of what is called determination.

Determination is a change in cell state, such that a cell and its progeny acquire one of several possible pathways of development. Determination is thus related to choice, whereas cell maturation does not involve choice. Prior to determination, a cell or its progeny might develop into, for example, cartilage or muscle. Following determination, only one line of differentiation is in general possible. Again, cells may become determined as leg cells as distinct from flank, or wing, cells. Or, in insects, cells may become determined with respect to a particular segment. In mammalian development there is an early determinative event by which cells become committed either to trophoblast or inner cell mass. Determination is usually assayed by classical techniques of transplantation or by the new powerful technique of clonal analysis. The important point is that the underlying molecular basis of determination is not known and there are no overt signs of change at the time of determination. Thus the assumption is made that at the time of determination, an important series of molecular changes has occurred, which is remembered by the cell. It is a very important question as to whether the changes associated with determination are similar to those associated with cell differentiation. While it is generally assumed that they are, there is no evidence to support this view.

Determination is also different from cell differentiation since it can be concerned with spatial organisation. For example, at quite an early stage of development, before the somites have formed, the character of the different vertebrae is specified. If the presumptive somite material is grafted to another site, say from the neck region, then neck vertebrae will still develop at this new site (Kieny, Mauger & Sengel, 1972). Thus, at a very early stage, these presumptive somite cells have a character which distinguishes them from adjacent regions, even though they will give rise to the same cell types. We have termed this type of determination which reflects a difference in intrinsic character of apparently similar cells, non-equivalence (Lewis & Wolpert, 1976), and it is an essential feature of pattern formation.

Pattern formation is the spatial organization of cellular differentiation and cell differentiation may be the same in different spatial patterns. Cell differentiation in the arm and leg may reasonably be assumed to involve similar, if not identical, cellular differentiation of muscle, cartilage, and tendons and yet their spatial organization is quite different. I wish to argue that this reflects non-equivalence among some of the cells, and that pattern formation involves the specification of a cell state termed positional value, which makes them non-equivalent. This is an aspect of cell differentiation not yet open to study *in vitro*. I shall illustrate this largely with respect to limb morphogenesis.

LIMB DEVELOPMENT

Our model for pattern formation in vertebrate limb development is based upon the idea of positional information (Wolpert, 1971, 1978). This suggests that cells are assigned positional values in a three-dimensional co-ordinate system and that the cells interpret these positional values by differentiation appropriate to their genome and developmental history. It is suggested that position is specified in a zone at the tip of the growing bud, called the progress zone, and different mechanisms have been proposed for specification along the three different axes. For the proximo-distal axis position it is assumed to be specified by the length of time – possibly measure by the number of cell divisions – the cells spend in the progress zone (Summerbell, Lewis & Wolpert, 1973). For the antero-posterior axis there is good evidence that position is specified by a graded signal from the polarising region, which is at the posterior margin of the bud (Tickle, Summerbell & Wolpert, 1975: Tickle, 1980). The ectoderm seems to specify the dorso-ventral axis (MacCabe, Errick & Saunders, 1974; Patou, 1977). It is a central idea that the cells early on acquire positional values, and that these positional values are remembered by the cell and passed on to its progeny. Moreover, these positional values programme not only the cells pathway of differentiation, but also other properties such as its growth. Once positional values are specified, cell behaviour is largely autonomous.

Consider, for example, the proximo-distal axis. The elements are laid down in a proximo-distal sequence and are derived from the cells of the progress zone. The cells in this zone appear to be structurally undifferentiated mesenchyme cells and they give rise to the cartilaginous elements, tendons and connective tissue of the limb. As the limb bud grows out, so the intrinsic character of the cells in the progress zone changes, and they give rise to more distal structures. Thus, a young progress zone, if grafted elsewhere will give rise to an almost complete limb, whereas an older one, will give rise only to distal structures. For example, if a young progress zone is replaced with a young progress zone, elements will be missing and the pattern may be just humerus – hand. For the opposite case – a young progress zone replacing an older one – elements are repeated: the resulting limb comprising humerus, radius and ulna, humerus, radius and ulna, wrist, hand (Summerbell & Lewis, 1975). Thus, the development of the cells in the progress zone is autonomous and dependent on how long they have been in the progress zone. Rubin & Saunders (1972) showed that this change in character is a property of the mesenchyme and not the ectoderm by exchanging the ectodermal shell between limb buds of different ages.

These changes with time, make early progress zone cells nonequivalent with later ones. This, we suggest reflects a change in positional values. This non-equivalence is further reflected in the growth of the cartilaginous elements. Each cartilaginous element along the limb is roughly the same length when it is laid down. The cartilaginous rudiments of the long bones elongate a great deal but the elements in the wrist hardly do so at all (Lewis, 1975; Summerbell, 1976). The cartilage cells in the wrist do not undergo the typical hypertrophy characteristic of the long bones. Moreover, this difference in growth seems to be tissue autonomous, and is maintained both in culture and when grafted elsewhere in the limb (Holder & Wolpert, in preparation). It is thus suggested that the growth of the cartilage is specified by its positional value. Indeed it can be argued that this positional value specifies the different times of ossification of the different cartilaginous elements (Holder, 1978) as well as the character of their growth plates (Wolpert, 1981).

Non-equivalence is shown again by grafts from hindlimb bud to forelimb (Saunders, Gasseling & Cairns, 1959). A small block of cells from the presumptive thigh region grafted to the tip of the wing bud forms a toe, and even induces the overlying ectoderm to form scales instead of feathers. Thus, at an early stage there is a difference between wing and leg mesenchyme. Like the vertebrae, this reflects their different positional values with respect to the main body axis.

There is also a positional value associated with the antero-posterior axis. When a polarising region is grafted to an anterior position in a wing bud, there is a mirror-image duplication of all the elements — cartilage, muscle, and tendons, across the midline. We have suggested that this results from the polarising region specifying new positional values (Tickle, 1980).

LIMB REGENERATION

Limb regeneration in amphibia provides further strong evidence for non-equivalence. When a newt limb is amputated, only regions distal to the cut regenerate. There is no signal from the stump which specifies that only missing parts should be regenerated, since the same distal parts are regenerated from a distal cut surface, resulting in mirror-image duplication of structures already present (Butler, 1955). A novel model for limb regeneration has been put forward which is based on the polar co-ordinate model for epimorphosis (French, Bryant & Bryant, 1976). This model assumes that in the amphibian limb positional values are specified in terms of polar co-ordinates and that provided there is a complete set of circular values, distal transformation – that is regeneration – will occur. Evidence in support of this model is reviewed by Bryant (1978). There is some evidence that in the amphibian limb the positional values are carried in the dermis and muscle and not in the cartilage, and the histological basis of this has been examined by Tank & Holder (1979).

MUSCLE DEVELOPMENT

The above discussion has emphasised the importance of non-equivalence in the development of the cartilaginous elements. For the development of muscle the evidence is that, by contrast, they are all equivalent. From an early stage, the presumptive muscle cells are a separate lineage in the limb bud. Transplantation studies have shown that a small number of muscle cells migrate into the region of the presumptive limb bud from the adjacent somites (reviewed by McLachlan & Wolpert, 1980). This small group of cells gives rise to all the muscles of the limb, but not to the muscle connective tissue. The evidence for equivalence comes from the observation that the source of these cells has no effect on the muscle pattern. Thus, a normal pattern of wing musculature will develop when cervical or leg somites are grafted adjacent to the presumptive wing. This is quite unlike the cartilaginous elements which develop according to their origin (see above). It may be argued that the muscle cells acquire 'wingness' when they migrate into the bud, but this seems an unlikely explanation. They are also absent from the progress zone and if they do acquire positional values this must be by some other means. The preferred explanation is that the muscle cells are all equivalent and that the pattern of muscles arises as a result of their interaction with the connective tissue cells (McLachlan & Wolpert, 1980). One possibility is that they have a higher affinity for the presumptive muscle connective tissue cells. They would thus initially accumulate as dorsal and ventral muscle masses. The later development of the muscle involves splitting of these masses. This could arise if the pattern of affinity within the connective tissue altered, or the connective tissue altered the rates of muscle growth or fusion (Shellswell, 1980). There is a differential pattern of collagen synthesis in the muscle, which shows differences between epimysium, perimysium and endomysium. The epimysium synthesis predominantly Type I, the perimysium Type III, and the endomysium Type IV, but this pattern develops quite late (Shellswell, 1980).

DETERMINATION OF CARTILAGE AND MUSCLE

There are no conclusive studies to show that the differentiation of muscle and cartilage *in vitro* is anything other than cell maturation. That is, cells that are already determined as muscle or cartilage cells complete this process *in vitro*. This view has been repeatedly stated by Holtzer (1978). However, there have been claims that it is possible to direct early mesenchyme cells to differentiate as either muscle or cartilage in culture, by metabolites such as nicotinamide, which alter the NAD concentration (Caplan, 1977). However, these experiments fail to distinguish between selection of determined populations and real changes in the direction of differentiation. *In vivo* we have found that nicotinamide analogues deleteriously affect a variety of cell types but do not alter the pathway of differentiation (McLachlan & Wolpert, 1980).

Quite persuasive evidence that early limb bud cells can have their fate changed comes from the observations of Nathanson, Hifler & Searls (1978) who found that clonally derived muscle cells can be transformed into cartilage with bone matrix.

The *in vivo* studies show that in normal development the important determinative step is not in relation to whether the cell will develop into muscle or cartilage, for the muscle cells have a quite different lineage (see above). Whether or not muscle cells can, under special circumstances, be persuaded to change their differentiated state, is of great interest, but not really relevant here. The important determinative event is whether the cells will form cartilage, or perichondrium, or dermis, or connective tissue. Very little attention has been given to these cell types either *in vivo* or *in vitro*. There is however evidence that the connective tissue/cartilage decision is probably labile up to about stage 23–25. Searls (1973) and Cioffi (1975) showed that when tissue from presumptive soft tissue regions was grafted into the presumptive cartilage at early stages it developed as cartilage, and at later stages, when core tissue was grafted to more peripheral regions, ectopic cartilage formed.

Solursh and co-workers (Solursh, 1981) have examined the development of chick cartilage *in vitro*. Their technique makes use of dense cultures which form cartilaginous nodules. From stage 20 onwards their cultures give cartilage, but even stage 17–19 buds give cartilage if cAMP is added. From stage 25 on, cartilage develops without going through an aggregation phase characteristic of earlier stages. They find no striking differences between periphery and core regions though the development of cartilage from peripheral regions is slower. These studies thus suggest that most of the cells in the early limb bud can give rise to cartilage *in vitro*. Kosher, Savage & Chan (1979) too have claimed that this is true of the bud in organ culture if the ectoderm is removed. It is thus difficult at this stage to relate these studies to the events occurring *in vivo*, where only about 5% of the cells give rise to cartilage (Lewis, 1977). Solursh has also found that stage 19 mesenchyme or tumour cells inhibit cartilage formation by stage 23–24 mesenchyme and suggest that some sort of homotypic interaction is involved. This may mean that a certain density of cartilage cells is required. This latter idea fits quite well with our observation on the effect of X-irradiation. It is always the smallest elements that are lost first, such as the radius and digit 2 (Wolpert, Tickle & Sampford, 1979) and suggests that a minimum density is required for cartilage formation.

In vivo the cartilage of the long bones undergoes a further sequence of development not seen in culture. The cells hypertrophy and eventually die. This hypertrophy is an important feature of growth since it adds to the dilation of the cartilage (Wolpert, 1981). This hypertrophy does not occur in the wrist elements and this may reflect different positional values, and also why the wrist grows so little.

THE PERICHONDRIUM AND CELL SHAPE

Studies of cell differentiation *in vitro* are not easily related to pattern formation, as just pointed out. Moreover, such studies tend to ignore cells whose phenotype is not easily recognised in culture, such as the perichondrium. In addition, they will miss mechanisms which involve mechanical interactions between tissues since there is increased evidence that cell shape can influence cell behaviour (Folkman & Tucker, 1980).

The perichondrium is the layer of cells surrounding the cartilaginous rudiment and it is rarely given much attention by those who work either *in vitro* or *in vivo*. In fact I believe it to be the main factor determining the shape of a cartilaginous rudiment (Wolpert, 1981). The growth of the cartilage involves both cell division, matrix secretion and cell hypertrophy. These bring about an increase in volume, and this dilation generates a hydrostatic pressure in the cartilage which can be regarded as a visco-elastic fluid. This pressure is resisted by the perichondrium and the shape is determined by how the perichondrium deforms. Thus, cartilage growth can only be understood in terms of the mechanical properties of the perichondrium. It may be that the development of the perichondrium is specified in the same way as suggested for the cartilage, that is by positional value. However, an alternative mechanism would involve the stretching of the cells by the developing cartilage.

One of the characteristic features of a developing cartilaginous rudiment is the presence of whorls of flattened cells at the edge. Gould, Selwood, Day & Wolpert (1974) have suggested that this pattern of cells arises from the pressure exerted by the formation of matrix, which dilates the tissue. Thus, when viewed in cross-section, matrix is first seen in the centre of the future cartilaginous element, the cells there become less dense, and the adjacent cells begin to flatten. It is possible that the development of the perichondrium is directly related to this stretching. Perhaps only mesenchyme that is so stretched forms the perichondrium. While it is possible that the perichondrium is specified independently, it is clearly attractive to find that it is specified mechanically by the development of the cartilage. Cartilage nodules in culture are described as having a perichondrium (von der Mark & von der Mark, 1977) and it seems unlikely that this perichondrium is specified independently. Of course it is not known whether this perichondrium is similar to that found *in vivo*.

NON-EQUIVALENCE AND POSITIONAL VALUE IN INSECT DEVELOPMENT

Some of the best evidence for non-equivalence and positional value comes from studies on intercalary regeneration in insects. The cells in the insect imaginal disc and in the cockroach leg can be regarded as having positional values specified as a polar co-ordinate system (French, Bryant & Bryant., 1976). Whenever non-contiguous positional values are placed adjacent to one another, intercalation of positional values occurs until a smooth set of positional values is present. There are many interesting, important and controversial aspects of this model, but here the crucial point is that intercalary regeneration provides at present, a unique assay for positional value. The intercalary regeneration is in no way affected by the 'differentiated' character of the cells involved – they are all epithelial and make different kinds of cuticle. It is only their positional value that matters.

Moreover, the interaction is a local phenomenon and does not depend on global properties. This is clearly shown by Bohn's (1970) experiment in which normally non-adjacent regions of the cockroach tibia are placed together. Depending on the sequence of the graft, the tibia can grow back to its normal shape or much longer. Again, French (1980) has shown that intercalary regeneration will occur if there is a discrepancy in the circular positional values, even though the tissues are from different proximo-distal levels. This means that cells which lie on a meridian have the same circumferential positional value at all levels along the leg and irrespective of what structures they give rise to.

Compartments, too, illustrate non-equivalence in a rather elegant manner. Thus, the anterior and posterior cells in the developing wing form quite discrete groups of cells, and a sharp boundary forms between them. It is thought that this results from activation of the *engrailed* gene in the anterior compartment, whereas it is not turned on in the posterior one. Morata & Lawrence (1978) have suggested that this phenomenon involves in part a change in surface properties, which affects cell adhesiveness. In this connection there is substantial evidence in two insect systems for gradients in adhesiveness, which run in paralled with the proposed sets of positional values (Nardi & Kafatos, 1976; Nübler-Jung, 1977).

POSITIONAL VALUE

There is now, as argued above, substantial evidence that there is a cell state, characterised by positional value. This evidence is all from *in vivo* studies. It will be very important to study positional value *in vitro* but as yet this is not possible since we have no assay for positional value *in vitro*. All the assays are from *in vivo* studies. How are we to distinguish, *in vitro*, cartilage cells or connective tissue cells from different digits? Until this is done we cannot ask even simple questions about the nature of positional value, such as its stability and clonal properties. In the case of the polarising region from the chick limb bud, we do have some very preliminary information. Polarising activity is lost when the cells are placed in culture for a period longer than 24 hours.

Even though we do not see how to answer them, one can still indulge in asking questions. Is the acquisition of positional value similar to either determination or differentiation? That is a particularly difficult question since we do not even know the molecular events underlying either, though it is widely assumed that differentiation involves specific gene activation. We should, at this stage, be willing to consider that positional value is specified in some other manner.

It is important to recognise a striking quantitative discrepancy between the number of different cell states related to positional value and the number of different cell states related to cell differentiation. In the limb mesoderm, for example, there are only a few different cell types such as cartilage, fibroblasts, muscle cells. Certainly not more than about 10. Now, while we do not know how fine grained the co-ordinate system of positional values is, there must be at least an order of magnitude more different positional values in the limb than differentiated cell states.

This has a further implication. Holtzer (1978) has claimed that determinative decisions are always binary. However, since the cell states that he is assaying *in vitro* are those related to cell differentiation – muscle, cartilage, and so on – the cell states associated with positional value are excluded. There is no reason to believe that *in vivo* only binary decisions are made, since the specification of positional value must be taken into account.

SUMMARY

The main difference between differentiation *in vitro* and *in vivo* is the loss of spatial organisation *in vitro*. This has important implications since many determinative events – an important aspect of differentiation – have a spatial basis. Moreover, one type of differentiation, the acquisition of positional values, cannot yet be assayed *in vitro*. Positional value reflects a cell's positional history and may be the basis of pattern formation. Positional value is reflected in non-equivalence between cells: that is, cells that undergo similar cell differentiation, are not the same and may have other properties which are different, such as their growth programme. This can be illustrated with reference to limb morphogenesis and insect development. Thus, much of what is studied *in vitro*, is cytodifferentiation related to cell maturation, rather than the crucial issue of changes in pathways of differentiation. Similar sorts of arguments can be made with respect to processes involving changes in form where mechanical coupling may be an important aspect of cell behaviour which can affect cell differentiation.

I am grateful to the Medical Research Council for supporting this work, and to Miss Maloney for helping prepare the manuscript.

REFERENCES

BOHN, H. (1970). Interkalare Regeneration und segmentale Gradienten bei ein Extremitaten von *Leucophaea*-Larven (Blattaria). I. Femur und Tibia. *Roux Archiv. Entwicklungsmechanik Organen*, **165**, 303–41.

BRYANT, S. V. (1978). Pattern regulation and cell commitment in amphibian limbs. In *The Clonal Basis of Development*, ed. S. Subtelny & I. M. Sussex, pp. 63–82. New York: Academic Press.

BUTLER, E. G. (1955). Regeneration of the urodele forelimb after reversal of its proximo-distal axis. *Journal of Morphology*, **96**, 165–282.

CAPLAN, A. I. (1977). Muscle, cartilage and bone development and differentiation from chick limb mesenchymal cells. In *Vertebrate Limb and Somite Morphogenesis*, ed. D. A. Ede, J. R. Hinchliffe & M. Balls, pp. 199–215. Cambridge University Press.

CIOFFI, M. (1975). Determination and stability during differentiation in the avian limb. Ph.D. Thesis, University of London.

FOLKMAN, J. & TUCKER, R. W. (1980). Cell configuration, substratum and growth control: In *The Cell Surface: Mediator of Developmental Processes*, ed. S. Subtelny & N. K. Wessells, pp. 259–75. New York: Academic Press.

FRENCH, V. (1980). Positional information round the segments of the cockroach leg. *Journal of Embryology and Experimental Morphology*, **59**, 281–313.

FRENCH, V., BRYANT, P. J. & BRYANT, S. V. (1976). Pattern regulation in epimorphic fields. *Science*, **193**, 969–81.

GOULD, R. P., SELWOOD, L., DAY, A. & WOLPERT, L. (1974). The mechanism of cellular orientation during early cartilage formation in the chick limb and regenerating amphibian limb. *Experimental Cell Research*, **83**, 287–96.

HOLDER, N. (1978). The onset of osteogenesis in the developing chick limb. *Journal of Embryology and Experimental Morphology*, **44**, 15–29.

HOLTZER, H. (1978). Cell lineages, stem cells and the 'quantal' cell cycle concept. In *Stem Cells and Tissue Homeostasis*, ed. B. L. Lord, C. S. Potten & R. J. Cole, pp. 1–28. Cambridge University Press.

KIENY, M., MAUGER, A. & SENGEL, P. (1972). Early regionalization of the somitic mesoderm as studied by the development of the axial skeleton of the chick embryo. *Developmental Biology*, **28**, 142–61.

KOSHER, R. A., SAVAGE, M. P. & CHAN, S. (1979). *In vitro* studies on the morphogenesis and differentiation of the mesoderm subjacent to the apical ectodermal ridge of the embryonic chick limb bud. *Journal of Embryology and Experimental Morphology*, **50**, 75–97.

LEWIS, J. H. (1975). Fate maps and the pattern of cell division: a calculation for the chick wing bud. *Journal of Embryology and Experimental Morphology*, **33**, 419–34.

LEWIS, J. (1977). Growth and determination in the developing limb. In *Vertebrate Limb and Somite Morphogenesis*, ed. D. A. Ede, J. R. Hinchliffe & M. Balls, pp. 215–28. Cambridge University Press.

LEWIS, J. H. & WOLPERT, L. (1976). The principle of non-equivalence in development. *Journal of Theoretical Biology*, **62**, 479–90.

MACCABE, J. A., ERRICK, J. & SAUNDERS, J. W. (1974). Ectodermal control of the dorsoventral axis of the leg bud of the chick embryo. *Developmental Biology*, **39**, 69–82.

McLachlan, J. & Wolpert, L. (1980). The spatial pattern of muscle development in the limb. In *The Development and Specialization of Muscle*, ed. D. F. Goldspink, pp. 1–17. Cambridge University Press.

Morata, G. & Lawrence, P. A. (1978). Cell lineage and homeotic mutants in the development of imaginal discs of *Drosophila*. In *The Clonal Basis of Development*, ed. S. Subtelny & I. M. Sussex, pp. 45–62. New York: Academic Press.

Nardi, J. B. & Kafatos, F. C. (1976). Polarity and gradients in lepidopteran wing epidermis. II. The differential adhesiveness model: gradients of a non-diffusible cell surface parameter. *Journal of Embryology and Experimental Morphology*, **36**, 489–512.

Nathanson, M. A., Hifler, S. R. & Searls, R. L. (1978). Formation of cartilage by non-chondrogenic cell types. *Developmental Biology*, **64**, 99–117.

Nübler-Jung, K. (1977). Pattern stability in the insect segment. I. Pattern reconstitution by intercalary regeneration and cell sorting in *Dysdercus intermedius*. *Wilhelm Roux Archiv. N*, **83**, 17–40.

Patou, M. -P. (1977). Dorso-ventral axis determination of chick limb bud development. In *Vertebrate Limb and Somite Morphogenesis*, ed. D. A. Ede, J. R. Hinchliffe & M. Balls, pp. 257–67. Cambridge University Press.

Rubin, L. & Saunders, J. W. (1972). Ectodermal-mesodermal interactions in the growth of limb buds in the chick embryo: constancy and temporal limits of ectodermal induction. *Developmental Biology*, **28**, 94–112.

Saunders, J. W., Gasseling, M. T. & Cairns, J. M. (1959). The differentiation of prospective thigh mesoderm grafted beneath the apical ectodermal ridge of the wing bud in the chick embryo. *Developmental Biology*, **1**, 281–301.

Searls, R. L. (1973). Newer knowledge of chondrogenesis. *Clinical Orthopaedics*, **96**, 327–44.

Shellswell, G. B. (1980). Cellular events in the early development of skeletal muscles. In *Development in Mammals*, 4, ed. M. H. Johnson, pp. 137–60. Amsterdam: Elsevier/North-Holland.

Solursh, M. (1981). Histogenic mechanisms in *in vitro* limb chondrogenesis. In *Current Research Trends in Prenatal Cranio-facial Development*, ed. R. M. Pratt & R. L. Christiansen, pp. 315–27. New York: Elsevier/North-Holland.

Summerbell, D. (1976). A descriptive study of the rate of elongation and differentiation of the skeleton of the developing chick wing. *Journal of Embryology and Experimental Morphology*, **35**, 241–60.

Summerbell, D. & Lewis, J. H. (1975). Time, place and positional value in the chick limb bud. *Journal of Embryology and Experimental Morphology*, **33**, 621–43.

Summerbell, D., Lewis, J. H. & Wolpert, L. (1973). Positional information in chick limb morphogenesis. *Nature, London*, **244**, 492–6.

Tank, P. W. & Holder, N. H. K. (1979). The distribution of cells in the upper forelimb of the axolotl. *Journal of Experimental Zoology*, **209**, 435–42.

Tickle, C. (1980). The polarizing region in limb development. In *Development in Mammals*, 4, ed. M. H. Johnson, pp. 101–36. Amsterdam: Elsevier/North-Holland.

Tickle, C., Summerbell, D. & Wolpert, L. (1975). Positional signalling and specification of digits in chick limb morphogenesis. *Nature, London*, **254**, 199–202.

von der Mark, K. & von der Mark, H. (1977). Immunological and biochemical studies of collagen type transition during *in vitro* chondrogenesis of chick limb mesoderm cells. *Journal of Cell Biology*, **73**, 736–47.

Wolpert, L. (1971). Positional information and pattern formation. *Current Topics in Developmental Biology*, **6**, 183–224.

Wolpert, L. (1978). Pattern formation and the development of the chick limb. In *The Molecular Basis of Cell-Cell Interaction*, pp. 547–9. New York: Alan R. Liss, Inc.

Wolpert, L. (1981). Cartilage morphogenesis in the limb. In *Cell Behaviour*, ed. R. Bellairs, A. S. G. Curtis & G. Dunn. Cambridge University Press (in press).

Wolpert, L., Tickle, C. & Sampford, M. (1979). The effect of cell killing by X-irradiation on pattern formation in the chick limb. *Journal of Embryology and Experimental Morphology*, **50**, 175–98.

INDEX